TM114 MIL

INTRODUCTION TO PLASTICS AND COMPOSITES

MECHANICAL ENGINEERING

A Series of Textbooks and Reference Books

Editor

L. L. Faulkner

*Columbus Division, Battelle Memorial Institute
and Department of Mechanical Engineering
The Ohio State University
Columbus, Ohio*

INTRODUCTION TO PLASTICS AND COMPOSITES

MECHANICAL PROPERTIES AND ENGINEERING APPLICATIONS

EDWARD MILLER

California State University, Long Beach
Long Beach, California

Marcel Dekker, Inc.   New York • Basel • Hong Kong

ISBN: 0-8247-9663-2

The publisher offers discounts on this book when ordered in bulk quantities. For more information, write to Special Sales/Professional Marketing at the address below.

This book is printed on acid-free paper.

MARCEL DEKKER, INC.
270 Madison Avenue, New York, New York 10016

Current printing (last digit):
10 9 8 7 6 5 4 3 2 1

PRINTED IN THE UNITED STATES OF AMERICA

Preface

Our knowledge of plastics has grown at a phenomenal rate during the past two decades. Formerly associated with cheap, unreliable items, plastics are now recognized as a respectable member of the class of materials that can be employed under extreme load and environmental conditions. This improvement in properties is directly related to the large number of technical publications concerning the properties of plastics. As would be expected, due to the relative newness of these materials and the vast array of new polymers, blends, copolymers, and additives, the literature is highly fragmented and therefore difficult to read. This text is designed to present an orderly array of the principal topics related to the mechanical properties of plastics. The objective is to permit an engineer or graduate student to obtain, without undue confusion and effort, the background required to understand the current literature.

Recent results reported in the literature are included. However, data and theories are presented not in an attempt to summarize all the latest research but rather to give clear examples of the aspects of the properties of plastics that are currently understood and the theories that are widely viewed as being usable in engineering applications and analysis of the behavior of plastics. Theories that are still at the stage where engineering application is unlikely in the near future, because of either excessive complications or lack of common agreement on the axioms of the theory, are mentioned only briefly. It is hoped that this selection will focus the reader's attention on ideas that the reader can confidently apply in analyzing the behavior of plastics.

Chapter 1 describes chemical bonding between atoms and the poly-merization processes that generate the most common polymers. The mech-anical properties are functions of these processes, and such a discussion is necessary to understand the arguments in the remainder of the text.

The mechanical properties are controlled by the molecular configuration, the extent of crystallization, and the amount of glassy material. Chapter 2 considers the arrangement of the atoms on the long-chain molecules of the polymer and how the arrangement of the molecules changes during crystal-lization and the glass transition. Chapter 3 describes the most common types of additives used and the problems and benefits associated with their usage.

The number of polymer types available is continually increasing, and detailed summaries of the properties of all the polymer types, blends, and additives would require several volumes in themselves. The most commonly used polymer types are discussed in Chapter 4, with indications of their most common uses and general properties. This compilation is designed so that the reader can gain insight into which plastic could be considered for application to a particular product. Chapter 5 describes the most common methods for manufacturing plastic products. These discussions are related to how individual techniques affect the resultant mechanical properties of the finished product and emphasize the interrelation of manufacturing techniques and mechanical properties. Chapter 6 deals with the most common types of test data for plastics, and comparison with metal behavior is provided. Chapter 7 discusses the rapidly growing application of fracture mechanics to plastics. The J-integral approach and the analysis of fatigue data employing stress intensity analysis are described.

Viscoelastic behavior is described throughout the book but is considered in detail in Chapter 8. The pseudoelastic approximation, the Boltzmann superposition theorem, and the heredity integral method for analysis of the behavior of viscoelastic materials under long-term or varying loads are discussed. Chapter 9 describes the various mechanical models for plastics and the behavior of these models under dynamic loading. The discussion is then expanded to include the complex modulus and the experimental techniques to determine the behavior of polymers under varying loads. Chapter 10 describes the behavior of laminae and laminates of composites, employing the well-established techniques of matrix analysis. The discussion is aimed at providing a clear, concise presentation, which the reader can follow to perform laminate analysis.

Edward Miller

Contents

Contents

INTRODUCTION
TO PLASTICS
AND COMPOSITES

1

The Chemical Structure of Plastics

1.1. INTRODUCTION

The industrial plastics which are so widely utilized in all aspects of manufacturing consist of large molecules which are primarily covalently bonded. These giant molecules are usually chainlike in dimensions, although side chains, circular molecules, and interconnections of the chains by various mechanisms may occur. It is the large dimensions of these molecules which generate the unique properties of the plastics as compared with other materials such as metals and ceramics. The underlying molecular structures of industrial plastics are related closely to those of organic living organisms, and many plastics are produced by modifying the structure of plant molecules.

Chemists are extremely adept at developing new organic compounds which have interesting properties as new plastics, and during the past 30 years many new and improved plastic structures were commercially introduced. However, the cost of developing the technology and manufacturing equipment for a new plastic is becoming prohibitive, and more effort is currently being devoted to improving properties of existing plastics by combining several plastics with known properties to develop new grades with improved properties.

It is difficult, if not impossible, to understand the behavior of plastic products under conditions of changing temperature, stress, manufacturing conditions, or environment without an understanding of the underlying structure of the plastic. This chapter will provide a brief summary of those aspects of the chemical bonding in polymers which affect the mechanical behavior.

1.2. PRIMARY CHEMICAL BONDING IN MOLECULES

A brief review of the simpler concepts of atomic bonding will assist in understanding the particular bonding in plastics.

Ionic Bonding

Ionic bonding primarily involves the ionization of one atom with consequent loss of an electron and the transfer of that electron to a different atom, whose affinity for electrons is higher. The classical example is the formation of LiF. The formation of the LiF molecule from the individual unbound atoms can be conceived as consisting of the following steps:

1. Ionization of Li atom to Li:

    $$Li - e \rightarrow Li^+$$

    The ionization energy (energy required for this reaction to occur) is $+123.5\ kcal/mole$.
2. Ionization of F atom to F^-:

    $$F + e \rightarrow F^-$$

    The electron affinity of F (energy required for electron absorption) is $-95\ kcal/mole$.
3. Coulombic attraction of the two ions:

    $$Li^+ + F^- \rightarrow Li^+F^-$$

The energy required to ionize the two atoms is defined as the value for the formation of two ions infinitely far from other atoms. When these two ions are attracted to each other, energy is released. The Coulombic attractive forces are eventually balanced by repulsion forces when the atoms approach so closely that their electronic shells interpenetrate excessively. The resultant total energy change of the system due to the attraction of these ions is therefore:

$$E = -\frac{e^2}{r} + \frac{b}{r^{12}} = -120\ kcal/mole$$

where $-e^2/r$ is the attractive energy and b/r^{12} is the repulsive term. The bond is stabilized by the Coulombic attraction of the two dissimilar charges. The total energy evolved when the ionic bond is formed is the sum of these three terms: $E_{\text{total}} = 123.4 - 95 - 120 = -91.6 \, \text{kcal/mole}$.

Covalent Bonding

Plastics are manufactured from relatively small groups of low atomic number elements. The major components are carbon and hydrogen, with some plastics containing oxygen, nitrogen, fluorine, chlorine, and sulfur. The electron energy levels permissible for electrons surrounding the nuclei of various atoms have been analyzed by quantum mechanics. Two electrons of opposite spin can occupy each level. In the diagrams in Fig. 1.1, upward and downward arrows signify electrons occupying an energy level, the different arrow directions indicating different spins. For the low atomic number elements with which we are concerned, the permissible energy levels are indicated by the principal quantum number n, which also primarily determines the average distance of the center of electronic charge from the nucleus. The second quantum number l affects the energy of the electron slightly but is the main determinator of the shape of the electronic distribution of charge around the nucleus. A common terminology is to ascribe the letter s to $l = 0$, since transitions involving this state produce sharp spectra, and p to $l = 1$, since these are the principal spectral lines observed.

The distributions of energy levels permissible are shown for the first few elements in Table 1.1. The spatial distribution of charge around the nucleus is spherically symmetric for s electron, and directed along the three coordinate axes for the three p states.

Table 1.1. Quantum Numbers for Electronic Energy States

n	l	Number of states	Number of electrons	Designation
1	0	1	2	$1s$
2	0	1	2	$2s$
2	1	3	6	$2p$
3	0	1	2	$3s$
3	1	3	6	$3p$
3	2	4	8	$3d$

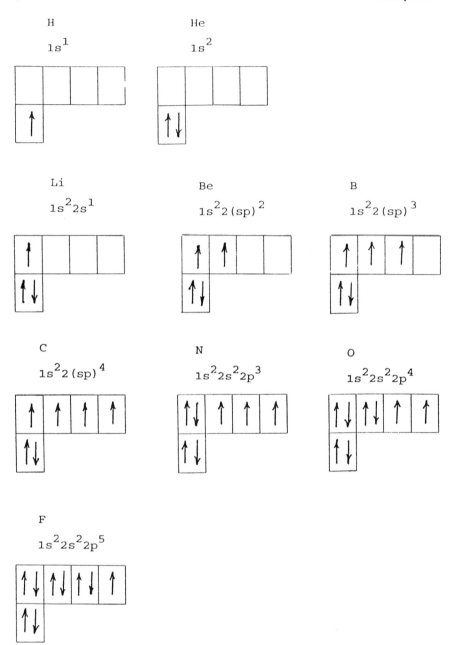

Figure 1.1. Electronic energy levels for principle elements in plastics in their hybridized states.

Hybridization of Electronic Orbitals

The s and p states with the same principal quantum number are close in energy, and a lowered total energy may occur during bonding if electrons from these two levels redistribute their charge by entering a hybrid state. Mathematically, these new hybrid energy levels are linear combinations of the wave functions of the ground level s and p states. For our purposes it is sufficient to recognize that these new distributions lower the energy of the bonded element by producing a greater number of bonds and by generating electron distributions which have increased electron concentrations in one direction, permitting stronger bonding in that direction. This change in electronic distribution as the sum of the different energy level wave functions is called *hybridization*.

Bonding in polymers is principally covalent bonding, in which an electron of opposite spin from each of the interacting atoms overlap regions of high electron density. Since the electrons are close to both atoms, each atom can be considered to fill its outermost electronic shell, stabilizing the structure.

The fundamental atom upon which virtually all polymers are constructed is carbon. When the carbon atom has formed only single bonds with neighboring atoms, the valence electrons are in a hybridized state, sp^3, and the valence is four (four outermost electrons capable of bonding with four other electrons). The direction of highest density of these electrons is best described by considering the carbon atom as located in the center of a tetrahedron, the direction of bonding toward the four corners of the tetrahedron. The resultant angle between the bonds is 109°28′. Long-chain structures (polymers) are developed by carbon atoms bonding to other carbon atoms. A representation of such a chain is shown in Fig. 1.2. This figure illustrates the structure of polyethylene, in which each carbon atom is bonded to two other carbon atoms and two hydrogen atoms.

The significant features resulting from the fixed bond angles of a carbon atom are:

1. The plastic molecules are neither linear nor planar.
2. The hydrogen atoms and carbon atoms bonded to one carbon are not in a planar configuration.

Double Bonds

Two atoms may form more than one bond between them. Bonds form by overlap of orbitals of electrons of opposite spins. Double bonds can be described as being formed by overlap of two sp^3 orbitals, but it is more common to assume that one p orbital on each carbon atom remains

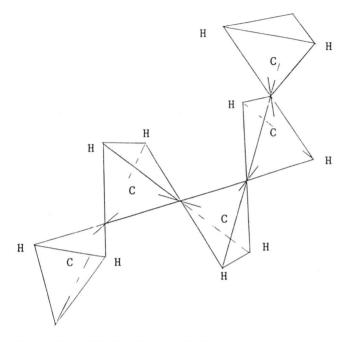

Figure 1.2. The bonding in a chain molecule.

unhybridized. This overlaps another *p* orbital on a second carbon atom side by side, instead of end to end, to form what is called a π-bond. The electronic density in this bond lies equally both above and below the two atoms. This hybridization improves the overall density of electronic charge which overlaps and thereby lowers the energy of the resultant molecule. The electronic configuration shifts to that shown in Fig. 1.3, in which the arrangement is such that the electrons form three sp^2 bonds such as those in beryllium and one *p* bond. However, because of the difference between the two bonds, a double bond is not as strong as two single bonds and may be easily broken if another attacking atom approaches the bonded pair. The concept of breakage of double bonds to generate reactive electrons for further bonding is utilized widely in the production of plastics and will be discussed further later. A second important difference between a single bond and a double bond is the restricted rotation and movement possible in a long-chain molecule at the double bond as compared with single bonds. The mobility and flexibility of the chain and the plastic are therefore decreased by the existence of double bonds in the polymer chain molecules.

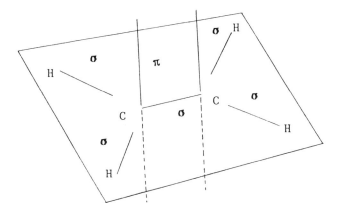

Figure 1.3. A double bond between two carbons atoms. Bonding in ethylene: the *s*-electrons have maximum overlap along the line through their centers (σ-bonding); the *p*-electrons have the greatest overlap when parallel to each other (π-bonding).

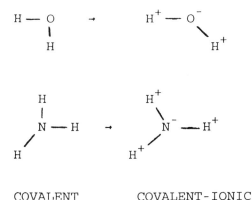

COVALENT   COVALENT-IONIC

Figure 1.4. Distortion of bonds due to the ionic contribution to the bond in water and ammonia.

Hybridized Covalent-Ionic Bonds

Just as the individual electronic levels can combine to form a new more stable bond, bond strengths can be increased and the energy of a structure lowered by having an ionic contribution to a covalent type of bond. The most well-known example of this type of mixed bonding is the water molecule. Since oxygen can bond covalently with two *p* electrons, the molecule should form with bonds displaced by 90°, as shown in Fig. 1.4.

However, since oxygen has a higher electron affinity than hydrogen, the electrons tend to shift toward the oxygen, so that the hydrogen becomes slightly positively charged and oxygen negatively charged. This tends to localize the electrons closer to the oxygen, producing a small negative charge on the oxygen and a positive one on the hydrogen. The repulsive force between the positively charged hydrogens opens the bond angle to the observed value of 105°. Similarly, the bond angles in ammonia (NH_3) should be 90° if the bonds were purely covalent. The observed value is 108°, also as a result of the repulsion between the positively charged hydrogen atoms (see Fig. 1.4).

Secondary Bonds

Besides ionic and covalent bonds, which form strong connections between atoms, other mechanisms exist that contribute to the bond strength. These are discussed below.

Dipole Moment of Partially Ionic Bonds

Since a charge separation is produced when an ionic contribution to the bond exists, a dipole is produced. The *dipole moment* is defined as the product of the charge and the distance between the centers of the positive and negative charges. Clearly, the greater the ionic contribution, the greater the dipole moment. The dissolving power of water for ionic materials is the result of the inherent dipole moment of the partially ionic water molecule. The positive hydrogen attracts negative ions and the negative oxygen attracts positive ions, dissociating and dissolving ionic materials. Whereas water is an excellent solvent for such materials, it is a poor solvent for nonpolar materials, and oils and other organic chemicals, which are largely nonpolar, tend to form into separate layers which float on the water. The dipole moment of an NaCl molecule (essentially completely ionic) is 9×10^{-18} esu-cm. Although hydrogen has a slightly greater attractive tendency for electrons than does carbon, the dipole moment of methane (CH_4) is zero, due to this small electronegativity difference and the symmetry of the molecule, in which the hydrogen atoms symmetrically surround the carbon. If one of the hydrogen atoms is replaced by a chlorine atom to yield CH_3Cl, the dipole moment increases to 1.9×10^{-18}, due to the attraction of the chlorine for electrons, introducing a strong ionic contribution to the Cl—C bond. This additional ionic contribution increases the strength of the bond between the two atoms.

The physical properties are a function of the strength of the bonds within the molecule and the attractive forces between neighboring molecules. The introduction of an ionic contribution and the resultant dipole increases the

Table 1.2. Effect of Adding a Polar Molecule

Compound	Dipole moment (esu)	Boiling temperature (°C)	Melting temperature (°C)
CH_4	0	-161.4	-182.6
CH_3Cl	1.9×10^{-18}	-24	-97.7

strength of the bonds between neighboring molecules. The negatively charged atom on one molecule attracts the positively charged atom on the adjoining molecule. The physical properties of CH_4 and CH_3Cl are compared in Table 1.2. Note the significant increase in boiling point and melting point as a result of the addition of the Cl atom, due to attraction between neighboring charged atoms.

Induced Dipoles

When a charged atom approaches another atom, it displaces the electrons on the originally neutral atom: a negatively charged atom repels the electrons and a positively charged atom attracts the electrons on the surrounding atoms. When this charge displacement occurs, the originally neutral atom has an induced dipole moment, and an attractive force is generated between the two charged atoms. When a chlorine atom is replaced by a hydrogen atom within a molecule, the attractive force between neighboring molecules is increased, even if the neighboring molecule does not have a permanent dipole moment. Figure 1.5 schematically illustrates the production of an induced dipole.

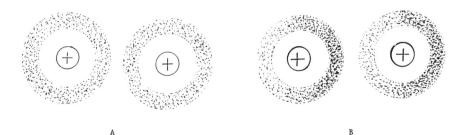

A   B

Figure 1.5. Induction of a dipole. (A) Two atoms before induction. (B) Induced dipole as a result of electronic rearrangement.

Polarizability

The magnitude of the induced dipole moment is proportional to the external field produced by the charged atom. The constant of proportionality is called the polarizability. Physically, the polarizability is the ease of distortion of the electronic distribution around the nucleus, and it tends to increase with increasing size of the atom, since the electronic distribution is larger, more diffuse, and less affected by the attraction to the nucleus itself.

The Hydrogen Bond

A hydrogen atom consists of only one electron and one proton. If the electron is redistributed away from the nucleus during bonding, the proton becomes exposed and has a very strong polarizing force on other atoms. Hydrogen is unique in its ability to polarize other molecules, forming strong secondary bonds. Bonding between hydrogen and other atoms which have ionic charges, such as chlorine, fluorine, or oxygen, is exceptionally strong.

Van der Waals Forces

Even neutral atoms have fluctuating dipole moments as a result of electronic motion. Although on the average this dipole moment is zero, the momentary dipole can induce a dipole on another atom by polarization. Attraction will occur between the atoms when they approach each other, and the electronic distribution around the atoms becomes disturbed by their mutual interaction. The bond generated in this manner is weaker than that generated by a permanent dipole.

Comparison of Interatomic Bond Strengths

These different bond types have significantly different strengths of attraction between the atoms. Table 1.3 gives the typical energy values required to break the bonds.

Table 1.3. Interaction and Bond Energies

Type of interaction	Energy (kcal/mole)
Primary covalent	50–200
Hydrogen bond	5–10
Induced dipole	1–5
Van der Waals	0.5–2

1.3. POLYMERS

Each atom within a molecule is capable of forming an induced dipole or van der Waals interaction with an atom from another molecule. Larger molecules therefore attract each other more strongly than smaller molecules. This is well illustrated by comparing the boiling and melting points of various straight chain hydrocarbons with increasing chain length. This series (the alkanes) can be described by the general formula:

$$H-[\underset{\underset{H}{|}}{\overset{\overset{H}{|}}{C}}]_n-H$$

An example is pentane, with $n = 5$:

$$H-\underset{\underset{H}{|}}{\overset{\overset{H}{|}}{C}}-\underset{\underset{H}{|}}{\overset{\overset{H}{|}}{C}}-\underset{\underset{H}{|}}{\overset{\overset{H}{|}}{C}}-\underset{\underset{H}{|}}{\overset{\overset{H}{|}}{C}}-\underset{\underset{H}{|}}{\overset{\overset{H}{|}}{C}}-H$$

The boiling and melting points of a compound are good indicators of the bond strength between molecules of that compound. The molecular kinetic energy increases with temperature, and the boiling point is that temperature at which the kinetic energy is sufficient to overcome the bond energy holding the molecules together. Table 1.4 lists the boiling and melting points for selected alkanes. The increase of boiling point with increasing chain length is evident, and the compounds change at room temperature with increasing chain length from being gaseous to liquid and then solid as interatomic forces increase. Polyethylene, a widely used plastic, is a member

Table 1.4. Boiling and Melting Points of Alkanes

Name	Number of C atoms	Melting point (°C)	Boiling point (°C)
Methane	1	−184	−161
Ethane	2	−172	−88
Pentane	5	−131	36
Hexadecane	16	20	287
Pentatriacontane	35	80	331
Polyethylene	>1000	>110	Decomposes before boiling

of the alkane group, and its reasonable mechanical properties are a result of the large number of interacting carbon atoms in the chain.

All plastics are composed of long-chain molecules, their strength and other properties being derived from this structure. The exact chain size is not constant for different molecules within a sample; a chain size distribution always exists. The melting point and chain length of polyethylene therefore cannot be precisely specified.

The Mer

Although plastic molecules are long chains, they are all composed of small basic units which are repeated to generate the chain. The *mer* is the simplest chemical structural unit which is repeated in the chain. For polyethylene, the mer is specified as that of the alkane series above (CH_2).

Degree of Polyermization

The number of repeating mers which produce the chain is the degree of polymerization (DP). Since different molecules may have different chain lengths, only an average number can be specified. Later sections will deal with details of how the averaging may be performed. A synonym for a plastic is *polymer*, and the two terms will be used interchangeably in the remainder of this text.

1.4. CHEMICAL BONDING IN POLYMERS

Thermal Vibrations

The atoms and the molecule as a whole are not at rest in any specific configuration. Thermal energy causes vibration of the atoms along and perpendicular to the directions of the chemical bonds and rotation around the bond directions. Among the atomic and molecular motions which are present and increase with increasing temperature are the following:

1. Atomic vibrations within the carbon backbone chain increase the average interatomic distance and the total chain length.
2. Vibrations perpendicular to the bond direction can cause small variations in bond angles.
3. Rotation of the carbon atoms around the bonds can occur, causing whipping and twisting of the chains.

The vibrational amplitude increases with increasing temperature. As a result of this kinetic energy, the interatomic distance between the atoms within the molecule is continuously changing with time, the average inter-atomic distance increasing with increasing temperature. The molecular coiling due to the motion of the individual atoms within the chain also increases with increasing temperature.

Increasing the Strength of Polymers

We are now able to view the overall bonding characteristics of polymer molecules. Within the long chain of the molecule, bonding is covalent. These primary chemical bonds are of high stability and strength, and rupture of the bonds within the molecule requires large amounts of energy (see Table 1.4). Interactions with neighboring molecules involve significantly weaker forces, and failure of a polymer may be more likely by separation of neighboring molecules rather than by rupture of the molecule itself. However, the strength of the attraction between the neighboring molecules can be increased in several ways. Starting with polyethylene (PE) with a tensile strength of 800–3000 psi as a reference, the intermolecular forces can be increased as follows:

Increasing Molecular Length

Increasing the number of intermolecular attractions increases the total force of attraction. The mechanical strength of a polymer generally increases with increasing DP, because mechanical failure is associated with the separation of molecules from one another, which requires breaking the attractive forces between all the atoms within the molecules. However, increasing the DP indefinitely does not continue to produce increased strength. The chain may become so long that individual parts of the chain may behave independently, with chain separation occurring in isolated regions. In addition, as the chain length increases, the strength of the large number of secondary bonds between molecules eventually approaches that of the covalent bonds within the chain, and chain breakage becomes a source of failure. Figue 1.6 shows the general dependence of strength, impact strength, and modulus of elasticity on molecular length. With increasing molecular weight these properties approach a value independent of increased chain length for some polymers, whereas in others the properties tend to continue to increase slightly with increasing length up to the limits of DP available for experi-mentally prepared samples. This typical variation can be given by an equation of the form:

$$\frac{1}{\text{property}} = a + \frac{b}{\text{DP}}$$

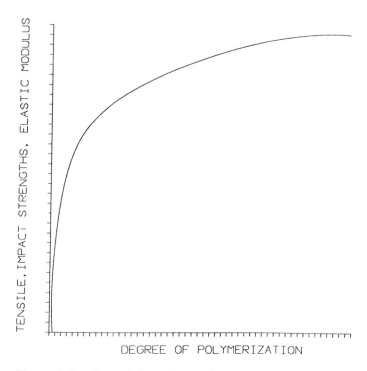

Figure 1.6. General dependence of mechanical properties on molecular chain length.

where a and b are constants that depend on the specific polymer and the property under consideration.

Adding Polar Atoms

Polyvinyl chloride (PVC) is structurally related to PE by replacing every other hydrogen by a chlorine atom:

$$
\begin{array}{ccccccc}
\text{H} & \text{Cl} & \text{H} & \text{Cl} & \text{H} & \text{Cl} & \text{H} \\
| & | & | & | & | & | & | \\
-\text{C} & -\text{C} & -\text{C} & -\text{C} & -\text{C} & -\text{C} & -\text{C}- \\
| & | & | & | & | & | & | \\
\text{H} & \text{H} & \text{H} & \text{H} & \text{H} & \text{H} & \text{H}
\end{array}
$$

This replacement increases the induced dipole attraction between neighboring molecules and also somewhat stiffens the chain, because of the chlorine atom's larger size. As a result, the tensile strength of PVC is 7000–9000 psi, greater than that of PE by a factor of 3.

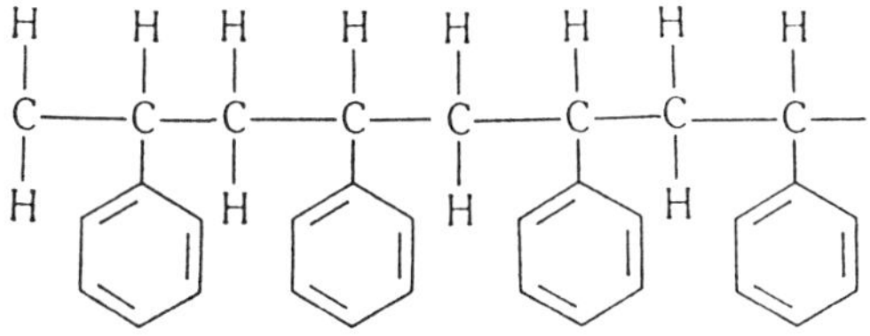

Figure 1.7. Molecular structure of polystyrene.

Steric (Spatial) Hindrance

The molecular twisting and rotations produced by thermal vibrations of the atoms produce a flexible polymer. The hydrogen atoms on the chain molecules of polyethylene are very small compared with the carbon atoms, and motion of the chains is controlled primarily by the flexing of the C—C bonds. Other atoms which may replace the hydrogen, such as Cl, are much larger in atomic radius, and their attachment to the chain may produce some interference with the chain motion. If large bulky groups of atoms replace the hydrogen, significant restriction in the motion of the chains will occur, similarly restricting the flexibility of the polymer. If the molecules are made stiffer, they are more likely to align closely with neighboring molecules.

The common model for the molecular configuration of the flexible plastic molecules is a bowl of spaghetti. To approximate the correct molecular dimensions, the spaghetti strands should be 200 times longer than those usually eaten. An even better model, to include the effect of thermal vibrations, is a bowl of wriggling worms! A model for stiff plastic molecules is a collection of pencils in a box. Obviously, the stiff pencils would tend to lie much closer to each other than the wriggling worms. The secondary forces decrease inversely with the sixth power of interatomic distance, and therefore such alignment and decrease of interatomic distance would greatly increase the intermolecular forces and the strength of the plastic.

Structurally, polystyrene (PS) can be related to PE by replacing every other hydrogen by a benzene ring (see Fig. 1.7), producing a very stiff molecule. As a result of this molecular stiffness, PS has a tensile strength in the range 6000–8000 psi.

1.5. POLYMERIZATION

Polymerization is the process by which the long-chain molecules are produced. Two main categories into which plastics are customarily divided are thermoplastics and thermosets.

Thermoplastics

As described previously, the long-chain molecules are bonded by primary covalent bonds within the chain. The individual chains are attracted to each other by secondary forces between the dipoles induced or present on the individual chains. The molecules are therefore capable of individual motion, and on heating the solid polymer can melt, becoming sufficiently free flowing to permit mold filling. The ability to manufacture parts by a molding process makes thermoplastics the most commonly used type of polymer.

Thermosets

If primary covalent bonds are established between all the chains, the entire sample becomes one molecule. In many thermosets, the structure can be visualized by considering the material as still composed of separate chain molecules with a small number of primary bonds interconnecting these chain molecules. These few covalent bonds between the chain now add to the induced dipoles to strengthen the intermolecular forces. In other thermosets, the number of interconnections is so large that the identification of any long chain becomes impossible. Since the covalent bonds cannot be broken without degrading the properties of the polymer, these materials do not melt on heating, but they char and decompose if heated excessively. The covalent bonds increase the strength of these polymers, but the ease of manufacture by melting and rapid mold filling with molten plastic is lost.

Monomers

Polymers are manufactured starting with small organic molecules, which can be gases, liquids, or solutions of the starting molecules dissolved in a suitable solvent. These starting small molecules are the monomers. Although the monomer is chemically similar to the mer (the basic structural unit of the polymer), chemical reactions occur to generate the solid polymer from the monomer. These reactions modify the monomer, so that the monomer is not identical to the mer.

1.6. ADDITION OR STEPWISE POLYMERIZATION

The addition polymerization process most commonly utilizes a monomer which has a double bond between two carbon atoms. Bonds are generated by electron overlap between neighboring bonded atoms, and, as discussed previously, the electrons in the π-bond do not overlap as fully as the other bonding electrons. A double bond, although stronger than a single bond, can therefore be decomposed to a single bond by various mechanisms.

Initiators

As discussed previously, a double bond between two carbon atoms is relatively reactive and can be broken by attack with a reactive unpaired electron. For the double bond to be attacked, it must be relatively exposed to approach. A double bond at the end of a molecule is therefore easily attacked, whereas a double bond within a molecule is screened from exposure and is less likely to be chemically degraded. The double bond illustrated below can easily be approached by a reactive atom;

$$CH_2\!\!=\!\!CH_2\!\!-\!\!CH_2\!\!-\!\!CH_3$$

The double bond illustrated next is less likely to be unaffected by a reactive atom:

$$CH_3\!\!-\!\!CH\!\!=\!\!CH\!\!-\!\!CH_2\!\!-\!\!CH_3$$

An initiator is a molecule which itself is rather unstable and decomposes into smaller parts containing unpaired electrons. A part of a molecule containing an unpaired electron is called a *free radical*. For example, hydrogen peroxide (H_2O_2 or $H\!\!-\!\!O\!\!-\!\!O\!\!-\!\!H$) contains the rather unstable oxygen–oxygen bond. This bond decomposes readily to generate two free radicals ($H\!\!-\!\!O\!\cdot$).

$$H_2O_2 = 2(H\!\!-\!\!O\!\cdot)$$

The unpaired electron reacts with the double bond in ethylene ($CH_2\!\!=\!\!CH_2$), breaking it and bonding to one electron of the pair:

$$H\!\!-\!\!O\!\cdot + CH_2\!\!=\!\!CH_2 \rightarrow \cdot CH_2\!\!-\!\!CH_2\!\!-\!\!OH$$

The product now contains an electron which is unpaired, and this reaction product is itself a free radical that can further attack another ethylene molecule, decomposing its double bond and generating a longer chain which is also a reactive free radical:

$$CH_2\!\!=\!\!CH_2 + \cdot CH_2\!\!-\!\!CH_2\!\!-\!\!OH \rightarrow \cdot CH_2\!\!-\!\!CH_2\!\!-\!\!CH_2\!\!-\!\!CH_2\!\!-\!\!OH$$

As this product is still a free radical, the process continues, with the chain continually growing in length and possibly forming very large molecules.

Hydrogen peroxide is rather a strong initiator, decomposing too readily for long-term storage. The two most common initiators are benzoyl peroxide and azobisisobutyronitrile:

Benzoyl peroxide:

$$\bigcirc - \overset{\overset{O}{\|}}{C} - O - O - \overset{\overset{O}{\|}}{C} - \bigcirc \longrightarrow 2 \left[\bigcirc - \overset{\overset{O}{\|}}{C} - O \cdot \right]$$

Azobisisobutyronitrile:

$$\begin{array}{ccc} CH_3 & CH_3 \\ \backslash & | \\ CH_3 - C - N = N - C - CH_3 \\ / & | \\ CN & CN \end{array} \longrightarrow 2 \left[\begin{array}{c} CH_3 \\ | \\ CH_3 - C \cdot \\ | \\ CN \end{array} \right] + N_2$$

Reaction Rate of Addition Polymerization

The simplest example of polymerization is the formation of polyethylene. The polymerization reaction was shown in the previous section. The starting material, ethylene ($CH_2{=}CH_2$), is a gas at the reaction temperature. Several different processes exist, operating at different temperatures and pressures and utilizing different catalysts. The reaction does not generally go to completion. In the high-pressure process, less than 30% of the monomer is reacted, the unreacted monomer being removed and recycled. The reaction is exothermic, because weaker π-bonds are converted to stronger σ-bonds, and cooling must be provided to maintain proper conditions and to prevent safety hazards. Different processing parameters generate polymers with different chain length distributions, average chain lengths, and number and length of side chains.

If we wish to generalize the addition polymerization process, the reaction can be considered first to involve the decomposition of the initiator into two free radicals. The resultant free radical is designated by the symbol R·. The second step in the reaction is the approach of the free radical to a double bond and the separation of the double bond, forming a longer free radical. This process continues by breaking other double bonds to generate the long-chain molecule. The steps are outlined below, where R· represents half of the initiator molecule and M represents a CH_2 group:

1. *Initiation*: The initiator decomposes to produce free radicals:

    $$R{\cdot}{\cdot}R = 2R{\cdot}$$

2. *Propagation*: Reaction proceeds between the free radicals and available double bonds. The result of the reaction is an increased chain length, the resultant chain itself being a free radical, capable of further reaction.

$$R\cdot + CH_2{=}CH_2 \rightarrow R{-}CH_2{-}CH_2\cdot$$

$$R{-}CH_2{-}CH_2\cdot + CH_2{=}CH_2 \rightarrow R{-}CH_2{-}CH_2{-}CH_2{-}CH_2\cdot$$

etc.

3. *Chain termination*: In principle, it might seem that any actively growing chain would continue to grow until all the monomer molecules had reacted. However, mechanisms for growth termination exist. The principal ones are:

   a. *Combination.* Two growing chains whose active ends come in contact will generate a stable single bond, combining the unpaired electrons on both free radicals and producing a longer stable molecule.

$$R{-}M{-}M{-}M{-}M{-}M\cdot + \cdot M{-}M{-}M{-}M{-}R$$
$$\rightarrow R{-}M{-}M{-}M{-}M{-}M{-}M{-}M{-}M{-}M{-}R$$

   b. *Transfer.* The unpaired reactive electron may transfer from one molecule to another, ending the growth of that molecule.

$$R{-}M{-}M{-}M\cdot + M{-}M \rightarrow R{-}M{-}M{-}M{=}M + M\cdot$$

   c. *Disproportionation.* The unpaired reactive electron transfers to another growing chain, re-forming a double bond on that molecule and stabilizing both molecules.

$$R_1{-}CH_2{-}CH_2\cdot + R_2{-}CH_2{-}CH_2\cdot$$
$$\rightarrow CH_2{=}CH_2 + R_2{-}CH_2{-}CH_2{-}R_1$$

As a result of the random termination of chain propagation, polymer products always contain a range of chain lengths. Special processing may minimize this variation, but if monodisperse samples (single chain length molecules) are desired, some fractionation process must be used to isolate selectively the molecules with the desired chain size.

Effect of Initiator Concentration

To give some picture of the effect on the rate of reaction of the concentration of initiator, let us consider a polymerization in which termination is primarily by combination. Since the reaction of a free radical and a double bond is very rapid, the rate of formation of new chains is proportional to the number of chains initiated. This, in turn, is dependent on the concentration of the initiator.

The rate of chemical reactions is proportional to the concentration of reacting species. If

V_{init} is the rate of initiation of new polymer chains,
K_{init} is the rate constant for the initiation, and
[R] and [M] are the concentrations of initiator and monomer, respectively,

then:

$$V_{init} = K_{init}[R][M]$$

The most common termination process is the interaction of two growing chains. The rate of termination (V_{term}) for this process is given by

$$V_{term} = K_{term}[M\cdot]^2$$

where K_{term} = rate constant for the termination reaction
$[M\cdot]$ = concentration of growing polymer chains
V_{term} = rate of termination

Since two chains must collide to produce a termination, the rate is proportional to the square of the growing chain concentration.

At steady state the rates of initiation and termination are equal and

$$[M\cdot] = \left\{ \left(\frac{K_{init}}{K_{term}} \right)([R][M]) \right\}^{1/2}$$

The concentration and rate of formation of growing chains during the polymerization are proportional to the square root of the initiator concentration. The exact dependence on monomer concentration has been found to vary with the solvent and the exact mechanism of initiation.

The rate at which monomer molecules attach to a growing chain is given by the reaction

$$M\cdot + M = M\!\!-\!\!M\cdot$$

The velocity of this propagation reaction (V_{prop}) is the number of monomer molecules added to growing chain per second.

$$V_{prop} = K_{prop}[M][M\cdot] = K_{prop}\left(\frac{K_{init}[R]}{K_{term}} \right)^{0.5} [M]^{1.5}$$

Since

$$V_{init} = \text{number of chains formed per second}$$

$$V_{prop} = \text{number of monomers added per second}$$

then

$$DP = \frac{\text{number of monomer molecules added}}{\text{number of chains formed}}$$

$$= \frac{V_{prop}}{V_{term}}$$

$$= \frac{K_{prop}(K_{init}[R]/K_{term})^{0.5}[M]^{1.5}}{(K_{init}[R][M])}$$

$$= K_{prop}\left\{\left(\frac{1}{K_{init}K_{term}}\right)\left(\frac{[M]}{[R]}\right)\right\}^{0.5}$$

This equation demonstrates that the degree of polymerization is inversely proportional to the square root of the initiator concentration. Since increased molecular length generally improves mechanical properties, the initiator concentration should be kept as low as practical to have reasonable reaction times.

Table 1.5 gives examples of common plastics produced by stepwise polymerization. In summary, the general characteristics of the addition polymerization mechanism are:

1. Termination of a particular chain's growth is random, and a distribution of chain lengths must always exist in any plastic product. When the DP or some other measure of molecular size is specified, it therefore can be only an average value for the sample.
2. The growth is by addition to the end of the growing chains. Only about 10^{-8} of the total amount of material is therefore reacting at any time.
3. The monomer concentration in the material decreases slowly throughout the process. In commercial manufacture of polyethylene, only about 30% of the starting monomer is reacted. The polymerized material is separated and the unreacted monomer recycled. In small-batch polymerization processes, such as in the fiberglass processing, a small amount of unreacted monomer may still exist in the final product, because a free radical may never encounter all the monomer molecules.

Table 1.5. Examples of Stepwise Polymerization Products

Name	Monomer	Mer	Polymer
Polyethylene (PE)	H H │ │ C═C │ │ H H	H │ —C— │ H	H H H H H H │ │ │ │ │ │ —C—C—C—C—C—C— │ │ │ │ │ │ H H H H H H
Polyvinyl chloride (PVC)	H Cl │ │ C═C │ │ H H	H Cl │ │ C—C │ │ H H	H Cl H Cl H Cl │ │ │ │ │ │ —C—C—C—C—C—C— │ │ │ │ │ │ H H H H H H
Polystyrene (PS)	H H │ │ C═C │ H	H H │ │ C—C │ H	H H H H H H │ │ │ │ │ │ C—C—C—C—C—C— │ │ │ H H H
Polyvinyl fluoride (PVF)	H F │ │ C═C │ │ H H	H F │ │ C—C │ │ H H	H F H F H F │ │ │ │ │ │ —C—C—C—C—C—C— │ │ │ │ │ │ H H H H H H
Polytetrafluoroethylene (PTFE)	F F │ │ C═C │ │ F F	F │ —C— │ F	F F F F F F │ │ │ │ │ │ C—C—C—C—C—C │ │ │ │ │ │ F F F F F F

4. The unpaired electron is very unstable and growth of the chains is very rapid, so large chains form immediately. As a result, a longer reaction time increases the yield of the reaction, but the average chain length of the grown chains is relatively independent of time.

5. The rate constants vary exponentially with temperature. Increasing the temperature by 10°C tends to increase the reaction rate by a factor of 2.

6. The catalyst concentration affects the final DP, with larger concentrations producing more active chains, resulting in a smaller average DP.

7. Chain termination may also occur by contact with impurities. Cleanliness of materials is therefore important.

8. The rate of polymerization can be decreased by utilizing less catalyst or by diluting the monomer by dissolving it in a nonreactive dilutent.

Inhibitors and Retarders

Addition polymerization may proceed spontaneously because of bond breakage as a result of thermal vibrations or the presence of undesired components or impurities which generate free radicals. To lengthen the shelf life, chemicals are added which will react with any undesired free radicals that form within the monomer. These additives tend to generate free radicals which will be insufficiently active to initiate chain propagation but are capable of bonding and destroying any unwanted chain-initiating free radicals that form randomly. The difference between inhibitors and retarders is largely one of degree. Inhibitors tend to be sufficiently effective that they completely suppress any polymerization. Eventually they are used up by combination with reactive components, and then chain propagation by the undesired processes proceeds at its normal rate. This produces an incubation period prior to polymerization. Retarders are less effective but slow down the polymerization rate continuously. Figure 1.8 shows the effect of these additions on the rate of polymerization.

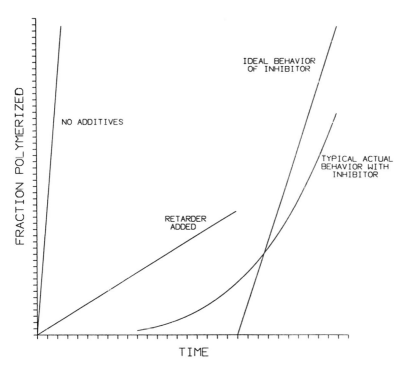

Figure 1.8.   Effect of inhibitors and retarders on the rate of polymerization.

Oxygen is a powerful inhibitor which reacts rapidly with free radicals, and the surfaces of fiberglass–polyester products frequently do not polymerize due to contact with the oxygen in air. To prevent this inhibition, wax may be added to the starting materials. During polymerization, the wax floats to the surface and prevents oxygen contact. The partially reacted surface may also be washed away with acetone or another solvent which will not affect the underlying polymerized material.

Other Stepwise Mechanisms

Mechanisms other than the free radical mechanism are also utilized. Ionic compounds such as BF_3 and $AlCl_3$ are also used. In a greatly oversimplified fashion, the mechanism can be considered to consist of the charged ions of the compound separating, attracting the electrons in the double bond, and decomposing it. The other part of the ionic compound also remains attracted to the growing chain. Other methods involve anionic or cationic polymerization with organometallic and acid initiators, respectively. Termination is never by combination, and in many cases the termination reactions are totally absent, resulting in much more uniform molecular lengths. Since the molecules are always theoretically available for further growth, these polymers are also called living polymers.

1.7. CONDENSATION POLYMERIZATION

Condensation polymerization involves the reaction of two different small organic molecules, each of which has a reactive or functional group capable of reacting with the functional group on the other type of monomer molecule. Unlike addition polymerization, the reaction of the two functional groups involves the formation of a small molecule such as water or nitrogen which must be removed from the reaction mass. In an addition reaction process, no extraneous molecules are produced.

Another name for condensation polymerization is step polymerization, to emphasize the method by which the reaction proceeds. For a condensation polymerization to proceed, there must be reactive centers (functional groups) on each of the molecules which react. There are always at least two different types of functional groups which react and join the two starting molecules together.

Polyesters

As an example of a condensation reaction, an organic acid and an organic base (alcohol) react in a manner similar to an inorganic acid and base to generate water and a new molecule. This molecular result of the reaction of an acid and alcohol in organic chemistry is called an ester. The water produced in this reaction must be removed during the reaction processing, either by applying a vacuum or by heating the reaction mass. The formation of an ester is the organic equivalent of neutralization of an acid and a base:

Acid + base → salt + water

$HCl + NaOH = Na^+Cl^- + H_2O$

alcohol + acid⟶ester + water

$$\underset{\overset{|}{H}}{\overset{\overset{H}{|}}{H-C-OH}} + \underset{}{\overset{\overset{O}{\parallel}}{H-C-O-H}} \longrightarrow \underset{\overset{|}{H}}{\overset{\overset{H}{|}}{H-C-O}} - \overset{\overset{O}{\parallel}}{C} - H \ + \ H_2O$$

The resultant ester is a longer molecule than the starting materials but is still small. To generate a large molecule, this process must be performed repeatedly. Such repetition is possible if each of the monomers has two functional groups:

diacid + dialcohol ⟶ still reactive product

$$\underset{\overset{|}{H}}{\overset{\overset{O}{\parallel}}{HO-C}} - \underset{}{\overset{\overset{O}{\parallel}}{C}} - C - OH + \underset{\overset{H\ H}{}}{\overset{H\ H}{HO-C-C-OH}} \rightarrow \ \underset{\overset{H}{}}{\overset{\overset{O}{\parallel}}{HO-C}} - \underset{}{\overset{\overset{O}{\parallel}}{C}} - \underset{\overset{H}{}}{\overset{\overset{}{}}{C}} - O - \overset{\overset{H\ H}{}}{C} - C - OH$$

       diacid   dialcohol   still reactive   still
                       (glycol)   reactive

Note that the small polyester molecules that are produced by this rection still have functional groups (OH or COOH) which are capable of reacting with another diacid or dialcohol to continue the polymerization. This process can continue indefinitely to produce a linear polyester.

The two different functional groups may be on the same molecule:

$$\underset{\overset{|}{H}}{\overset{\overset{H}{|}}{H-O-C}} - \overset{\overset{O}{\parallel}}{C} - OH$$

Polyesters can also be generated by the reaction of a glycol (a molecule with two OH groups) and an ester:

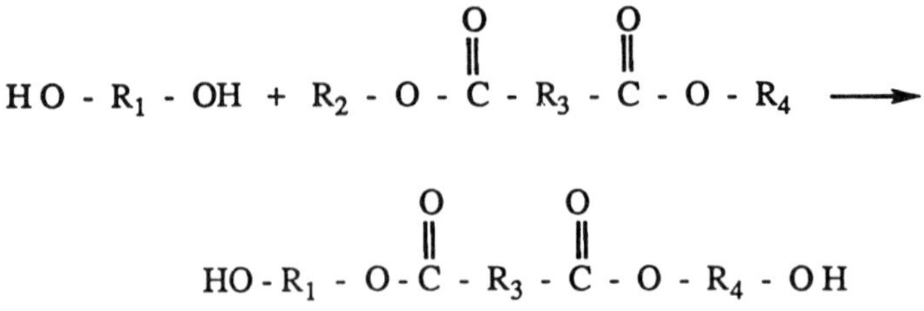

$$HO - R_1 - OH + R_2 - O - \overset{\overset{\textstyle O}{\|}}{C} - R_3 - \overset{\overset{\textstyle O}{\|}}{C} - O - R_4 \longrightarrow$$

$$HO - R_1 - O - \overset{\overset{\textstyle O}{\|}}{C} - R_3 - \overset{\overset{\textstyle O}{\|}}{C} - O - R_4 - OH$$

R_1 TO R_4 = any non-reacting group

Polyamides

The condensation polymerization which generates the nylon family of plastics is that between an acid and an amine. As in the production of polyesters, to generate a long-chain molecule both monomers must be bifunctional. Water is generated during the reaction and must be removed from the final product.

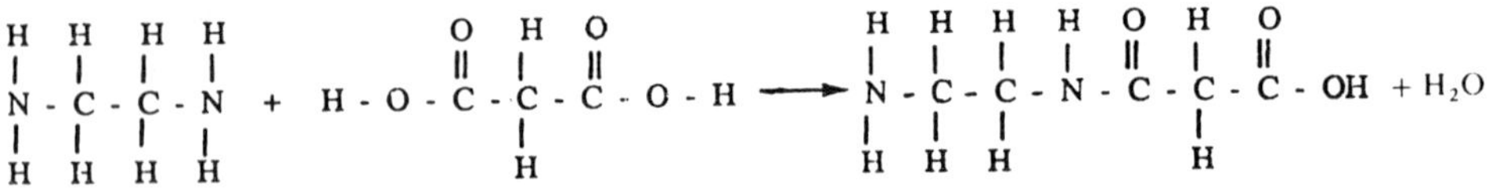

Many other condensation reactions are possible, all following the general concepts outlined above. It is not necessary to restrict this process to linear polymers. Thermosetting polymers may be produced by starting with trifunctional molecules:

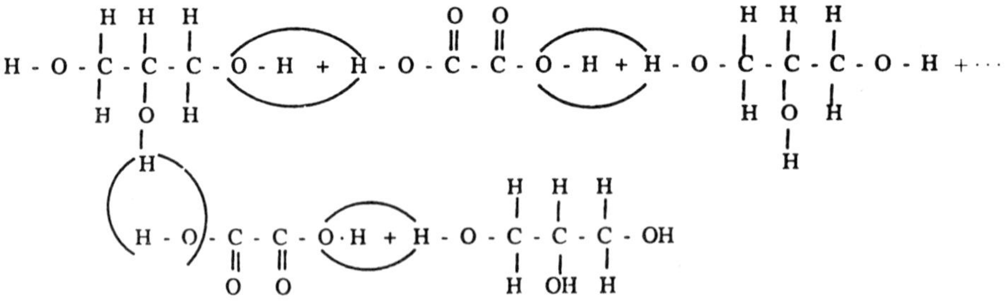

General Characteristics of Condensation Polymerization

In addition polymerization, only the growing chains are reactive. In condensation polymerization all molecules contain functional groups and are active. As a result, the monomer molecules react initially, and all the monomer tends to disappear at an early stage, being replaced by somewhat longer-chain molecules. The chain length therefore increases gradually as condensation proceeds. To generate an average chain length sufficiently great to produce satisfactory mechanical properties, the reaction must proceed to near completion. Long reaction times and increased temperatures are therefore common for the condensation polymerization process.

Kinetics of Condensation Polymerization

As an example, let us consider the production of a polyester. The reaction that proceeds is that of an OH group with a COOH group. Experimentally it is found that this reaction is catalyzed by acids: the greater the acid concentration, the greater the rate of reaction. With this experimental information, the rate of reaction will be proportional to the concentration of reacting groups and a rate constant, which will be also proportional to the concentration of acid. Since the rate of reaction equals the rate of loss of acid molecules,

$$\frac{d[\text{COOH}]}{dt} = -k[\text{OH}][\text{COOH}]$$

where [ ] represents the concentration of the component.

Since the reaction constant is proportional to the acid concentration,

$$k = k_1[\text{COOH}]$$

Then

$$\frac{d[\text{COOH}]}{dt} = -k_1[\text{OH}][\text{COOH}]^2$$

Assuming that the reaction starts with an equal number of acid and alcohol functional groups, then any time [OH] = [COOH] and

$$\frac{d[\text{COOH}]}{dt} = -k_1[\text{COOH}]^3$$

If the concentration of acid molecules at the start of the reaction is C_0, then integrating this equation from the start of the reaction at $t = 0$ to any time t:

$$\frac{1}{[\text{COOH}]^2} - \frac{1}{C_0^2} = 2k_1 t$$

The concentration of acid reactive groups therefore varies with the inverse of the square root of the reaction time.

Extent of Reaction with Time

To measure the total extent of reaction, let p be the fraction of functional groups reacted:

$$p = \frac{(C_0 - [COOH])}{C_0}$$

or:

$$[COOH] = C_0(1 - p)$$

Note that when $p = 0$ the concentration of acid is its initial value, and when $p = 1$ the reaction is complete and the concentration of acid has fallen to zero. Substituting this into the equation above:

$$\frac{1}{[C_0]^2(1 - p)^2} - \frac{1}{C_0^2} = 2k_1t$$

$$2k_1tC_0^2 = \frac{1}{(1 - p)^2} - 1$$

or

$$2p^2k_1C_0^2t = 1 - (1 - p)^2$$

$$p^2[2k_1C_0^2t + 1] - 2p[2k_1C_0^2t + 1] + 2k_1C_0^2t = 0$$

and solving for p:

$$p = 1 - (2k_1C_0^2t + 1)^{-0.5}$$

A plot showing the general dependence of the extent of completion of the condensation polymerization versus time is shown in Fig. 1.9. The curve indicates a rapid initial increase in polymerization, but long times are required for the reaction to proceed to the stage where the material is principally reaction product.

An approximation to the degree of polymerization can be obtained by relating the DP to the ratio of the original number of molecules of acid to the number present at time t. Although this assumes all molecules have equal length, it should still give a useful average molecular length.

$$DP = \frac{C_0}{C}$$

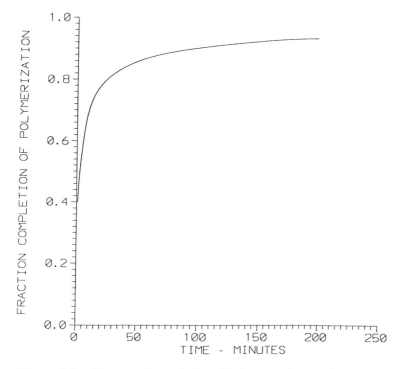

Figure 1.9. Fraction of completion of polymerization vs. time.

since

$$p = 1 - \frac{C}{C_0}$$

$$DP = \frac{1}{(1 - p)}$$

The variation of DP with fractional extent of reaction p is shown in Fig. 1.10.

This illustrates the requirements for almost complete reaction in condensation polymerization. If the reaction is 97% complete, the DP is only 33. Reaction completion of 99.9% is required to obtain a DP of 1000. Note the difference from the stepwise reaction, in which long chains are obtained at small fractions of reaction completion, with the unused monomer being recycled.

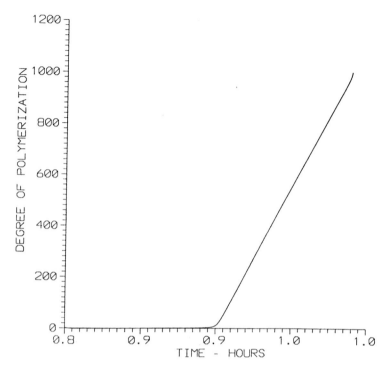

Figure 1.10. Variation of degree of polymerization with time for a condensation reaction.

Thermoset Formation

Consider molecules which have more than two functional groups per monomer. The result of the polymerization can be a thermoset.

Let:

f = number of functional groups per molecule

N_0 = number of monomer molecules initially present

N = total number of monomer molecules present at time t

Then:

$N_0 f$ = number of functional groups initially present

$N f$ = number of functional groups present at time t

p = number of functional groups reacted/number of functional groups originally present

Since two functional groups are lost for each bond formed, the fraction of functional groups lost or the fractional completion of reaction is

$$p = \frac{2(N_0 - N)}{N_0 f} = \frac{2}{f} - \left(\frac{2}{f}\right)\left(\frac{N}{N_0}\right) = \frac{2}{f} - \frac{2}{(\mathrm{DP} \cdot f)}$$

and

$$\mathrm{DP} = \frac{2}{(2 - fp)}$$

If the starting material has more than two functional groups, complete reaction is not necessary to produce a thermoset material (an infinite DP). Interesting results can be obtained by proper choice of the starting monomers. If the starting material consists of 60% bifunctional and 40% trifunctional monomers, then

$$f = 0.6 \cdot 2 + 0.4 \cdot 3 = 2.4$$

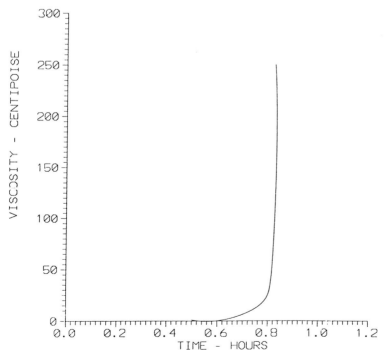

Figure 1.11. Viscosity of a polymerizing solution vs. time.

and the degree of polymerization becomes infinite when $p = 0.83$. For a trifunctional material ($f = 3$), only 66% completion of reaction is necessary. For quadrifunctional monomers, only 50% completion is necessary.

Because, as shown in the previous section, the chain length does not become large until the extent of completion of the reaction is very large, it is possible to permit partial reaction to occur with no danger of an insoluble mass forming. One useful technique is to choose a catalyst which accentuates the reaction of the end groups of the trifunctional molecule, with the third functional group in the center of the molecule unreacted, so that linear molecules can be produced. The relatively small molecules are in the form of a viscous but spreadable polymer which can then be further reacted to produce a thermoset when ready. The viscosity of a solution in which polymerization is occurring increases as the DP increases. Figure 1.11 shows the viscosity as a function of reaction completion for this mixture. The molecular weight remains relatively low for extended periods of time and then suddenly increases. The point of rapidly increasing molecular weight produces large increases in viscosity, and this point is termed the *gelation point*. It is possible to stop the reaction prior to this point to obtain a low-viscosity liquid which can be readily painted or spread on a fibrous material and then permit the reaction to proceed to form a permanent thermoset molded product. Figure 1.11 shows the incubation period in which the solution remains relatively fluid, followed by the sudden increase in viscosity when gelation begins.

2
Molecular Configurations

2.1. MOLECULAR CONFIGURATIONS

Isomerism

Although the mer gives the arrangement and types of atoms within the long-chain molecule, due to the tetrahedral orientation of the four covalent bonds that carbon forms, different spatial configurations are possible with the same chemical formula. Such different forms are called *stereoisomers*. On a two-dimensional page, the forms of polyvinyl chloride (PVC) are represented as follows:

Isotactic (Cl atoms all on the same side of the C atoms):

$$
\begin{array}{cccccc}
\text{H} & \text{H} & \text{H} & \text{H} & \text{H} & \text{H} \\
| & | & | & | & | & | \\
-\text{C}-\text{C}-\text{C}-\text{C}-\text{C}-\text{C} \\
| & | & | & | & | & | \\
\text{H} & \text{Cl} & \text{H} & \text{Cl} & \text{H} & \text{Cl}
\end{array}
$$

Syndiotactic (Cl atoms on alternate sides of the C atoms):

$$
\begin{array}{cccccccc}
\text{H} & \text{H} & \text{H} & \text{Cl} & \text{H} & \text{H} & \text{H} & \text{Cl} \\
| & | & | & | & | & | & | & | \\
-\text{C}-\text{C}-\text{C}-\text{C}-\text{C}-\text{C}-\text{C}-\text{C}- \\
| & | & | & | & | & | & | & | \\
\text{H} & \text{Cl} & \text{H} & \text{H} & \text{H} & \text{Cl} & \text{H} & \text{H}
\end{array}
$$

Atactic (Cl atoms randomly arranged on either side of the C atom):

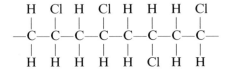

Strong attractions between neighboring atoms on different molecules require close alignment of the molecules. Any irregularity in the structure prevents such alignment and weakens the product. For polypropylene (PP), as an example, the atactic polymer is commercially uninteresting, having low strength as a result of the inability of the molecules to align closely to promote strong intermolecular attractive forces. The greater the fraction of the polymer which is isotactic, the greater the mechanical properties. Figure 2.1 shows the variation of strength with *isotactic index*. This is one measure of the fraction of the material which is isotactic. It is experimentally determined by measuring the fraction of the polymer soluble in *n*-heptane.

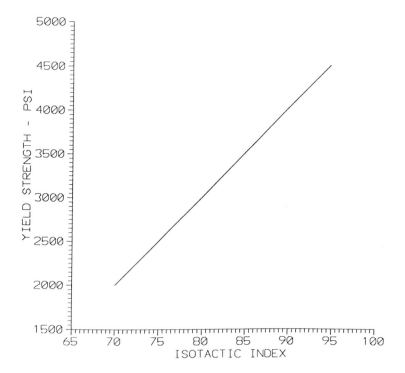

Figure 2.1. Variation of yield strength of PP with isotactic index.

The atactic molecules cannot align as easily and tend to form amorphous disordered regions. The isotactic portion of the sample tends to crystallize. This crystalline portion is more tightly bound and is impervious to attack by the solvent. The amorphous atactic regions are less strongly bound and have empty or free space available to infiltration by the solvent, and they are therefore soluble. Although other variations in molecular configurations affect solubility, the isotactic index therefore gives a reasonable value of the fraction of the isotactic isomer in the sample.

Branching

Whereas the simplified discussion of polymerization indicates that only long chains are produced, the free radicals may attack and remove hydrogen atoms along the chain, producing side chains or branches at irregular intervals. In polyethylene, the occasional side chains hinder perfect alignment and resultant crystallization. Two general classes of polyethylene are low density (LDPE) and high density (HDPE). The principal difference is that LDPE has more small side chains (two to six carbon atoms long), which hinder crystallization, resulting in a product approximately 60% crystalline. HDPE has far fewer side chains and is therefore 95% crystalline. This produces a significant difference in the density of the two products: LDPE has a density of 0.916 g/cm^3, whereas HDPE has a density of 0.97 g/cm^3.

Copolymers

The development cost for new polymers is becoming prohibitively high. In copolymerization, utilizing known polymers two different monomers are polymerized into one chain molecule. The long chain therefore contains two or more different mers. These mers can be arranged on the chain in various configurations. The properties generated by these molecular designs will be discussed later, in the section on specific plastics. Various configurations are possible:

1. *Random copolymer*: The different mers are arranged at random in the molecule.
2. *Alternating copolymer*: The differing mers alternate along the molecule.
3. *Block copolymer*: Blocks of one mer exist along the molecule. The lengths of the blocks of the various components are not the same, and the lengths of all the blocks of one mer are also somewhat randomized.
4. *Graft copolymer*: The long chain consists of one mer, with long side chains of a different mer. Graft copolymers are particularly useful as interfacial binding materials in a two-phase material, capable of

having the main chain dissolve in one material and the graft dissolve in an immiscible second phase.

Blends

Copolymers contain different mers that are covalently bonded within the molecule, and require polymerization practice different from that applied for the individual polymers. Blends are mixtures of independently polymerized plastics. These are then mechanically mixed, resulting in the individual chain molecules containing only one mer type. Polymers tend to be relatively immiscible, and most of the blends are two phase, with small regions of each polymer. The most widely used blends are those containing mixtures of rubbery particles and a stronger plastic, incorporating the strength of the stronger phase and the crack propagation resistance and impact resistance of the rubbery phase. Combinations of mechanical properties unavailable in a single plastic can be obtained by this technique.

However, the properties of the resultant mixture can be very poor if the two polymers are totally incompatible, and a sharp interface occurs between the two materials, producing an easy path for crack propagation. Other problems can arise, such as easy diffusion of vapor and solvents along these boundaries. For a blend to be useful, either of two conditions must apply:

1. The polymers must be so similar that the interaction between the different types of molecules tend to produce an overall chemical attraction. In the process of mixing of small molecules, some repulsion may exist and a solution may still form, as the increase in entropy (disorder) when the solution forms overcomes the repulsion between atoms. For polymers, the entropy of mixing of the two materials is low due to the small number of molecules present (as a result of the large molecular size), and the enthalpic term resulting from the intermolecular attractions or repulsions is the controlling factor. As a result very few commercially available blends are single phase.
2. The polymers may form two phases, and some mechanism for producing strong interfacial bonding between the two materials exists. This is accomplished by utilizing graft or copolymers, so that some tie molecules can protrude between the two phases, providing strong attractive forces. In most blends, the morphology of the two phases is such that one phase is the background matrix and the other phase exists as discrete globules within the matrix phase. The size of the globules and their shape affect the mechanical properties, and different processing conditions may produce quite different resultant properties by changing the configuration of the two phases.

Interpenetrating Networks

It is possible that a blend of two polymers has a structure in which both phases are continuous. The common analogy is that of a wet sponge, in which the sponge material and the water are both continuous and interpenetrate each other. The usual mechanism for generating an inter-penetrating network polymer blend is to cross-link one of the polymer constituents while it is mechanically mixed with the other. Interpenetrating network polymers are therefore by definition thermosetting.

2.2. MOLECULAR WEIGHT

The molecular weight of a molecule with a clearly defined chemical formula can be readily computed. Polymers, however, have an assortment of molecules with widely different molecular lengths, and, as indicated above, some variations in chemical structure may occur from one molecule to another.

A specific molecular weight cannot therefore be assigned to a polymer, since a distribution of molecular lengths and weights occurs. A typical distribution is shown in Fig. 2.2. The distribution affects the mechanical and physical properties of the polymer. Two samples with the same average chain length but with different distributions around that average will have significantly different properties. Some evaluation of the distribution is therefore necessary. Experimental procedures for determining the distribution are discussed in the next section. However, complete molecular weight distribution analysis is very complex, time consuming, and expensive. As a result, it is much more common to obtain several average values which will give some information about the distribution.

Average values are experimentally calculated from physical property measurements. Some properties are a function of the average length of the molecules and others a function of the average weight of the molecules; the viscosity of a solution of a polymer is a complex function of both. Therefore, three common average values are reported.

Number Average Molecular Weight

The number average molecular weight corresponds to the usual definition of average of any quantity. The fractions of molecules with a given molecular weight are summed:

$$M_n = \frac{\sum n_x M_x}{\sum n_x}$$

where n_x = number of molecules having molecular weight M_x.

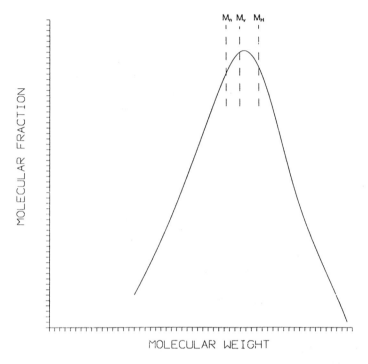

Figure 2.2. Molecular weight distribution and molecular weight averages.

Weight Average Molecular Weight

In this average, the weight fraction of molecules with a given molecular weight is summed:

$$M_w = \frac{\sum w_x M_x}{\sum w_x}$$

where w_x = weight of molecules having molecular weight M_x.

Viscosity Average Molecular Weight

Determination of the viscosity of a solution of the polymer in a suitable solvent is a relatively simple experimental procedure, and for PVC this procedure is routinely performed to evaluate an average molecular weight. However, viscosity depends on the weight and number of molecules in a complex manner. The viscosity average molecular weight determined by these measurements can be related to the molecular weight by the formula

$$M_v = \left(\frac{\sum n_x M_x^{(1+a)}}{\sum n_x M_x} \right)^{1/a}$$

where a depends on the material but is close to $a = 0.5$ for many polymers.

Comparison of Measurements of Average Molecular Weight

For a monodisperse sample, in which all molecules have exactly the same length, all three of these values are identical. In general, the higher the M_n, the greater the stiffness, the strength, and the chemical resistance and the lower the ductility of the polymer. The greater the variation in molecular size and the greater the spread in the distribution, the greater the difference in these values. In the computation of the weight average molecular weight, the molecules having a larger mass are counted more heavily than in the number average molecular weight, and therefore M_w is always larger than M_n. M_v numerically is always between M_n and M_w. For this reason, the ratio M_n/M_w is utilized as a measure of the dispersion or breadth of the distribution. Samples with the same M_n may have different mechanical properties if the distribution shape is different.

For many properties, a monodisperse material gives the optimum properties. Lowering the M_n improves the flow characteristics, permitting easier processing. The effect of molecular weight on toughness varies with the polymer: for most tough polymers, toughness increases with increasing molecular weight. For acrylics, which generally have low toughness, the toughness tends to decrease with increasing molecular weight.

Examples

Example 2.1.   Consider a monodisperse sample of polyethylene (PE) in which all the molecules have the same chain length with a degree of polymerization (DP) of 1000. Find the molecular weight.

$$M_n = \frac{100(14 \cdot 1000)}{100} = 14{,}000 \text{ g/mole}$$

$$M_w = \frac{(100 \cdot 14{,}000) \cdot (14{,}000)}{(100 \cdot 14{,}000)} = 14{,}000 \text{ g/mole}$$

$$M_v = \left[\frac{(100 \cdot 14{,}000^{1.5})}{(100 \cdot 14{,}000)} \right]^2 = 14{,}000 \text{ g/mole}$$

Note for a monodisperse polymer all three molecular weight averages are the same.

Example 2.2. Consider a sample of PE which is 50 mole % DP = 1000, 50 mole % DP = 1500.

For DP = 1000:

Molecular weight of a molecule = $14 \cdot 1000 = 14{,}000$

For DP = 1500:

Molecular weight of a molecule = $14 \cdot 1500 = 21{,}000$

$$M_n = \frac{(50 \cdot 14{,}000 + 50 \cdot 21{,}000)}{100} = 17{,}500 \text{ g/mole}$$

$$M_w = \frac{(50 \cdot 14{,}000) \cdot (14{,}000) + (50 \cdot 21{,}000) \cdot (21{,}000)}{(50 \cdot 14{,}000 + 50 \cdot 21{,}000)} = 18{,}200 \text{ g/mole}$$

$$M_v = \left[\frac{(50 \cdot 14{,}000^{1.5} + 50 \cdot 21{,}000^{1.5})}{(50 \cdot 14{,}000 + 50 \cdot 21{,}000)} \right]^2 = 18{,}030 \text{ g/mole}$$

Note that the viscosity average molecular weight is intermediate in value between the number average and weight average molecular weights.

Example 2.3. A sample of 20 g of PE with a DP of 10,000 is blended with 80 g of PP with a DP of 20,000. Find the average molecular weights.

Number average molecular weight = fraction of molecules having a given molecular weight · molecular weight

To find the fraction of molecules with a specified molecular weight, we need to calculate the number of molecules in each part of the blend. Let A = Avogadro's number = $6.02 \cdot 10^{23}$ molecules/mole.

Mer weight of PE = 14
Molecular weight of PE = $14 \cdot 10{,}000 = 140{,}000$

$$\text{Number of PE molecules} = \frac{20 \text{ g} \cdot A}{(140{,}000)} = 1.428 \cdot 10^{-4} A$$

Mer weight of PP = 42
Molecular weight of PP = $42 \cdot 20{,}000 = 840{,}000$

$$\text{Number of PP molecules} = \frac{80 \cdot A}{(840{,}000)} = 0.952 \cdot 10^{-4} A$$

$$\text{Fraction of PE molecules} = \frac{1.428}{(1.428 + 0.952)} = 0.6$$

$$\text{Number average molecular weight} = 0.6 \cdot 140,000 + 0.4 \cdot 840,000$$

$$= 420,000 \text{ g/mole}$$

$$\text{Weight average molecular weight} = \text{weight fraction} \cdot \text{molecular weight}$$

$$= 0.2 \cdot 140,000 + 0.8 \cdot 840,000$$

$$= 700,000$$

Viscosity average molecular weight

$$= \left[\frac{(0.6 \cdot 140,000^{1.5} + 0.4 \cdot 840,000^{1.5})}{(0.6 \cdot 14,000 + 0.4 \cdot 84,000)} \right]^2 = 653,000$$

2.3. DETERMINATION OF MOLECULAR WEIGHT

End Group Analysis

The chain is composed of repeating mers. The ends of the chain molecule are chemically different. In a linear polyester each chain has an unreacted alcohol or acid group at the end. The number of such groups can be determined by titration. In an addition polymerization, the molecules are ended either with a fragment of the initiator or with a double bond in the case of disproportionation. Either one of these is detectable by chemical reaction or by infrared analysis. From the number of end groups and the weight of the sample, the average molecular weight can be obtained directly. For this technique to be applicable, the chemical nature of the end group must be known. For large molecules, when the molecular weight exceeds 10,000, the number of end groups becomes such a small fraction of the total material that these methods become insensitive.

Osmotic Pressure

A schematic diagram of the equipment for osmotic pressure measurements is shown in Fig. 2.3. A semipermeable membrane (such as cellophane) through which the small solvent molecules can pass readily but through which the larger polymer molecules cannot pass separates two containers. A solution of the polymer in an appropriate solvent is placed in one side, and the pure solvent is placed in the other. The solvent diffuses through to equalize its chemical activity on both sides. Equilibrium is reached when the balance of the pressure generated by the additional height of fluid produced by the diffusion balances the chemical force tending to produce diffusion.

The height rise of the fluid is a function of the concentration, density

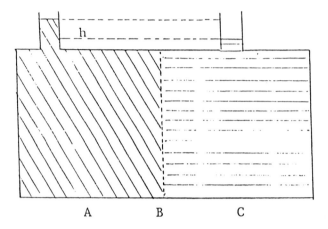

Figure 2.3. Osmometry apparatus: (A) solution of polymer in solvent; (B) pure solvent; and (C) semipermeable membrane.

of the solution, and number average molecular weight:

$$h = \frac{RTc}{\rho M_n} + Bc^2 + \cdots$$

At infinite dilution, the height rise is dependent only on the molecular weight:

$$\lim_{c \to 0} \frac{h}{c} = \frac{RT}{\rho M_n}$$

To analyze the data, the fluid height/concentration is plotted versus concentration and extrapolated to infinite dilution as shown in Fig. 2.4.

Osmotic pressure measurements cannot be performed on samples with excessively low molecular weights, as the molecules are then small enough to diffuse through the membrane. If the M_n is too large, the method becomes insensitive. As for the most methods, the range of M_n which can be evaluated is 50,000 to 1,000,000. ASTM D3750 shows the complete calculation procedure required.

Freezing Point Depression

The freezing point of a solvent is lowered when any solute is placed in a dilute solution. The lowering of the freezing point is independent of the chemical composition of the solute, depending only on the M_n. As for osmotic pressure, the relationship between M_n and the measured value is simplified if the data are extrapolated to infinite dilution. Under these conditions,

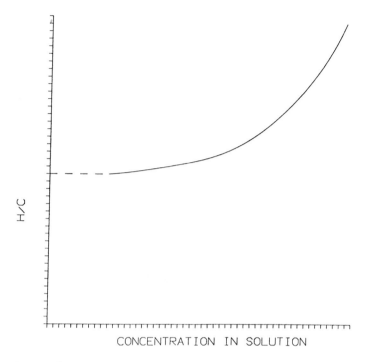

CONCENTRATION IN SOLUTION

Figure 2.4. Extrapolation of osmotic height data to zero concentration.

the two are related by

$$\lim_{c \to 0} \frac{\Delta T_f}{c} = \frac{RT^2}{\rho \Delta H_f M_n}$$

where ΔH_f = heat of fusion of the solvent
ρ = density of solution
T = temperature in K

The freezing point must be measured very accurately; a 1% solution of a polymer lowers the freezing point by 0.001°C.

Boiling Point Elevation

Similarly, the boiling point of a solvent is raised slightly by a solute. The relationship at infinite dilution is

$$\lim_{c \to 0} \frac{\Delta T_B}{c} = \frac{RT^2}{\rho \Delta H_v M_n}$$

where ΔH_v = heat of vaporization of the solvent

Light Scattering

Light is passed through a solution of the polymer, and the intensity of light scattered at right angles to the incident beam is measured. Although the relationship between light scattering and M_w is known theoretically, the equipment is usually calibrated with known substances. Dust or impurity particles scatter light very effectively compared with the polymer molecules, so the solutions must be carefully prepared and filtered or ultracentrifuged to remove such impurities. Molecular weights in the range 10,000 to 10,000,000 can be obtained. ASTM D4001 shows the complete calculation of molecular weight from the experimental data.

Viscosity

The viscosity of a solution of a thermoplastic in an appropriate solvent increases with concentration and the molecular weight of the plastic. Measurement of the viscosity therefore requires finding an appropriate solvent, and for many polymers the solvents are difficult to handle and pose threats of toxicity. The most accurate viscosimeters are the type in which the time of flow of a known volume of solution through a capillary tube is determined. The time required for a solution to flow through the viscosimeter is proportional to the kinematic viscosity (the ratio of viscosity to density). The Brookfield type of viscosimeter, in which a paddle rotates within the solution, it also used. It is more commonly used to determine the effect of shear rate on the viscosity of the solution, rather than determination of molecular weight.

For the flow type of viscosimeter, the flow time of a solution of known concentration is compared to that of the pure solvent. The relative viscosity (η_{rel}) is defined as:

$$\eta_{rel} = \frac{t}{t_0}$$

where t = time for a known volume of solution to flow through the viscosimeter

t_0 = time of flow for the same volume of pure solvent

The viscosity number (the viscosity divided by the concentration or the logarithmic viscosity number (the natural logarithm of the relative viscosity divided by the concentration) is then plotted vs concentration. The *intrinsic viscosity* is the extrapolation to zero concentration of the viscosity number, and is related to the viscosity average molecular weight. Techniques are available which give a good approximation to the intrinsic viscosity from one reading taken at a concentration of 0.5 gr/ml or less. Theoretically, the intrinsic viscosity is related to the viscosity average molecular weight by the equation:

$$\eta_{intrin} = K'M_v{}^a$$

Since exponent a and the constant K' are not known precisely for most systems, this presents problems in accurately defining the molecular weight. In practice, this method is used more commonly for comparison purposes in quality control rather than actual molecular weight determinations, and it is particularly commonly applied to processing PVC. The procedure is described in ASTM D1243.

Fractionation

Determination of the molecular weight distribution requires lengthy and careful experimental procedures. A complex sample can be separated into individual portions with a narrow molecular weight distribution by controlled precipitation from a solution. The solubility of polymers decreases with increasing molecular weight. If a nonsolvent for the polymer is added to the solution and agitated vigorously, some polymer will precipitate. This can be decanted, filtered, and dried. Further separation into other fractions can be performed by further additions of the nonsolvent. A 1-g sample of polymer can be separated into 20 fractions. The molecular weight of each fraction can be determined by one of the methods described above. Other techniques can also be used, such as evaporating the solvent or lowering the temperature of the solution.

Gel Permeation Chromatography

This is the most common and precise method for determining molecular weight distributions and is also suitable for determining the average molecular weight. It requires that the polymer be dissolved in a solvent; tetrahydrofuran is the standard solvent which is specified by ASTM in D3593. The dissolved sample is injected into a long column packed with a solid porous substance through which pure solvent continually passes. The injected sample must be a maximum of 70 microliters in volume. A satisfactory column may contain glass beads, impregnated with a gel of crosslinked polystyrene. Column dimensions are 0.5–5 cm inner diameter, 10–150 cm long. Shorter molecules tend to become adsorbed within the gel, and their passage down the column is retarded. Longer molecules tend to be flushed down the column more rapidly, and separation increases with column length. The time of passage through the column is calculated by determining the refractive index of the liquid exiting the column. The system must be calibrated with multiple samples of polystyrene (PS) of known molecular weight. Some of the other methods described above should be used for confirmation of the absolute molecular weight. Since PS is the standard for this technique, the procedure for PS has been specified independently in ASTM D3536. Although the analysis requires careful procedures and complex calculations, much of it has been computerized.

2.4. CRYSTALLIZATION OF POLYMERS

Gases

Although we will not discuss gaseous materials in detail, it is convenient at this point to quickly compare the gas phase to the solid and liquid. In a gas, the atoms or molecules move independently, the average kinetic energy of the molecules being directly proportional to the absolute temperature of the gas. The molecules are far apart, the average distance between them being much greater than the molecular diameter or length. Chemical attractive forces exist between molecules but are insufficient to hold the molecules together due to the large kinetic energy of the molecules at elevated temperatures. Because of the large molecular weight of plastics, the tendency to vaporize is quite small at room temperature. At elevated temperatures the polymers tend to decompose, liberating smaller fractions of the molecules and oxidation and decomposition products.

Liquids

At temperatures well above the melting point of the polymer, the molecules have sufficient kinetic energy to move relatively independently of each other. Liquids and solids are very incompressible compared with gases. Doubling the pressure on a liquid (from atmospheric to two atmospheres) would cause the volume to decrease by about 0.01%, whereas for a gas the volume would be halved (if an ideal gas is assumed). For the liquid to be incompressible, the molecules must be in contact. The kinetic energy of the molecules is insufficient to break the strong chemical bonds attracting the molecules to each other. The molecules are quite mobile at temperatures close to the melting point and are in a random configuration, with the molecules moving erratically as a result of interaction and transfer of kinetic energy at collisions. Because of the rapid bending and twisting of the molecules, each molecule effectively occupies a volume far larger than the volume of all of its atoms alone. One of the more important parameters which defines the properties of polymers is the *free volume*, the effective empty space within the material:

Free volume = total volume of the polymer − volume of all atoms or expressed as a percent:

$$\text{Free volume} = \frac{\text{total volume} - \text{atomic volume}}{\text{total volume}}$$

Liquid Crystals or Thermotropic Liquids

Liquid crystals are polymers which have extremely stiff backbones, so that the molecule can be considered rodlike, with relatively little flexibility. As a result, in the liquid state the molecular flexing of most polymers is lacking and the rodlike molecules align into highly ordered arrangements parallel to each other. The primary difference between the solid and liquid form is the molecular mobility: in the liquid state the molecules can still move readily relative to each other, while maintaining their parallel alignment. This order produces an anisotropic liquid!

When solidified, the stiff molecules remain aligned, and nearest neighbor molecules remain bonded by the strong secondary forces. These materials form liquids which are highly ordered at temperatures just above the melting point. As the temperature is raised further, the order diminishes, and at sufficiently high temperatures, a random liquid, similar to other liquid polymers, is formed. In ordered form, the molecules behave as stiff, rodlike particles. These tend to align parallel to each other. This configuration is distinctly different from the random, kinked and intertwined molecules found in the disordered liquid form.

The development of such materials is receiving significant attention, since they have properties which are different from those of other forms of the polymers. Because the liquid is highly ordered, the heat evolved on crystallization is relatively low. This permits rapid molding, as heat transfer requirements are minimized, with molding cycle times typically one-half those of other polymers with good mechanical properties. Due to the orientation of the molecules in the liquid, they have excellent melt processing characteristics; the rodlike molecules can flow past each other with a minimum of entanglement, resulting in low liquid viscosities. Such polymers therefore require lower molding pressures and processing energy. Molecular alignment is enhanced by shear stresses in the liquid, as shear stresses tend to generate some molecular motion, resulting in increased ordering and alignment of the molecules. Therefore liquid crystals are shear thinning, the viscosity decreasing with increasing flow velocity. Due to this alignment of the molecules, flow in small channels such as the runners in injection molding and thin mold parts occurs readily, permitting the molding of thinner cross sections or parts with complicated shapes.

The mechanical properties of these materials are excellent. They have tensile strengths in the range 100–250 MPa and moduli of 10–25 GPa.

However, due to the molecular orientation, the resultant parts are highly anisotropic, with high strength in the flow (molecular alignment) direction. Shrinkage during the molding process is also controlled by this alignment, being virtually negligible in the flow direction (less than 0.001 cm/cm) and five times greater (0.005 cm/cm) in the transverse direction. Thermal expansions are also anisotropic, being less than $10 \cdot 10^{-6}$ in the flow direction and five times greater in the transverse direction. Because such large differences in shrinkage can induce warping, molding with several gates may be desirable to reduce the large differences in directional properties. Due to the strong intermolecular bonding, these materials maintain their mechanical properties at high temperatures. This is of importance in the electronics industry, since soldering of parts attached to a liquid crystal substrate is possible. Due to the large amount of orientation they exhibit as-molded properties similar to those of fibrous reinforcements. Other advantages are the reduction in residual stresses in the finished product and reduction in flash and flash removal costs, since lower total injection pressures are required. Liquid crystal polymers tend to be among the most expensive polymers, costing up to $45/kg, but the processing ease and improved properties often overcome this initial expense.

Such thermotropic materials are currently being studied for blending with the flexible molecule plastics, as result of their very promising mechanical properties. However, as they have such high orientation and the other polymers have very different molecular characteristics, blending is thermodynamically unlikely. As a result, current investigations employ extrusion or drawing of a molten blend of the thermotropic material with a thermoplastic. As a result of its lower viscosity, the thermotropic material develops into fibers in a matrix of the thermoplastic. Because of the orientation of the molecules during the flow, the thermotropic material forms elongated fibers in the direction of extrusion or flow. One principal advantage over the use of fibrous reinforcing materials is that such fiber–polymer mixtures have high viscosities, making molding difficult. Conversely, these blends have viscosities lower than that of the matrix thermoplastic. An additional advantage of these blends is avoidance of the excessive equipment wear caused when the fiber–polymer mixture is extruded. Currently, the lack of interfacial bonding of the thermotropic material to the matrix polymer significantly lowers the strength of these blends.

Crystalline Solids

Solids are defined as materials with an internal repeating order of molecules or atoms. Industrially utilized plastics cannot be manufactured with all the material in crystalline form; instead, plastics are either essentially completely amorphous (randomly arranged) or partially crystalline. The difficulty in producing crystalline polymers is readily apparent, since the long molecules of a polymer are difficult to align in perfect order. Such uniform ordering and crystallization is impossible in thermosetting plastics, because thermosets form in a random configuration and realignment of the chains cannot occur without breaking the covalent bonds which interconnect the atoms. Thermoplastics have completely separated molecules and can crystalline by rearrangement and ordering of the molecules. For these molecules to fit into a uniform arrangement, the molecular chains cannot contain irregularities. Chemical factors tending to produce easily crystallizable polymers are:

1. Simple, relatively branch-free chain molecules so that chain alignment is not hindered. Side branches, an atactic arrangement, and random molecular attachments all tend to prevent crystallization. A uniform degree of polymerization of the various molecules also encourages greater crystallization.
2. The presence of polar groups on the molecules, which aid in increasing intermolecular forces.
3. Lack of foreign molecules, which would hinder uniformity of molecular alignment. Impurities, unreacted monomer molecules, and blends of different molecules would all be obstacles to alignment.
4. Molecular stiffness, which aids alignment and increases crystallization. Side groups which are small enough not to prevent close approach of the chain molecules but which accentuate steric hindrance aid crystallization.

The unit cell is defined as the smallest volume which can generate the entire solid by repeated translation in the three Cartesian coordinates. The unit cell of a polymer usually contains several mers and therefore contains up to several hundred atoms, whereas in metals the unit cell usually contains only a few atoms. The crystal structure of polyethylene has been studied extensively. The polyethylene molecules align with the c-axis, with the normal bond angle between atoms, so that the chains are not linear, but the molecule does lie completely in one plane. Polytetrafluoroethylene (PTFE), although similar to PE, has a distinctly different crystal structure. Although the two materials can be considered similar, with hydrogen atoms completely replaced by fluorine atoms in PTFE, the larger size of the F atom compared with the H atoms prevents the chain molecule from lying in one plane as it does for PE. Rather, a twisting or helical structure is required which allows the F atoms to avoid making contact with each other, as would be the case if the molecule were planar.

Crystallization

The transformation from the liquid to crystalline forms is a first-order thermodynamic reaction. Above the melting point, the free energy of the liquid is lower than that of the solid, whereas at the equilibrium melting point the free energies of the solid and liquid phases are equal. Discontinuities occur in the other thermodynamic functions, with the enthalpy, entropy, and heat capacity of the two phases being different. These functions are shown in Fig. 2.5.

Crystallization requires the formation of a small nucleus of the required ordered atomic arrangement by chance motion of the polymer molecules and subsequent diffusion of additional molecules to align themselves with the molecules already existing in the nucleus. Inorganic and organic molecules with low molecular weights generally crystallize completely. However, at the crystallization temperature polymer liquids are very viscous. The molecules are mechanically entangled, and branching and atacticity hinder perfect

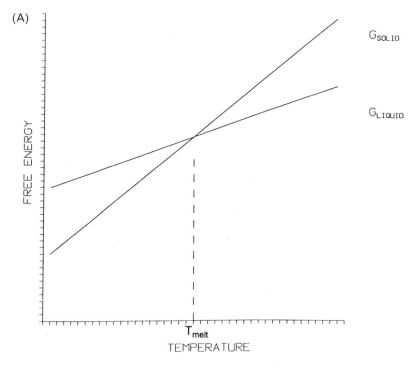

Figure 2.5. Thermodynamic properties of the liquid and solid near the melting point. (A) The free energies of the solid and liquid do not change discontinuously at the melting point. Supercooled liquid and superheated solid are possible.

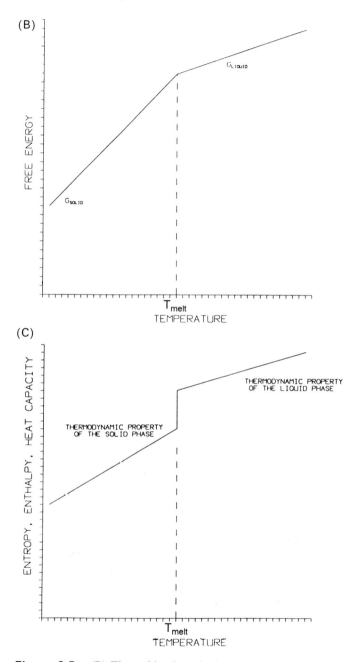

Figure 2.5. (B) The stable phase is the one which has the lowest free energy. (C) Enthalpy, entropy, and heat capacity all show a discontinuity at the melting point.

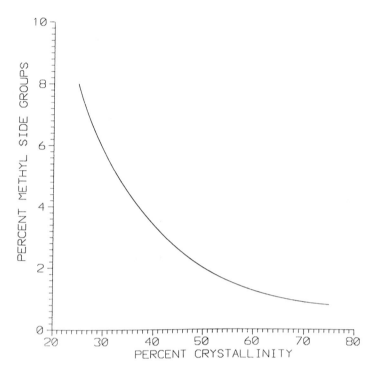

Figure 2.6. Effect of side chains on the crystallinity of PE.

alignment. Long-chain polymers have therefore never been observed to form completely crystalline material in industrial materials; instead, they form a very fine mixture of crystalline and disordered regions. Figure 2.6 shows the effect of side chains on the crystallinity of PE. Note the very rapid decrease in crystallinity with relatively few side chains. These side chains are only several carbon atoms long.

On cooling below the melting point, the resultant material may partly crystallize, but part of the material that has not been able to fold correctly within the crystal unit cell may still retain the amorphous characteristics of the liquid. The exact spatial relationship between the crystalline and amorphous regions depends on the mechanism of solidification. If the sample contains crystalline regions, molecules must fold to fit within the correct unit cell configuration. The chain molecule may completely end with the crystalline region, or portions may extend outside into the amorphous region that separates and surrounds the small crystalline regions. Some molecules extend through the amorphous region into another separate crystalline region, and others may exist entirely in the amorphous regions.

An important classification of polymers is with respect to the fraction of crystalline material:

1. *Crystalline*: No polymer is 100% crystalline. Crystalline polymers contain large fractions of crystalline material, typically being 60–90% crystalline.
2. *Semicrystalline*: Large fractions of amorphous material exist. The crystalline portion may range from 20 to 60%.
3. *Amorphous polymers*: Many commercially useful polymers contain either no measurable crystalline regions or crystalline regions that are very small fractions of the total weight.

Crystallization from Solutions

The size and arrangement of the crystals formed depend strongly on the method of preparation of the plastic. The first single crystal of PE was made by preparing a dilute solution of PE in a solvent and then slowly precipitating the PE from the solution by cooling the solution. Alternatively, crystalline solids can be precipitated by adding a second liquid which dissolves in the solution but in which PE is less soluble. In a dilute solution, the molecules of the polymer tend to remain independently dissolved and free to translate and rotate independently of each other, permitting slow and equilibrium solidification. The crystals formed by such precipitation are extremely small, appearing as plates (lamellae) only approximately 10 μm in length and 100 Å thick. The density of these crystals is less than that calculated from the crystal structure, indicating that some imperfections or amorphous material must exist along with the crystallized material. Analysis of the X-ray diffraction patterns of these crystals and all crystals obtained for polymers reveals a remarkable folding of the molecules that permits them to exist within such a small crystalline volume. Rather than lying along and entirely within the widest dimension of the crystal, the molecule chain appears to lie as shown in Fig. 2.7, folding and refolding repeatedly within the crystal. For small crystals grown carefully from solutions, the folds appear to be tight (the size determined by the permissible bond angles), and the molecules reenter the crystal adjacent to the previous portion of the molecule.

For other crystals developed under conditions of more rapid growth, the molecules develop very loose folds, and a significant fraction of the molecule may not exist within the crystalline region. If the polymer is dissolved in a concentrated solution, the individual molecules tend to intertwine and entangle within the solution as a result of molecular motion, and when crystallization occurs, the molecules become trapped in more than one small crystal, with portions of the molecules between the crystals being irregular or amorphously arranged. Such interconnecting molecules

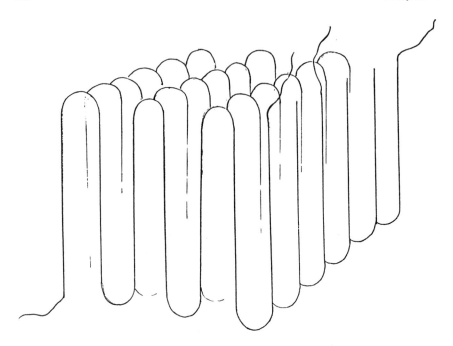

Figure 2.7. Molecular folding in crystals.

are called *tie molecules* and in the extreme can connect the crystals into one large, loose mass.

Crystallization from Melts

When a polymer liquid cools, the randomly arranged, coiled molecules must realign to form the crystalline solid. In the melt, solidification is hindered by the high viscosity of the molten polymer, chain entanglements, and the irregular structure of the polymer molecules. Observation of such solidification reveals that many small crystals nucleate at the same location within the liquid and grow outward from the nucleation center. The combination of all these crystals within one spherically growing bundle is called a spherulite.

A spherulite is composed of many crystals, interconnected by tie molecules, with amorphous material separating the crystals. The individual crystals are similar in dimensions to those formed during crystallization from a solution and are about 100 Å thick, with the molecular chains being folded within the crystal. Due to their small size and relative platelike appearance, these crystals are called lamellae. The intercrystalline regions have dimensions

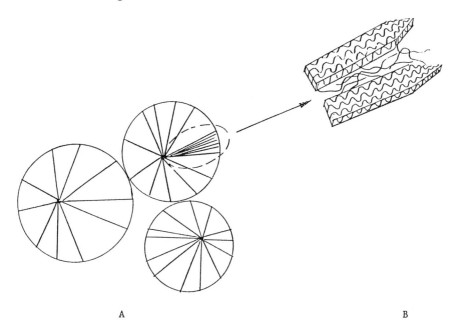

A   B

Figure 2.8. Spherulitic structure. (A) Idealized appearance of spherulites under polarized light. (B) Small, shardlike crystallites within the spherulites.

of the same order and contain tie molecules and other material which cannot fit into the crystalline regions. As a result, portions of the molecules which contain side chains, unusual atoms, atactic regions, impurity molecules, and unincorporated catalyst are concentrated in this interlamellar region. The spherulite is therefore a combination of crystalline lamellae and amorphous interlamellar material.

Figure 2.8 shows schematically the structure of a spherulite. Note that the small crystallites are directed radially outward from the nucleation center. These crystalline ribbons are of the order of 10^{-6} cm thick. The spaces between the ribbons contain amorphous material, including all the impurities and other irregular molecules which do not fit neatly into the crystal structure. The sheaves of lamellae can be seen clearly at the magnification possible with an electron microscope.

The lamellar in the spherulites are always very thin, but their thickness depends on the temperature at which crystallization occurs. Some under-cooling must always occur, and the thickness of the lamella decreases with decreasing solidification temperature, with the greatest decrease occurring just below the thermodynamic melting point. Similarly, there is an increase

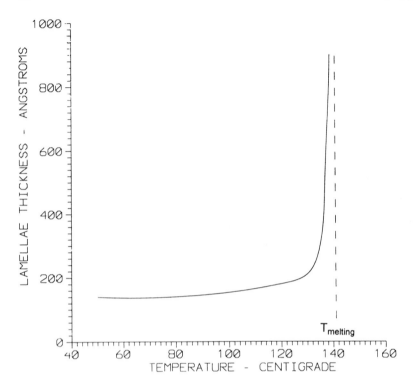

Figure 2.9.   Effect of annealing on the thickness of lamellae.

in the lamellar thickness if the solidified plastic is held at temperature close to the melting point (Fig. 2.9).

Crystallization During Processing

Most processing of thermoplastic materials involves heating the polymer above its melting point and causing flow under pressure into a mold or tool. During processing, since flow rates and cooling rates differ widely between different parts of the product, considerable variation in properties will result. This dependence on manufacturing conditions tend to be greater for polymers than for other materials. Solidification of polymers during or immediately after such flow generates a different molecular morphology than for the same polymer permitted to solidify from the quiescent liquid. If the flow rate or the shear stress generated during flow is not excessive, row structures of long stacks of lamellae form. The centers of these structures have long chains which are elongated in the direction of flow. However, the

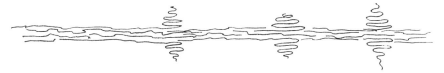

Figure 2.10. Shish kebab structure formed during solidification under flow conditions.

normal configuration of the crystallized molecules is folded, and such folding occurs to generate extensions around the central thread, resulting in a structure called a "shish kebab" structure (Fig. 2.10). In both the central linearly stretched molecular region and the folded outlying extensions, the molecules all tend to have their chain axes directed along the flow direction.

The greater the amount of shear stress generated by flow prior to solidification, the smaller the diameter of the extended regions around the central thread. Since the molecules are aligned in the flow direction, the elastic modulus of the material is increased in the flow direction, and decreased in the transverse direction. The individual elongated crystal regions are separated by amorphous zones. The elastic modulus in the transverse direction of this mixture of amorphous and crystalline material can be calculated by the rule of mixtures for composite materials:

$$\frac{1}{E_{transverse}} = \frac{V_{cryst}}{E_{cryst}} + \frac{(1 - V_{cryst})}{E_{amorp}}$$

where E_{cryst} = the modulus of purely crystalline material

E_{amorp} = the modulus of purely amorphous material

V_{cryst} = the volume fraction of crystalline material

The amorphous regions therefore primarily control the modulus in the direction transverse to the flow direction. For example, Fig. 2.11 shows the modulus in the transverse direction for a material with $E_{cryst} = 10^6$ psi and $E_{amorp} = 10^5$ psi. The importance of the amorphous regions is evident for fractions of amorphous material greater than 15%.

The Melting Point

The melting point of a material is the temperature at which the free energies of the liquid and solid phases are equal. Therefore, it is a thermodynamic parameter which should have a single value for any pure material. However, the existence of the extremely small lamellar structure of plastics introduces an additional energy term associated with the surface energy of the lamellae. Depending on the size and number of lamellae, this surface energy can be

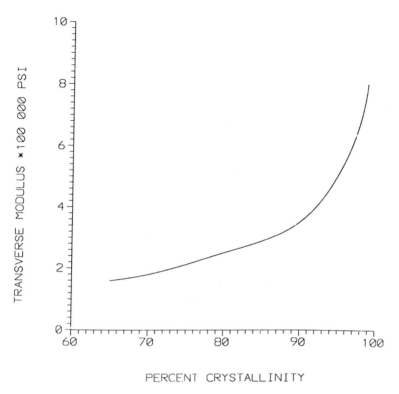

Figure 2.11. Transverse modulus as a function of the fraction of crystalline material in a polymer.

significant compared with the free energy difference between the solid and liquid phases. As a result, a specific melting point of a plastic cannot be defined independently of the lamellae size. The smaller the lamellae, the greater the surface energy and the lower the melting point of the sample.

This concept can be quantified. For a material with negligible surface energy, the free energy change on melting at any temperature is

$$\Delta G_m^\circ = G_l^\circ - G_s^\circ = \Delta H_m^\circ - T \Delta S_m^\circ$$

where the superscript $^\circ$ indicates the properties of the material with negligible surface area, and

$$\Delta G_m = \text{the change in free energy on melting}$$

$$G_l, G_s = \text{free energy of the liquid and solid, respectively}$$

$$\Delta H_m, \Delta S_m = \text{the change in enthalpy and entropy on melting}$$

At the equilibrium melting point T_{mp}, the free energies of the liquid and the solid are equal:

$$G_s^\circ = G_l^\circ$$

$$\Delta G_m^\circ = \Delta H_m^\circ - T\,\Delta S_m^\circ$$

$$0 = \Delta H_m^\circ - T\,\Delta S_m^\circ$$

Then

$$\Delta S_m^\circ = \frac{\Delta H_m^\circ}{T_{mp}^\circ}$$

or at another temperature T, considering that ΔH°, ΔS° change only slowly with temperature:

$$\Delta G_m^\circ = \Delta H_m^\circ\left(1 - \frac{T}{T_{mp}^\circ}\right)$$

Now, for a solid composed of small lamellae, with surface area A and surface energy per unit area $= \gamma$, the energy of the solid phase is

$$G_s = G_s^\circ + A\gamma$$

where G_s° = free energy neglecting surface energy. Therefore:

$$\Delta G_m = G_l - G_s = \Delta H_m^\circ\left(1 - \frac{T}{T_{mp}^\circ}\right) - A\gamma$$

The melting point of the material including surface area is again determined by the condition that $\Delta G_m = 0$:

$$\Delta H_m^\circ\left(1 - \frac{T_{mp}}{T_{mp}^\circ}\right) - A\gamma = 0$$

$$T_{mp} = T_{mp}^\circ\left(1 - \frac{A\gamma}{\Delta H_m^\circ}\right)$$

where T_m is the actual melting point of the material as a result of the increase in total energy due to the presence of the surface area associated with the lamellae. The greater the surface area or the smaller the individual lamellae, the lower the melting point of the material. This is an important concept. Because lamellae in polymers are very small and vary in size throughout the sample, a melting range is expected rather than a discrete melting point! The dependence of melting point on surface area is shown schematically in Fig. 2.12.

As a result of the changes in spherulite size possible with different rates of solidification, hysteresis is also exhibited between liquefaction and

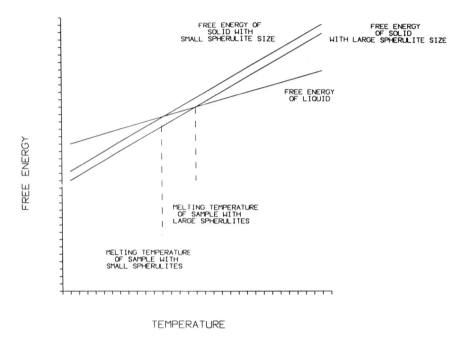

Figure 2.12.   Effect of surface energy on the melting point.

solidification. The melting range will therefore vary with the cooling rate experienced by the material. While some supercooling always occurs on solidification, the graph of Fig. 2.13 shows the melting point dependence on the temperature at which the sample was originally solidified. This can be rationalized by noting that the lower crystallization temperature produces smaller lamellae and a greater surface area. As discussed before, holding a polymer slightly below the melting point tends to thicken the lamellae. The initial melting temperature of a particular sample therefore depends on the rate of heating, as slow heating would develop larger lamellae and a higher melting point.

Nucleus Formation

A liquid may transform into a crystalline solid when cooled below its stable freezing point. This transformation proceeds by the initial formation of a small stable particle of the solid, called a nucleus, and subsequent growth of that nucleus to a much larger size. During the formation of the nucleus,

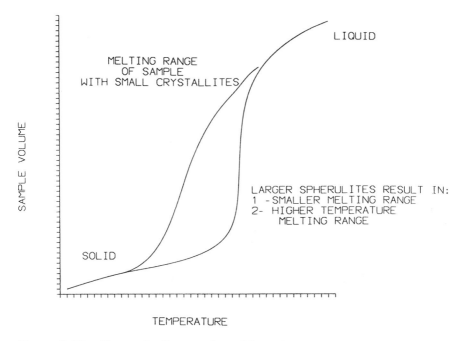

LIQUID

MELTING RANGE
OF SAMPLE
WITH SMALL CRYSTALLITES

SAMPLE VOLUME

LARGER SPHERULITES RESULT IN:
1 - SMALLER MELTING RANGE
2- HIGHER TEMPERATURE
    MELTING RANGE

SOLID

TEMPERATURE

Figure 2.13.   Hysteresis effects on the melting point.

the molecules shift from the random arrangement in the liquid to the ordered one in the solid. Obviously, there is no specific change at the melting point that causes sudden ordering. The molecules in the liquid are in constant random motion, and at temperatures both above and below the melting point small portions of the liquid momentarily form in the arrangement corresponding to that of the crystalline solid. Above the melting point the configuration is unstable and decomposes, the molecules developing some other random arrangement. Below the melting temperature, a cluster with the crystalline order should be stable and enlarge when other folds form within the molecule to expand the size of the crystal region or other molecules become attached.

However, the stability of the newly formed nucleus is controlled initially by the surface energy of the nucleus. For the nucleus to be truly stable, the energy of the system must decrease with its growth.

Let ΔG_{sl} = free energy evolved per cm^3 of liquid transformed to the solid. If, for simplicity, the nucleus is considered spherical with radius r, then the free energy change ΔG_v during formation of this volume of the nucleus is

$$\Delta G_{\text{v}} = \tfrac{4}{3}\pi r^3 \Delta G_{\text{sl}}$$

This equation considers only the energy change due to the difference in energy between the ordered solid and the disordered liquid. An additional factor must also be recognized. When the solid nucleus develops, a surface is generated separating the two phases. The atoms located on the surface of the nucleus cannot be in the configuration of atoms buried within the nucleus, surrounded by other molecules in the correct configuration. These surface atoms must be in higher energy states, and there is therefore always a driving force attempting to minimize the surface area of any system. For a given volume of a material, the shape with minimum surface area is a sphere. If the surface energy is independent of orientation within the solid, the material tends to generate that shape.

Since the surface energy of solid polymers and other materials generally depends on direction within the crystal, the generated crystals and nuclei are not spherical but rather develop shapes with planar facets. For simplicity of calculation, we will continue to consider an idealized spherical nucleus. If the energy per unit area of surface of the nucleus is γ, then

$$\text{Total surface energy} = 4\pi r^2 \gamma$$

The formation of a surface always increases the energy of the system; this term is therefore always positive.

The total free energy change when the nucleus forms is therefore

$$G_{\text{v}} = \tfrac{4}{3}\pi r^3 \Delta G_{\text{sl}} + 4\pi r^2 \gamma \tag{2.1}$$

Above the melting point the liquid must be lower in energy than the solid because it is stable, and ΔG_{sl} is positive. In this case the total energy increases on forming a small nucleus of the solid, since both terms are positive, and the nucleus will decompose. Below the melting point, the solid has a lower energy than an equal volume of liquid, and ΔG_{v} is negative. To visualize the physical significance of Eq. 2.1, consider a plot of these two terms as a function of the radius of the nucleus at different temperatures (Fig. 2.14). The surface energy term is always positive and is almost temperature independent. At the melting point the solid and liquid are in equilibrium and both are stable. ΔG_{sl} must therefore be zero. As the temperature drops below the melting point, this term becomes increasingly negative.

The total energy change produced when the nucleus forms is the sum of the volume and surface terms. If we consider sufficiently small radii, the volume energy term can be neglected, since $r^3 \ll r^2$ for small r. Conversely, at large radii the surface energy term is negligible ($r^2 \ll r^3$ for large r). The summation curve is also shown in Fig. 2.14. At small nuclear sizes, the energy

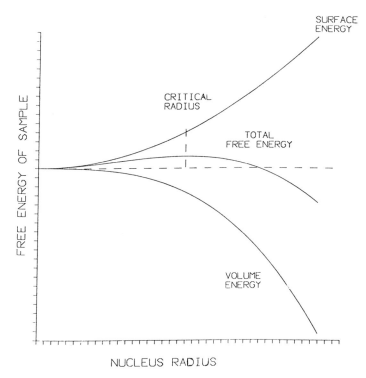

Figure 2.14.   Effect of surface and volume energy terms on total free energy change during solidification.

of the system increases when a nucleus of the solid forms, since surface energy is controlling. These small nuclei will therefore decompose rather than grow, despite the fact that the material is below the stable melting point of the polymer. A minimum size of nucleus (the critical nucleus) must therefore exist before the nucleus can be stable. Only growth of nuclei larger than this size will cause the energy of the system to decrease.

The lower the temperature, the smaller the critical nucleus, and the greater the number formed per second by random molecular motion. Eventually, the temperature may become so low that atomic motion is reduced to the point where molecules cannot move freely to generate new nuclei, and nucleus formation for the solid phase slows.

Growth of Nuclei

Nucleation is hindered by the need to generate new surface area. Once a nucleus larger than the critical size is formed, it will grow by further attachment of molecules from the disordered liquid. This process is dependent on the kinetic motion of the atoms in the liquid. At temperatures near the melting point the kinetic energy is high and the growth rate is large. As the temperature drops, the rate of growth decreases. At sufficiently low temperatures, the growth rate of the nucleus will become negligible.

Two independent steps are therefore required for the crystallization of a liquid: nucleation of the solid and growth of the nuclei. If either of these steps occurs slowly, the overall transformation rate will be low. The total solidification rate can therefore be expressed as

Total growth rate = rate of nucleation · rate of nuclear growth

The generalized result is shown in Fig. 2.15. The overall crystallization rate is low immediately below the freezing point, because nucleation rates are low. The solidification rate increases with decreasing temperature as nucleation

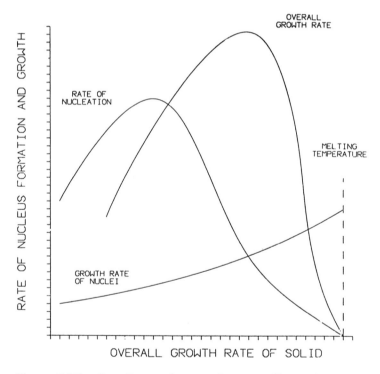

Figure 2.15. Overall growth rate of a crystalline polymer as a function of temperature. The overall growth rate is a function of the rate of nucleation and the growth rate of the nuclei.

rates increase and growth rates are still high, and then decreases as the rate of growth decreases at low temperatures. Therefore, even if a polymer has the uniform molecular structure required for crystallization, if the cooling rate is sufficient to pass through the temperature range where solidification occurs most rapidly without appreciable crystallization occurring, the material remains amorphous. Negligible crystallization will occur at the lower temperatures, and the polymer will remain at least partially amorphous at room temperature. The state of supercooled frozen random molecules is called the glassy state and is discussed in the next section.

Slow Continued Solidification

Conversion of the disordered liquid molecular configuration to the crystalline form may continue for long periods of time after the sample has been cooled to room temperature. Such slow transformation is common in Nylon 66. Some continued crystallization of the amorphous fractions occurs over periods of several years after molding. This crystallization produces dimensional changes. High-precision parts may be held for extended periods before use to stabilize their dimensions. Annealing at elevated temperatures induces crystallization and speeds the approach to maximum crystallinity, although 100% crystallinity can never be realized. It should be noted that some materials which do not crystallize (such as polycarbonate (PC)) also exhibit dimensional changes with time. This is due not to crystallization but to changes in free volume. This will be discussed later.

Determination of Extent of Crystallization

Density

The simplest experimental way to determine the extent of crystallization of a polymer is to determine the density of the solidified sample. The amorphous regions are less dense than the crystalline regions. The density of the crystalline material can be calculated from the crystal structure, and the density of the amorphous material can be obtained by measurements on samples cooled so rapidly that crystallization does not occur. The densities are most commonly obtained by density-gradient column measurements. A density-gradient column contains a series of immiscible liquids of different densities, the most dense liquid being at the bottom. The column is calibrated by dropping weights of known densities into the column. Each weight floats at the level at which its density equals the density of the fluid. Similarly, a small polymer sample dropped into the column floats at the location of the corresponding fluid density.

The percent crystallinity can be calculated by considering that the sample can be divided into the parts that are amorphous (subscript a) and crystalline (subscript c).

If ρ is the density, and V the volume of the composite, then the component contributions can be indicated as:

V_c = Volume of the crystalline fraction

V_a = Volume of the amorphous fraction

v_c = volume fraction of the crystalline fraction

v_a = volume fraction of the amorphous fraction

ρ_c = density of the crystalline fraction

ρ_a = density of the amorphous fraction

Then:

$$\rho V = \rho_c V_c + \rho_a V_a$$

Since $V = V_c + V_a$

$$\rho = \rho_c v_c + \rho_a v_a = \rho_c + \rho_a(1 - v_c)$$

$$v_c = \frac{(\rho - \rho_a)}{(\rho_c - \rho_a)} \tag{2.2}$$

If the weight fractions of the crystalline and amorphous fractions are w_c and w_a respectively, and the total weight of the sample is W, then:

$$w_c = \frac{W_c}{W} = \frac{\rho_c V_c}{\rho V} = \frac{\rho_c v_c}{\rho}$$

Substituting from Eq. 2.2 above for v_c:

$$w_c = \left(\frac{\rho_c}{\rho}\right)\left(\frac{\rho - \rho_a}{\rho_c - \rho_a}\right)$$

Differential Scanning Calorimetry

Differential scanning calorimetry (DSC) is a well-established technique in which a sample is either heated or cooled while its temperature is monitored. Simultaneously, a reference sample located within the same chamber is also monitored. The temperature difference between the two samples is recorded and plotted. Any difference between the two specimens is a result of heat generation or absorption within the specimen because of polymerization or a phase change. Thus, this technique is used for monitoring the polymerization process, as well as determining the extent of crystallization of a sample. If the temperature difference is plotted versus the sample temperature, the height along the y-axis is proportional to the rate of heat transfer to or from the specimen and the area under the curve is the heat transfer during the process. The degree of crystallization of a plastic can therefore be determined by comparing the heat evolved on crystallization to that which should be generated by 100% crystallization:

$$w_c = \frac{\Delta H}{\Delta H_f}$$

where ΔH = heat evolved during crystallization of the sample/unit weight

   ΔH_f = heat evolved during complete crystallization/unit weight

Figure 2.16 (Gupta and Salovey, 1990) shows the DSC plot on heating quenched poly ether ether ketone (PEEK). The quenching produces amorphous material. The peak on the graph at 178°C is a result of heat evolved during the conversion of the amorphous material to crystalline material on heating, and the dip at 336°C is due to melting of the crystallized material.

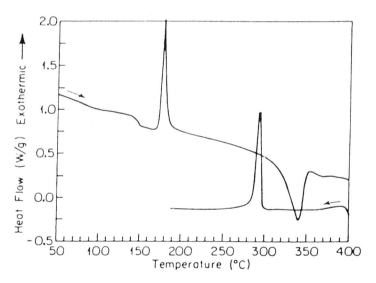

Figure 2.16.   DSC results for crystalline and amorphous PEEK during heating and cooling. During heating, crystallization begins to occur at about 178°C, absorbing energy, and melting of the crystalline material occurs at 336°C absorbing energy. Note that on cooling, solidification occurs well below the temperature at which melting occurred due to the requirements of nucleation of the solid. [From H. Gupta and R. Salovey, *PES 30*, 453 (1990). Reprinted by permission of the publishers.]

Rate of Crystallization

Crystallization can occur either from the melt on cooling below the melting point or from amorphous rubbery material on standing at temperatures below the melting point. In either case, data concerning the extent of crystallization with time are well fitted by the Avrami equation, originally developed for the solidification of metals. The derivation includes the impinging of various solidifying (crystallizing) regions within the sample, as well as the rates of nucleation and growth.

$$\frac{C(t)}{C(\infty)} = 1 - \exp(-kt^n) \tag{2.3}$$

or transposing,

$$\frac{[C(\infty) - C(t)]}{C(\infty)} = \exp(-kt^n)$$

where $C(t)$ = fraction of the sample crystallized at time t

$C(\infty)$ = fraction of the sample crystallized after an infinitely long time (complete crystallization need not occur)

k, n = constants which must be experimentally determined

Since many mechanical properties depend on the fraction of the sample which is crystalline, this form can be directly applied to the variation in properties during crystallization by replacing the concentration by the property directly in the equation.

Taking logarithms of both sides twice yields a linear type of equation which permits easy extrapolation of experimental data and the determination of the constant n from the slope of a plot of the left-hand side versus t:

$$\ln\left\{-\ln\left[\frac{C(\infty) - C(t)}{C(\infty)}\right]\right\} = \ln k + n \ln t$$

2.5. THE GLASS TRANSITION

Atomic Motions in a Polymer

The kinetic energy of the atoms bound within the long-chain molecules can result in the following:

  1. Independent atomic motions parallel and perpendicular to the molecular length. These motions occur within both the amorphous and the crystalline regions.

2. Random motion of the large portions of the chain within the amorphous region. Such motions, which do not result in total repositioning of the entire molecule, generate kinks within the molecule. Kinks of various sizes continually form and disappear. The free volume of an amorphous polymer decreases linearly with decreasing temperature as a result of the decrease in molecular motion which tends to separate the molecules. The availability of the free volume within the sample permits such motions, and decreasing temperature limits the kink formation both by diminishing atomic vibrational energy and by diminishing available free volume.

3. Random motion of the entire molecule within the amorphous region. While such motions may occur at random, stress-assisted motion will result in creep of the sample under applied load.

As the temperature is lowered, molecular motion decreases to such an extent that the long-range molecular motions producing chain migration or kinking can no longer occur. The temperature at which this transition occurs is called the *glass transition temperature* (T_g). The large-scale kinking above T_g generates rubbery patterns of behavior, while the stiffer material below the transition is the glassy form.

Changes in Volume with Temperature

Above the glass transition temperature the volume of an amorphous plastic decreases with decreasing temperature because of three effects:

1. A decrease in interatomic distance between atoms within the molecular chains.
2. A decrease in the free volume.
3. Changes in shape of the molecule as a result of molecular motion.

In the crystalline portion of a polymer the change in volume with temperature is due only to a decrease in the interatomic distances. Below the glass transition temperature molecular motion which results in kinking is effectively frozen, and the free volume therefore remains constant with continued decrease in temperature. Continued decrease in volume with decreasing temperature below T_g is therefore due only to the continuing contraction of the chains themselves. Figure 2.17 shows the change in volume as a function of temperature, indicating the change in slope at T_g due to the freezing in of the free volume at that temperature. The behavior of a crystalline polymer is also shown for comparison. At the melting point there is a discontinuous change in volume for the crystalline sample, but since the material is ordered above and below T_g, no change is observed at T_g. Conversely, a fully amorphous sample will exhibit no change at the melting

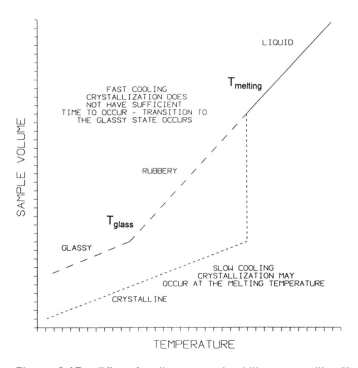

Figure 2.17. Effect of cooling rate on the ability to crystallize. Slower cooling rates may produce crystallization. If crystallization does not occur, the rubbery material will exhibit a glass transition. The crystalline material will not.

point but a marked reduction in thermal contraction coefficient at the glass transition temperature. If the material is partially amorphous, a smaller change will be observed at the freezing point, where some crystallization occurs, and some change will occur also at T_g, where the amorphous material becomes glassy. The greater the fraction of the sample that is amorphous, the greater the change at T_g and the smaller the change at the melting point. The exact temperature at which the glass transition occurs depends on the rate of cooling, higher cooling rates resulting in a slightly lower T_g.

The Glass Transition as a Thermodynamic Transition

A first-order phase transition is defined as one in which a discontinuity occurs in the first derivative of the free energy (G) with respect to temperature at constant pressure. At constant pressure:

$$\left(\frac{dG}{dT}\right)_{trans} = -S$$

where G and S are the free energy and entropy of the system, and T is the temperature of the transition. Therefore there is a discontinuity in the entropy (and other thermodynamic properties) at the transition temperature.

A second-order phase transition is defined as one in which the first derivative of the free energy is continuous, but the second derivative is discontinuous. Since at constant pressure:

$$\left(\frac{dG^2}{dT^2}\right)_{trans} = \frac{C_p}{T}$$

the heat capacity (C_p) therefore exhibits a discontinuity at the transition temperature.

Although there are some similarities between the glass transition and a second-order phase transition, the exact temperature at which the change in slope of the volume-temperature curve occurs depends on the cooling rate. The heat capacity versus temperature curve changes slope at T_g but does not become discontinuous. The glass transition cannot therefore be classified as a thermodynamic transition but is a function of the kinetics of the molecular mobility of the plastic.

Free Volume at T_g

Although no large-scale kinking or chain movement occurs below T_g, total motion cannot be frozen in at any temperature above absolute zero. The equilibrium free volume is a function of temperature only. When a sample is held below T_g for extended periods the molecular motion reduces the free volume to its equilibrium value, resulting in some contraction of the material. Figure 2.18 illustrates this behavior. The equilibrium free volume of the amorphous material should decrease along the line labeled "equilibrium." At T_g the free volume becomes fixed at a value v_g which depends slightly on cooling rate. However, for the usual range of cooling rates observed during processing of polymers, the v_g values do not change dramatically. If the material is held at a temperature T below T_g, the free volume will tend to decrease from v_g to its equilibrium value at temperature T, and the total volume will shrink slowly.

Data for polyetherimide (Yu et al., 1991) shown in Fig. 2.19 are for samples quenched to various temperatures from 220°C. T_g for this material is 215°C. The specific volume decreases with holding at temperatures below T_g as the free volume tends to decrease toward the equilibrium value at that temperature. For temperatures of 207°C and 196°C, the specific volume becomes constant with time, indicating that the equilibrium free volumes for those temperatures were reached in the time of the experiment. For lower holding temperatures the continuous decrease of free volume with time

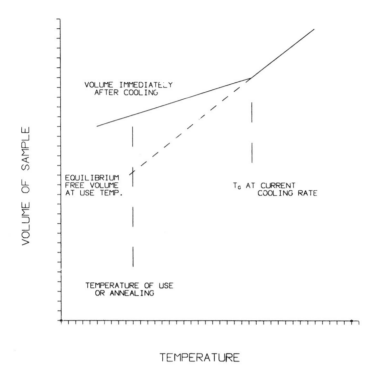

TEMPERATURE

Figure 2.18. The effect of holding below T_g on the free volume. Extended holding at temperatures below T_g will permit the free volume to approach the equilibrium value, decreasing the volume and increasing the density of the sample to the equilibrium value at that temperature.

indicates that equilibrium was not reached in the testing time. The data also show clearly that the equilibrium specific volume is smaller for lower testing temperatures.

Values of T_g can vary dramatically between samples of different molecular weights, number and types of side chains, and atactic/isotactic ratio. Furthermore, experimental determinations of T_g on the same sample by different methods often give different value. Values obtained by scanning differential calorimetry are frequently lower than those obtained by mechanical measurements. The values given in Table 2.1 are therefore only estimates, and the values for an individual sample may differ sharply from the values listed.

Crystalline regions may undergo transformations by changing to different crystal structures on cooling.

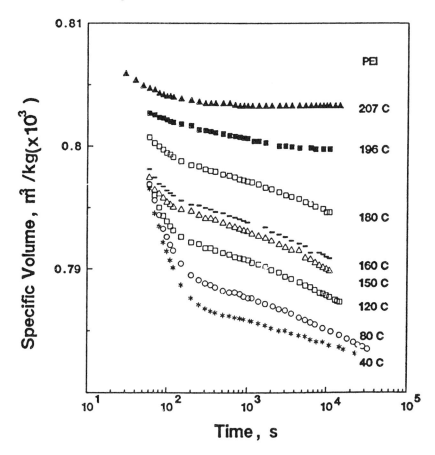

Figure 2.19. Approach of specific volume to equilibrium value during annealing of poly(ether imide). The specific volume of samples held at constant temperature shows an initial rapid decrease with time due to equilibration of the sample with the equipment. The slow change after that is due to the sample approaching its equilibrium free volume at that temperature. [From J. S. Yu, M. Lim, and D. M. Kalyon, *PES 31*, 145 (1991). Reprinted by permission of the publishers.]

Effect of Molecular Morphology

Because the glass transition results from freezing of molecular motion, the stronger the intermolecular forces between molecules, the higher the temperature at which the transition to a glass occurs. Table 2.1 gives the transition temperatures for selected polymers, and in general the higher the melting point the higher T_g, but no exact correlation between the two has

Table 2.1. Transitions in Common Polymers (°C)

Plastic	T_{mp}	T_g	Other transitions
Acetals	160 to 165	-15 to -75	-70
Polycarbonate PC	225	149	-120
Polyethylene PE	115 to 142	-80 to -110	~ -105
Polyethylene terephthalate PET (Dacron)	256	67	
Polypropylene PP	165 to 175	~ 5	$\sim -20, -80$
Polystyrene PS	230	95	-110
Polytetrafluoro- ethylene PTFE	325	115	-87
Polyvinylchloride PVC	240 to 250	70	-25

been developed. Dacron is primarily crystalline (95%), but the glass transition is observed in the amorphous regions.

The ends of the molecules are less restrained than the middle of the chain, and therefore the greater the molecular weight, the higher T_g. Generally, the dependence of T_g on molecular weight follows

$$T_g = T_g^\circ - \frac{A}{M_n}$$

Figure 2.20 shows the general variation of the volume versus temperature for samples of various DP values. The change in slope on decreasing temperature indicates the glass transition.

Effect of Copolymerization

Copolymerization tends to produce greater randomness of structure with reduced intermolecular bonding as compared with pure polymers. The extent of crystallization and T_g therefore tend to decrease as the percentage of the other polymer introduced is increased. Figure 2.21 shows the general

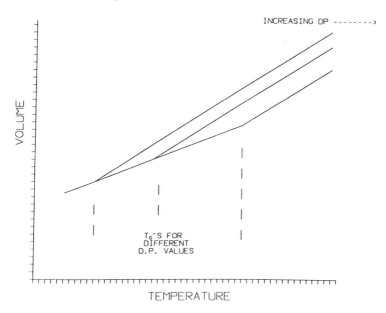

Figure 2.20. Dependence of T_g on DP. Increasing molecular length tends to decrease the free volume due to fewer chain ends, and increases the glass transition temperature accordingly.

decrease in crystallinity and the lowering of T_{melt} and T_g observed as a result of copolymerization. If a block copolymer is produced, two independent transitions may be observed as freezing of kink motion in the separate parts of the molecule occurs (Fig. 2.22).

Effect of Cross-Linking

Cross-linked polymers cannot crystallize, but kinking of the molecules between the cross-links occurs above T_g. Since thermosets are amorphous, all exhibit a glass transition. The T_g increases in a generally linear manner with increasing cross-link density.

Effect of Blending

If the two polymers which are blended are immiscible, the resultant blend has the characteristics of each of the components. Two distinct glass transition temperatures will be observed, each corresponding to the freezing of molecular motions in one of the components, similar to that shown for block copolymerization in Fig. 2.22. Less commonly, the blend may be

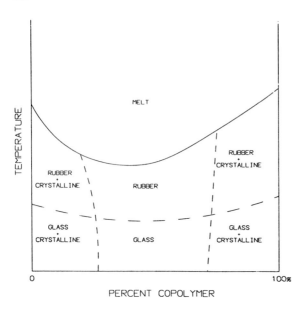

Figure 2.21. General dependence of structure on extent of copolymerization. Copolymerization decreases the extent of crystallinity, and decreases the melting temperature and the glass transition temperature.

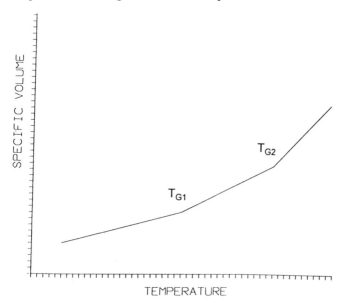

Figure 2.22. Specific volume vs. temperature for a block copolymer. Two glass transitions may be observed in the separate regions of the two blocks.

between miscible polymers and the resultant polymer has one glass transition, at a temperature between those of the two components. For such miscible blends Fox (1956) has proposed the equation for the glass transition temperature of the blend:

$$\frac{1}{T_{g(blend)}} = \frac{wt \% A}{T_{g(a)}} + \frac{wt \% B}{T_{g(b)}}$$

where $T_{g(a)}$ = glass transition temperature of component A, etc.

  wt $\%$ A = weight percent of component A, etc.

Other Transitions

The glass transition is the most easily identified and the most prominent change that occurs in polymers below their melting points. However, other molecular transitions occur in many polymers which are related to changes in the types of molecular motions possible. Generally, in amorphous polymers the transitions are due to the freezing of motion of small side chains, small polar groups on the main molecular chain, or small groups of two to four carbon atoms in the main chain. Since the transition results in further reduction in atomic mobility, when such a secondary transition occurs, there is a slight further decrease in the thermal expansion coefficient of the polymer at temperatures below the transition. Most common polymers exhibit these transitions below room temperature. Some of the transitions observed in common polymers are listed in Table 2.1.

Effect on Mechanical Properties

The glass transition temperature is an important concept in understanding the underlying molecular behavior of polymers. Because molecular motion becomes frozen at the glass transition, the mechanical properties in general change drastically around that temperature. Above T_g the material is amorphous and rubbery, generally exhibiting relatively low strength and large ductility and toughness. In the region around T_g the modulus tends to increase dramatically, reaching the higher value of the glassy material after this transition region. However, T_g is not always a reliable indicator of the temperature at which the material converts from a ductile rubbery material to a brittle one. The heat deflection temperature is determined by a simple mechanical test in which a beam of a sample is loaded and the temperature at which excessive distortion occurs under the load is found. The results of this test are not directly applicable to design of samples for use at elevated temperatures, because the rate of heating will not be the same as for an actual application. Nevertheless, the difference between the temperature at

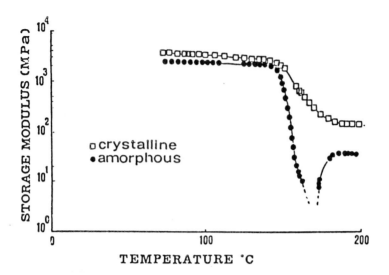

Figure 2.23. Changes in the modulus of PEEK for amorphous and crystalline material. The modulus of the crystalline materials at low temperatures is only slightly above that of the glassy material. The large drop in modulus of the amorphous material is due to the transition to the rubbery state. At higher temperatures crystallization of the rubbery material begins, increasing the modulus again. [From A. D'Amore, J. M. Kenney, and L. Nicolais, *PES 30*, 314 (1990). Reprinted by permission of the publishers.]

which a sample deflects under those test conditions and the glass transition temperature of some materials is startling. Acetals exemplify this difference. The T_g is below 0°C (see Table 2.1) but the heat deflection temperature is close to the melting point. Direct experimental determination of the applicability of a material for a specific temperature range of use is necessary.

The behavior of any polymer can be understood only by recognizing all the possible changes in molecular morphology. The modulus of PEEK was determined as a function of temperature by D'Amore et al. (1990) and is shown in Fig. 2.23 for partly crystalline and completely amorphous samples. The rapid drop in modulus above T_g due to the conversion of the amorphous regions from glassy to rubbery is evident. However, the amorphous material exhibits a much greater drop, since only part of the crystalline sample experiences the transition from glassy to rubbery material. The amorphous material then exhibits an unexpected increase in modulus just above T_g. This is a result of the increase in molecular mobility in this temperature range that permits crystallization to occur, with resultant improvement in modulus.

3
Solubility and Additives

3.1. SOLUBILITY

Predictions of Solubility

Predicting the solubility of two materials is an extremely important industrial problem that occurs throughout the chemical industry. The polymer industry requires solvents during the many stages of production of the polymers and during processing of the polymers, and utilizes the concepts of solubility to add components to the polymers to improve their properties. The rate of polymerization in a solvent is affected by the solubility of the resultant polymer in the solvent. Polymer blends require that the various components be compatible. In general, the solubility of plastics with each other decreases as the molecular weight increases. While small molecules such as inorganic and organic liquids are frequently soluble in any proportions, long polymer molecules are almost invariably insoluble in each other, and soluble blends are very difficult to obtain.

Other phenomena involving solubility of plastics besides blending concern the plastics industry: the dissolution of plastic in a solvent or the dissolution of a desirable additive in the plastic. During processing of block and graft copolymers, the type of solvents employed can affect the structure and final properties of the material. As a result, many procedures for predicting solubility have been developed. The most direct and commonly employed parameters are the Hildebrand and Hansen parameters, which will be discussed below. However, many other parameters and methods of predicting solubility have been proposed, and are more effective in some cases. Extensive lists of many of these parameters are available, so that predictions of solubility may be readily performed. The accuracy of prediction is not known for any given combination of solvent and solute, and these predictions can only be used as guides for experimental testing.

For any reaction, such as the solution of one material in another, to occur, the change in Helmholtz free energy (ΔA) of the reaction must be negative:

$$\Delta A = \Delta E - T(\Delta S) \tag{3.1}$$

where ΔE = change in internal energy of the system

ΔS = change in entropy of the system

Note that, in general, a solution is more disordered than the two pure separated components, so that the change in entropy is positive for the formation of most solutions. Since the term containing entropy has a negative sign, the disorder on formation of a solution always tends to produce solubility. However, the entropy of mixing of two materials depends on the number of molecules involved, the entropy change increasing with the number of molecules. If an ideal solution forms, the entropy change is dependent on the number of permutations of the different molecules within the system. For small molecules, this results in a large entropy change, since so many possible rearrangements may occur. For the large molecules in polymers, the number of permutations is orders of magnitude less, and the entropy change is much smaller and is frequently neglected in comparison with the change in internal energy in Eq. 3.1. For polymers, therefore, the determination of whether solution will occur rests primarily with the sign of the internal energy change.

Loosely following the reasoning of Hildebrand, to obtain a crude estimate of the possibility of solubility, consider two materials A and B. In each pure material, bonding of neighboring molecules occurs. Let

E_{A-A} = energy of a bond between two A molecules

E_{B-B} = energy of a bond between two B molecules

E_{A-B} = energy of a bond between an A and a B molecule

N = number of nearest molecules to which bonding occurs

A solution can be considered to form if we take an A molecule from material A and replace it with a B molecule. This will break N A–A bonds and N B–B bonds and form $2N$ A–B bonds. The energy change during this formation of a solution is therefore

$$\Delta E = 2NE_{A-B} - N(E_{A-A} + E_{B-B}) \tag{3.2}$$

If the net energy change in this reaction is zero ($\Delta E = 0$), a solution will form and the molecules have no preference about their nearest neighbors, forming a truly random or ideal solution. In this case, with $\Delta E = 0$:

$$E_{A-B} = \frac{E_{A-A} + E_{B-B}}{2}$$

If the net energy change is positive, then

$$E_{A-A} > \frac{E_{A-A} + E_{B-B}}{2}$$

and since the energy of the bonds between dissimilar molecules is greater than that between similar molecules, no solution will occur. With small molecules solutions may still occur because of the entropy effect of increased disorder, but in polymers, as discussed above, this is usually insufficient to effect solution. In this case the two polymers will remain as separate phases, and no solution will occur.

If the net energy change is negative, then

$$E_{A-B} < \frac{E_{A-A} + E_{B-B}}{2}$$

and bonding between dissimilar molecules is preferred, producing a strong interaction. This is uncommon in polymers.

The bond strength between neighboring molecules can be determined from the heat of vaporization of the material, since the energy required to convert the material into the gaseous phase equals the energy needed to break all the intermolecular forces. If:

ΔE_{vap} = energy required to vaporize one mole of material

computing the enthalpy change during vaporization:

$$H = E + PV$$

$$\Delta H_{vap} = \Delta E_{vap} + P(V_{gas} - V_{liq})$$

However, the volume of the liquid is significantly less than that of the gaseous phase:

$$\Delta H_{vap} = \Delta E_{vap} + P(V_{gas})$$

and if we consider the gaseous material to be an ideal gas, so that we may use $PV = RT$ for one mole of material:

$$\Delta E_{vap} = \Delta H_{vap} - P(V_{gas}) = \Delta H_{vap} - RT$$

This can therefore be used as a measure of the bonding energy within a material. The Hildebrand solubility parameter is defined as the bonding energy per unit volume:

$$HB = \left(\frac{\Delta H_{vap} - RT}{V}\right)^{0.5}$$

The square root comes from analogy with other thermodynamic functions of solutions. Hildebrand considered that solutions would tend to form, as discussed above (see Eq. 3.2), if the total bonding energies of the two materials forming the solution were similar, and he proposed the general equation for the energy change when a solution formed from two separate materials:

$$\Delta E = c(HB_1 - HB_2)^2 \cdot VF_1 \cdot VF_2$$

where $HB_{1,2}$ are the solubility parameters of the two components, $VF_{1,2}$ are the volume fractions of the two components, and c is a constant. Note that if HB_1 and HB_2 are equal, the criterion for complete solubility is satisfied, and as the two solubility parameters become farther apart in numerical value, the energy of mixing increases, making solution less likely.

General rules for the solubility of a polymer in a solvent in terms of the solubility parameters are as follows. If

$HB_{plastic} = HB_{solvent} \pm (2)$ Thermoplastics dissolve

Thermosets swell up to 20%

$HB_{plastic} = HB_{solvent} \pm (2 \text{ to } 5)$ Thermoplastics swell up to 10%

Thermosets swell slightly

$HB_{plastic} = HB_{solvent} \pm (>5)$ Thermoplastics slightly well

Thermosets hardly swell

For a solvent composed of several organic molecules, the solubility parameter can be found as the volumetric average of the solubility parameters of the two solvents (if the solvents dissolve in each other):

$$HB = HB_1 \cdot VF_1 + HB_2 \cdot VF_2$$

Hildebrand solubility parameters of solvents can be found from the heat of vaporization. Since polymers decompose before vaporizing, this technique is not applicable. The solubility parameter of polymers can be estimated as being equal to that of the solvents in which they readily dissolve. The solubility parameter can also be theoretically computed from the chemical formula, as the contribution to the bonding energy of each bond within the

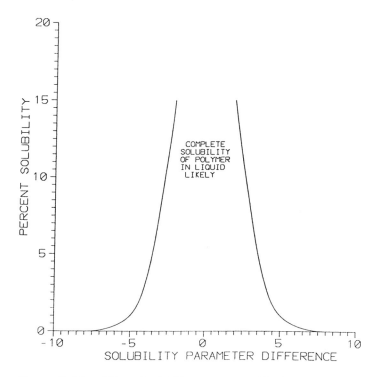

Figure 3.1A. Effect of solubility parameter of liquid environment on the solubility of a thermoplastic.

material can be estimated. Extensive tables of solubility parameters are available to assist in selecting a solvent for a particular plastic. One of the most extensive compilations is the *Handbook of Solubility Parameters and Other Cohesion Parameters* (CTC Press, Boca Raton, Florida, 1983). Modifications of the Hildebrand solubility parameter to adjust the parameter for the particular types of bonding in the materials have also been suggested, and these additions are also listed in the literature (see reference above). Schematic diagrams showing the effect on the solubility of the solvent in a thermoplastic and the swelling of a thermoset on the difference in solubility parameters of the solvent and polymer are shown in Fig. 3.1A and Fig. 3.1B.

Viscosity of solutions

In a good solvent, the molecules of a polymer spread out and are similar in appearance to the liquid or rubbery phase. In a poor solvent, a molecule tends to attract like molecules and remains in a balled configuration

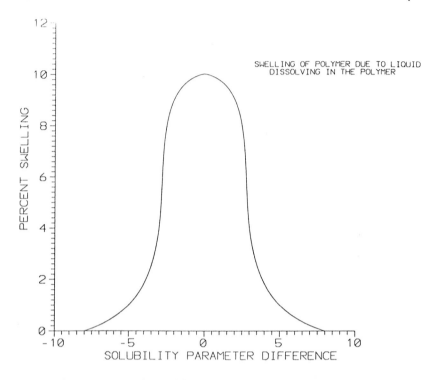

Figure 3.1B. Effect of solubility parameter of liquid environment on the swelling of a thermoset.

As a result, the viscosity of a polymer-solvent solution is higher for a good solvent, since the extended molecule will resist flow under an applied shear stress to a greater extent than a balled molecule. A schematic diagram showing the variation of viscosity with the difference in solubility parameter between the low molecular weight solvent and the solute polymer is shown in Fig. 3.2.

A small list of typical values of solubility parameters is shown in Table 3.1. It can be seen that water will not dissolve any of the polymers listed and that trichloroethylene should be the best solvent (of those in the list) for polystyrene (PS).

Hansen Solubility Parameter

The Hildebrand solubility parameter was derived assuming the solution formed by the two components was an ideal one, in which the interactions are sufficiently small so that no heat evolution or absorption occurs when the solution formed. This criterion is effective if the materials are bonded

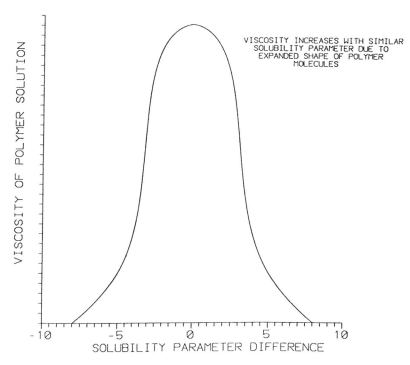

VISCOSITY INCREASES WITH SIMILAR
SOLUBILITY PARAMETER DUE TO
EXPANDED SHAPE OF POLYMER
MOLECULES

(y-axis) VISCOSITY OF POLYMER SOLUTION

(x-axis) SOLUBILITY PARAMETER DIFFERENCE
-10   -5   0   5   10

Figure 3.2. Effect of solubility parameter difference between solvent and solute on the viscosity of a solution.

primarily by dispersion forces, the weak interaction between the moving electrons of one atom and the nucleus of another atom. However, if there are strong dipole interactions, or hydrogen bonding occurs between molecules, the Hildebrand solubility parameter does not give an accurate prediction of solubility. These additional interactions should be included in an evaluation of the bonding forces between atoms. The total cohesive force holding the atoms together is the sum of the contribution of these forces:

$$\Delta E_t = \Delta E d_d + \Delta E_p + \Delta E_h$$

where ΔE_t = total energy of attraction

$\Delta E_d, \Delta E_p, \Delta E_h$ = contributions of the dispersion forces, dipole interactions, and hydrogen bonding to the total attractive force

Continuing in the same manner as for the Hildebrand parameter, a parameter (δ) relating to each of these terms is obtained by dividing each by the molar volume to establish an energy density.

Table 3.1. Solubility Parameters

Solvents	Solubility parameter $(\text{cal-cm}^3)^{0.5}$
Acetone	10
Bromobenzene	18
Chlorobenzene	17
Chloroform	11
Ethanol	13
Methyl chloride	10
Tetrahydrofuran	13
Trichloroethylene	9.4
Water	23
Xylene	9
Polymers	
Polyethylene (PE)	8
Polymethyl methacrylate (PMMA)	9
Polypropylene (PP)	8
Polystyrene (PS)	9.5
Polyvinyl chloride (PVC)	9.5

$\delta_d, \delta_p, \delta_h$ = dispersion, polar, hydrogen bonding parameters

The total Hansen solubility parameter (δ_t) is defined as:

$$\delta_t^2 = \delta_d^2 + \delta_p^2 + \delta_h^2$$

The Hildebrand parameter corresponds to δ_d. The Hansen parameter is very effective in some cases. Extensive lists of many of these parameters are available, so that predictions of solubility may be readily performed. The accuracy of predictions is not known for any given combination of solvent and solute, and these predictions can only be used as guides for experimental testing.

3.2. ADDITIVES*

The large number of polymers available suggests that the properties desired for a specific product should be readily obtainable without the necessity

---

* Sections 3.2 to 3.4 are based on the article "Property Modification by Use of Additives", E. Miller in *Engineered Materials Handbook—Engineering Plastics*, Vol. 2, 1988, Courtesy ASM International, Materials Park, Ohio.

of utilizing additional chemical compounds. However, most polymers have properties which can be greatly improved by additions of small amounts of other chemical components. Additives are utilized either to increase the ease of processing of the polymer or to improve the properties of the final product. Processing additives improve the processing characteristics of the polymer by:

1. Increasing the lubricity of the polymer, so that less energy is required for manufacturing and less mechanical degradation of the polymer occurs during processing.
2. Stabilizing the polymer so that less degradation occurs at the elevated temperatures encountered during processing.

Additives to improve properties include chemicals to improve flame retardation characteristics, color, light stability, weather resistance, impact resistance, density, static charge buildup, and mechanical properties. Additives may be present in small quantities or may be a large fraction of the total material. For example, a typical formulation for a colorless flame-retardant PVC sheet is:

53.5% suspension PVC
32.5% plasticizers
 8.0% extenders
 5.0% flame-retardant agents
 0.5% lubricants
 0.5% stabilizers

Additives need not be incorporated in large percentages. The composition of a clear transparent polystyrene molding would be typically:

97% polystyrene
 3% flame-retardant agents

Methods of Addition

The additives are usually combined with the polymer during the polymer production process. Additions are performed by several methods:

1. The additives can be mixed with monomers prior to polymerization. For this method to be satisfactory, the additives must either be soluble or be stabilized in an emulsion in the solution in which polymerization occurs and the additive must be stable under the conditions of polymerization. The additive must also have little effect on the kinetics of polymerization. Antioxidant additions to impact-resistant blends of polystyrene (PS) and PVC are frequently incorporated at this stage.

2. More commonly, the additives are introduced into the previously polymerized material in the form of powder, paste, or granules. Each type of polymer requires a different method of addition. Powders are mixed with polymers below the softening point of the polymer to prevent agglomeration of particles before complete mixing is effected. Granulated polymer particles are rather large for effective mixing with some additives. The additive-polymer mixture is therefore further ground to reduce the particle size. In PVC, the plasticizer is mixed at low temperatures with the solid to form the paste for production.

3. Additives may be added during processing. Colorants, blowing agents, and fillers are frequently added by the processor, as the exact conditions of processing will determine the amount and technique of additive addition.

3.3. ASSOCIATED PROBLEMS

To be useful, an additive should generally provide enhancement of a specific property or set of properties without significantly degrading the other properties of the plastic. Additives should not produce any discoloration of the polymer over the expected lifetime of the part. Since additives are complex chemicals, they can interact with each other and the bulk plastic in various ways, producing unexpected deleterious effects. Some of the common problems associated with the presence of additives in plastics are listed below.

1. Incomplete mixing during compounding will produce small hard particles approximately 500 µm in diameter scattered throughout the plastic part (flecks). These are small particles of inadequately dispersed additives.

2. Insufficient stabilizers (additives to prevent decomposition) or a nonuniform distribution of stabilizers in the compound may result in small burnt particles occurring throughout the part if temperatures during processing exceed the upper limit of the plastic's stability.

3. Decomposition of additive chemicals can generate water vapor or other gaseous products which produce small bubbles in the molded part. Bubbles may also be caused by additives which oxidize at the molding temperature or by moisture introduced with additives which have not been completely dried prior to compounding.

Bleeding and Blooming

Additives must be stable within the polymer compound. For such stability the additives must either:

1. Form a stable second phase with little tendency for migration. Examples of stable second phases which exist in polymers are solid filler particles, small pigment particles, or extended networks of impact modifiers. The stability of these second-phase particles is due to their low vapor pressure and low diffusion coefficients.

2. Form a stable solution within the polymer. High molecular weight molecules are in general not miscible with each other, and two phases tend to form if the additive is a long-chain polymer. Lower molecular weight additives may form solutions if the chemical bonding in the additive is similar to that within the polymer. The solubility of the additive should be sufficient at all temperatures, from that of processing to below room temperature, to ensure that the additive remains in solution over the entire operating temperature range. Decreases in solubility with temperature or insufficient solubility at room temperature will result in the additive being exuded. The continuous slow loss over a period of time due to insufficient solubility is called bleeding.

Blooming occurs when the additives form a supersaturated solution in the polymer at elevated temperatures. If the diffusion rate in the polymer is sufficiently large, the additive may migrate to the surface. When the polymer is cooled from the elevated processing temperature, the additive will then precipitate on the surface, resulting in a surface layer of the additive which appears as a tacky or sticky surface film. Since the diffusion coefficient can be orders of magnitude different from one polymer-additive system to another, the tendency for blooming varies widely between polymer-additive compounds. For example, the tendency for blooming of antioxidants is about ten times greater in low-density polyethylene (LDPE) than in high-density polyethylene (HDPE). The diffusion rate depends on the extent of crystallinity in the polymer, the extent of orientation, the spherulite size, and other morphological characteristics. These vary with processing temperature, pressure, and flow rate. The tendency for blooming therefore also depends significantly on the particular processing parameters.

Plate-out

Plate-out refers to the deposition of some additive-containing plastic on the surface of the molding or processing equipment during manufacture. There is a distinct difference between plate-out and blooming. Materials such as inorganic pigments and fillers, which do not exhibit blooming because they are not soluble in the plastic at any temperature, are carried along and appear in the plate-out. With calendered films, plate-out on the rolls causes surface defects and in extrusion dies can lead to variations in dimensions of the extrudate. Excessive cross-sectional variation or breaking away of sections from the main part can cause rejection rates to become very high. Plate-out is caused when some additive (such as the stabilizer or lubricant) is not satisfactorily compatible with the plastic. Phase separation of the additive takes place at the elevated temperatures during processing. Pigments and other additives dissolve in the insoluble additive phase and are carried along to the plate-out defect.

Chalking

Chalking describes the disintegration of the plastic's surface to such an extent that the pigment or filler particles present are exposed and form a dustlike coating on the product. Chalking is due to thermal or light-accelerated surface oxidation which affects the plastic but does not affect the more stable additive particles.

Toxicity

Plastic additives include a wide range of chemicals. Additives must have sufficiently low toxicity to be acceptable for use, both during processing and during the actual performance of the finished product. These requirements are particularly severe if the final product is to be utilized in the food, packing, textile, or toy industries and in other applications where extended handling by the user is likely.

In manufacturing, the greatest danger of toxic inhalation, ingestion, or skin contact arises before the additive is compounded with the polymer. Locations where large packages of additives are opened and emptied into hoppers are the areas of principal concern. Protective clothing must be worn by operators involved in these processes. Short-term exposure after they have been mixed with the polymer is much less hazardous.

Additive materials which may come in contact with food must meet special requirements. Of particular concern is the possibility that the additive is soluble in some foodstuff and can diffuse into it after extended contact. The Food and Drug Administration has specified that testing of such diffusion can be simulated by having the plastic compound come in contact with various standard solutions and determining the uptake of the additive by the simulating solution as a function of time. The testing procedure specifies that the plastic compound be held at $49°C$ until the change in the amount of additive absorbed by the test solution per unit time remains constant. Animal testing is required to determine the toxicity of new additives migrating from a plastic. Such testing is time consuming and expensive, and new additives are not readily approved for the food industry.

3.4. ADDITIVE TYPES

A discussion of the various additives is given below. It is not intended that a complete listing of the various chemical additives be included in this section, as new additives, changes in concentrations, and modifications of additives are being developed continually. Specific information can best be obtained from the manufacturers.

Heat Stabilizers

All polymers are susceptible to degradation by reacting with the oxygen present in the atmosphere. The severity of the oxidation and the severity of the effects of this oxidation on the properties depend on the conditions of use and the polymer type. Polymers with a high intrinsic resistance to oxidation would be likely to be chosen for parts designed for severe conditions. In addition, antioxidants are utilized routinely to further stabilize polymers against degradation.

The principal effects of oxidation are as follows.

1. Oxidation may cause a change in appearance of the plastic. Most polymers tend to be naturally yellowish in color as a result of absorption of light in the blue color range. Oxidation tends to increase this yellow color in most polymers.
2. Cross-linking may occur in thermoplastics as a result of oxidation. The surface of the sample becomes harder with increasing time of exposure to heat and air, leading to a loss of surface gloss. Chalking, a fine powdery residue of oxidized product and other additives which are broken free of the polymer during oxidation, may produce a fine coating on the surface.
3. Bulk properties may deteriorate. Impact strength, tensile strength, and elongation may all be degraded by oxidation.

Thermal Oxidation Processes

There are three general mechanisms for degradation of polymers:

1. Random degradation. Attack and decomposition occur at random throughout the polymer chain. As a result, degradation produces a polymer of significantly smaller molecular weight even in the early stages of degradation. Polyethylene (PE) exhibits this type of attack.

2. Chain unzipping. The ends of the chain are attacked, and complete degradation of the chain to monomer size is possible. Polymethylmethacrylate (PMMA) degrades by this mechanism.
3. Defect attack. Some polymers as manufactured contain defects in their long-chain structure which are characterized by weaker bonds that are readily attacked. PS prepared under certain conditions exhibits these defects.

All polymers react with ambient oxygen spontaneously. Oxidation occurs through a series of chemical reactions involving the formation of free radicals. Initiation of the reaction occurs by the generation of a free radical from the polymer chain:

$$RH \rightarrow R^{\cdot} + H^{\cdot} \qquad \text{(initiation of oxidation)}$$

Oxygen then reacts with the free radical, forming a peroxide radical which can attack other polymer molecules, perpetuating the reaction:

$$R^{\cdot} + O_2 \rightarrow ROO^{\cdot} \qquad \text{(peroxide formation)}$$

$$ROO^{\cdot} + RH \rightarrow ROOH + R^{\cdot} \qquad \text{(propagation)}$$

The hydroxyperoxide (ROOH) may also decompose to give other free radicals. The oxidation process may be terminated by several processes. The peroxide radical may decompose to yield oxygen and stable compounds, or the free radicals present may interact to form a stable constituent.

$$R + R^{\cdot} \rightarrow R\text{—}R$$

$$R^{\cdot} + ROO^{\cdot} \rightarrow ROOR \qquad \text{(termination reactions)}$$

$$ROO^{\cdot} + ROO^{\cdot} \rightarrow ROOR + O_2$$

To stop this thermal oxidation process the two free radicals, $R^{\cdot}$ and $ROO^{\cdot}$, which propagate the reaction must be eliminated. Under normal atmospheric conditions the concentration of $ROO^{\cdot}$ is far greater than that of $R^{\cdot}$. To decrease the rate of oxidation, therefore, the concentration of $ROO^{\cdot}$ must be reduced. Toward this end, two classes of antioxidants are produced. One group accelerates the chain termination process by containing an easily reacted hydrogen which reacts with the peroxide to end the propagation process. Most of these antioxidation compounds are substituted phenols.

Secondary antioxidants react with the hydroxyperoxides to prevent their decomposition to generate other free radicals. These antioxidants are principally sulfur-containing esters or esters of phosphorous acid. They behave synergistically with primary oxidants, increasing their effectiveness considerably. The concentration necessary for antioxidants to be effective is of the order of 1%.

Light Stabilizers

The energy available in the ultraviolet (UV) region of the spectrum is capable of causing many degradative reactions to occur in polymers. The ultraviolet radiation from the sun is partly absorbed by water vapor, oxygen, and carbon dioxide in the atmosphere. The remaining radiation is scattered by dust particles and clouds, although such scattering does not decrease the total amount of ultraviolet light reaching ground level. Approximately 6% of the total radiation that reaches the earth's surface is in the ultraviolet region of the spectrum and varies with daily weather conditions.

Damage occurs by absorption of energy from the ultraviolet radiation by the chemical bonds within the polymer. The light absorbed corresponds to the energy necessary to break the chemical bonds, changing and degenerating the chemical structure of the polymer. Table 3.2 gives the wavelength of the ultraviolet light which can interact with and destroy various types of bonds in the polymer.

To inhibit degradation as a result of ultraviolet radiation, light-stabilizing additives are incorporated into plastic parts for outdoor use. The concentration range usually employed is 0.05–2%. Other additives which prevent light from penetrating the polymer also act as light stabilizers. Among such additives are carbon black, pigments, and fillers.

Resistance to ultraviolet-generated degradation is a function of the polymer thickness, degree of orientation, extent of crystallinity, and presence of colorants, antioxidants, fillers, and other additives. As a result of scattering of light at the surface of the crystallite interfaces, the path length of the light through a partially crystalline polymer is significantly larger than in a purely amorphous polymer, and semicrystalline polymers are therefore generally more susceptible to damage by light.

The excitation of the molecule by adsorption of the photons in the ultraviolet range generates free radicals. The process is similar to and in

Table 3.2. Wavelength of Light Producing Bond Breakage

Bond	UV absorption wavelength (μm)
C—H	290
C—C	300
C—O	320
C—Cl	350
C—N	400

many cases difficult to separate from thermal oxidation. Although any bond can be activated and broken by ultraviolet light of sufficiently high frequency and energy, the main constituents which initiate photodegradation in most polymers are impurities, catalyst residues, hydroxyperoxides, carbon groups, and double bonds.

Degradation is initiated by the interaction of UV photons with a bond in the easily activated molecule, breaking the bond.

$$R\text{—}R + hv \rightarrow 2R^{\cdot}$$

$$R^{\cdot} + O_2 \quad \rightarrow ROO^{\cdot}$$

Chain propagation proceeds primary by reaction with oxygen as in the case of thermally generated oxidation:

$$ROO^{\cdot} + RH \rightarrow ROOH + R^{\cdot}$$

Also as in thermal oxidation, chain termination occurs by several possible reactions:

$$R^{\cdot} + R^{\cdot} \rightarrow R\text{—}R$$

$$R^{\cdot} + ROO^{\cdot} \rightarrow ROOR$$

$$ROO^{\cdot} + ROO^{\cdot} \rightarrow ROOR + O_2$$

Stabilization may be effected by several techniques:

1. Incorporate additives which absorb the light before it can come in contact with the degradation-initiating center. The harmful UV radiation is absorbed by the stabilizer, and the energy is dissipated as heat. These compounds must therefore have high UV absorption and heat and light stability. Such stabilizers are most useful in thicker articles, where sufficient depth exists for complete absorption of the ultraviolet radiation. The surface may not be well protected by this type of additive. The principal additives in this category are hydroxybenzophenones and hydroxyphenylbenzotriazoles.
2. Deactivate the excited state of the initiating center. Such reactions produce energy, which is generated either in the form of heat or as fluorescent radiation.
3. Transform the hydroxyperoxides generated into other stable forms which do not generate free radicals. Metal complexes of sulfur-containing compounds are very efficient in this regard and can be used at low concentrations effectively. Dialkyldithiocarbamates, dialkyldithiophosphates, and thiobisphenolates are used.

4.   Incorporate additives which react with the free radicals formed to stop the reaction from proceeding. Hindered amines most likely operate by this mechanism.

The exact mechanism by which commercial stabilizers perform is not precisely known, and they probably operate by more than one of these mechanisms simultaneously. Other additives can affect the light stability. Some colorants can lower the light stability by several orders of magnitude. This effect is most severe for yellow and red organic pigments. Conversely, phthaloblue and phthalogreen increase the UV stability by acting as UV absorbers, preventing the radiation from penetrating into the bulk of the product (process 1 above). Pigmentation affects the surface temperature of a product exposed to sunlight. Surface temperature differences of close to 20°C have been measured between white and black PVC products exposed to the same conditions, resulting in accelerated oxidation of the darker specimens.

Plasticizers

Plasticizers are additives which lower the glass transition temperature of the plastic and increase the free volume, thereby affecting the properties associated with these two parameters. As a result, these additives make the polymer more flexible, increase elongation to failure, lower the melting point, lower the elastic modulus, and increase the impact resistance. In addition to the effect on properties, plasticizers improve the processing ease. Low concentrations of plasticizers may produce some embrittlement. Some plasticizers must therefore be added in concentrations above a minimum value.

Bleeding of Plasticizers

In order to be effective, plasticizers must be capable of dissolving in the bulk polymer. As a result, their diffusion rates are rather high, and bleeding is a common problem associated with the use of large percentages of plasticizer. Even when bleeding does not normally occur, diffusion of the plasticizer from the bulk plastic to another sample in contact with it is possible. If two parts containing the same plasticizer are in contact, the plasticizer will tend to diffuse into the part which contains the lower concentration of plasticizer.

PVC is the polymer most affected by plasticizers, and their utilization is responsible for the tremendous range of properties and products prepared from PVC-based polymers. PVC plasticizers are subdivided into three groups:

1.   *Primary plasticizers* have solubility parameters sufficiently close to that of the base material that the plasticizer will not bloom or bleed

from the bulk material. They affect the properties sufficiently to gel the polymer within the processing temperature range.

2. *Secondary plasticizers* have solubility parameters that are not close to that of the bulk polymer and, if utilized alone, may bloom or bleed. When used in conjunction with a primary plasticizer, however, they form stable solutions with the polymer. They have a lower gelation ability than primary plasticizers.

3. *Extenders* have solubility parameters quite different from that of the polymer and will bleed extensively if used alone. Their low cost makes it advantageous to use them in conjunction with primary plasticizers. They remain in solution in the combined polymer–primary plasticizer compound and reduce the overall cost.

Solubility Temperature

A parameter which aids in the evaluation of plasticizers for PVC is the critical solubility temperature. The lower the value of this parameter, the better the gelation capacity of the plasticizer for PVC. This permits reduction in processing temperature and greater ease of mixing of the plasticizer and the PVC. Values for common plasticizers are shown in Table 3.3.

Lubricants

Lubricants are additives which improve the processing properties of plastics. Without lubricants, processing of PVC would not be possible, and the high production rates of other polymers could not be maintained. Processing is

Table 3.3. Critical Solubility Temperature for PVC Plasticizers

Plasticizers	Critical temperature (°C)
Tributyl phosphate	58
Dibutyl phthalate	90–95
Butyl benzyl phthalate	96–100
Tricresyl phosphate	101–105
Dioctyl phthalate	116–120
Dioctyl adipate	121–125
Dioctyl sebacate	161–165

From: W. Sommer, Plasticizers, in *Plastics Additives*, 2nd ed. R. Gachter and. H. Muller, eds. Cincinnati, Hanser, 1983, pp. 252–296.

normally performed close to the decomposition temperature of the plastic. Although increasing the temperature would decrease the viscosity and increase flow rates, this tends to degrade the material to an intolerable degree.

Lubricants are supplied to the compounder as a powder with a relatively low melting point or as a liquid. Most lubricants are added in the concentration range 0.5–1%. They function by:

1. Lowering the viscosity of thermoplastics over the processing temperature range. Although most often molded in the liquid state, the thermoplastic liquids have very high viscosities. It is desirable to lower the viscosity and thereby lower the energy required for production and to decrease wear on the equipment.
2. Reducing the external friction of the polymer melt moving relative to the parts of the molding equipment with which it makes contact. This external friction is a function of the tendency of the melt to adhere to the metal parts of the molding equipment. If adherence occurs, the surface of the finished product appears rough and torn. High gloss and smoothness of the finished products are a direct result of proper lubrication.

Requirements for Lubricants

1. No bleeding of the lubricant should occur during processing. Bleeding of lubricants is particularly serious, as the lubricants tend to dissolve other additives and transport them to the walls of the equipment. This causes a layer of the deposited additives to build up on the walls of the processing equipment, generating unsightly surface defects (plate-out) in the finished products.
2. The lubricants must be thermally stable at the elevated temperature of molding.
3. Since the lubricants' function ends when processing is complete, it is necessary that they have little effect on the mechanical properties of the polymer.

Impact Modifiers

Impact modifiers are rubbery high polymers added in large percentages and could be considered as parts of the bulk plastic rather than as additives. Figure 3.3 shows the relation between impact resistance and flexural modulus for impact-modifier polypropylene (PP). The resultant compound exhibits a structure more suitable for resistance to crack propagation and with more mechanisms for impact energy absorption without crack initiation. High molecular weight polymers do not generally form true solutions at any

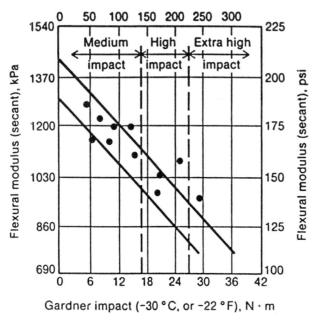

Figure 3.3. Relation between impact energy and stiffness of polypropylene. [From M. R. Rifi, H. K. Ficker, and D. A. Walker, *Mod. Plast.* 64 (2), 62 (1987). Reprinted by special permission from *Modern Plastics*, McGraw-Hill, Inc., New York, NY 10020.]

temperature up to their decomposition temperature. As a result, when the high molecular weight impact modifier is added to the base polymer, a heterogeneous mixture is produced, with distinct separate regions of the two polymers which are easily observable by microscopic examination. The mechanical properties of the mixture therefore depend on both the properties of the two phases and the morphology of the mixture. For impact improvement to be achieved, the added phase must be more impact resistant and have a lower modulus than the parent phase. Impact modifiers are amorphous materials. The negative effect of crystallinity of the modifier on the impact resistance is shown in Fig. 3.4.

Generally, the morphology that occurs on addition of an impact modifier is either of two types:

1. A network honeycomb dispersion of the rubbery phase in the matrix of the polymer. Examples are PVC/EVA (polyvinyl chloride/ethylene vinyl acetate) and PVC/CPE (polyvinyl chloride/chlorinated polyethylene). These mixtures are opaque, indicating that the size of the

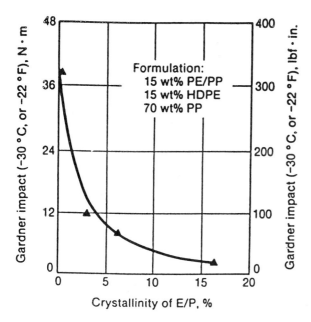

Figure 3.4.   Effect of crystallinity produced by additives on impact energy. [From M. R. Rifi, H. K. Ficker, and D. A. Walker, *Mod. Plast. 64* (2), 62 (1987). Reprinted by special permission from *Modern Plastics*, McGraw-Hill, Inc., New York, NY 10020.].

dispersed particles is of the same order of magnitude as the wavelength of light. The base polymer is present as a continuous matrix, and the other phase is dispersed as a honeycomb. This dispersion occurs during the melt processing. If the homogenization is extended excessively or if excessive shearing or working occurs during production, the honeycomb network breaks down into an agglomeration of rubbery particles in the plastic, and the effectiveness of the impact modifier is reduced. Manufacture of these products must be carefully controlled, as the melt temperature and shear forces must be maintained within a relatively narrow range. For PVC-based compounds, impact strength reaches a maximum for milling times of the order of 5 minutes and mold temperatures of about 190°C.

2. Individual particles of the rubbery phase in a matrix of the impact-sensitive matrix phase. Examples are PVC/PAE (polyvinyl chloride/polyaryl ether), PVC/ABS (polyvinyl chloride/acrylonitrile-buta-diene-styrene), and PVC/MBS (polyvinyl chloride/methacrylate-butadiene-styrene). Unlike the honeycomb-producing additives, these additives produce spherical rubber particles. The size of these

particles is controlled during the manufacturing of the polymer and not during compounding. If the compounding and molding are performed correctly, the rubbery particles are essentially spherical and uniformly dispersed in the matrix.

Spherical elastomeric particles are optimum in shape for impact resistance, as crack initiation and propagation are least likely to occur in this morphology in a two-phase structure. The particles are strongly bonded to the PVC matrix, and debonding and crack initiation at the interface are unlikely. The uniform dispersion is extremely stable under the usual operating conditions and particle agglomeration and growth do not occur. Due to this stability of the particles, high impact resistance is generated over a wide range of processing temperatures, pressures, and production rates.

PAE as an additive is cross-linked, and as a result the PAE particles are very resistant to breakdown after the cross-linking has occurred. During processing of the PVC/PAE compound, the PAE particles partially agglomerate, then disperse into small particles of fairly consistent size and consistent interparticle spacing. The particles are then very resistant to further dispersion. This desirable behavior permits PVC/PAE compounds to be processed over a wide variety of conditions without loss of impact strength.

Properties of Impact-Modified PVC

The modulus of elasticity, hardness, and heat distortion temperature of these materials are nearly those of rigid PVC. These materials have low mold shrinkage and little postmolding shrinkage. These compounds are transparent, and blow molded objects and packaging films are produced. Because butadiene contains double bonds which are very susceptible to degradation by UV radiation, it cannot be utilized in compounds for outdoor application.

Properties of Impact-Modified PS

Impact-modified PS plastics are manufactured by graft polymerization of styrene and acrylonitrile. Synthetic rubber is either dissolved in the monomer syrup or in the form of a latex when polymerization occurs in aqueous solutions. Properties of impact-modified polymers are shown in Table 3.4.

Fillers

Inert fillers are added primarily to lower the finished product's unit cost by replacing some of the high-cost plastic with lower-cost materials. They also cause increases in modulus of elasticity, flexural strength, hardness, and heat deflection temperature. The mechanical properties also tend to decrease less rapidly with increasing temperature than for unfilled plastics.

Table 3.4. Properties of Impact-Modified Plastics

Properties	Impact-resistant polystyrene (IPS)			Styrene-acrylonitrile copolymer (SAN)		Acrylic-styrene-acrylonitrile (ASA)
	Semi-impact resistant	Impact resistant	High-impact resistant	Not reinforced	Reinforced by 35 wt % glass fibers	
Density, g/cm³	1.04	1.04	1.04	1.08	1.36	1.07
Melt flow index, g/10 min	1.5–20	3–20	3–12	—	—	—
Impact strength						
At 20°C (68°F), kJ/m² (ft·lbf/ft²)	40–60 (2750–4100)	60–80 (4100–5500)	No break	20–25 (1400–1700)	10–18 (690–1250)	No break
At −40°C (−40°F), kJ/m² (ft·lbf/ft²)	35–50 (2400–3450)	50–70 (3450–4800)	70 (4800) No break	—	—	20–50 (1400–3450)
Notched impact strength						
At 20°C (68°F), kJ/m² (ft·lbf/ft²)	5–6 (340–410)	5–8 (340–550)	8–14 (550–960)	2–3 (140–210)	4–5 (270–340)	7–14 (480–960)
At −40°C (−40°F), kJ/m² (ft·lbf/ft²)	3–5 (210–340)	4–6 (270–410)	6–12 (410–820)	—	—	—

From: G. Menzel, High polymeric additives as impact modifiers for thermoplastics, in *Plastics Additives*, 2nd ed., R. Gachter and H. Muller, eds. Cincinnati, Hanser, 1983, pp. 361–396.

The effects of fillers on the properties of a plastic are largely dependent on the particle size distribution of the filler and the strength of the bond between the filler particle and the matrix plastic. Most of the inexpensive inorganic fillers are mined as natural ores and then ground to their final particle size. The particle size and shape are a result of random fracture and fretting of particles against each other in the grinding mill. A wide range of sizes and differently shaped particles are therefore present in any filler. ASTM D1210 specifies a method for determining particle size distributions. The particle size distribution can be shown in a plot of the weight fraction of material finer than a specified particle size versus size. Another specification which can be more easily utilized in mathematical analysis is the packing factor, defined as the maximum fraction of the total volume occupied by the solid filler particles. The greater the variation of particle sizes in the filler, the greater the packing factor, as the smaller particles can fit in the interstices of the larger particles with no increase in total volume of the sample. If two fillers with different packing factors are evaluated at the same volume percent filler, the one with the larger packing factor will have a wider range of particle sizes and should therefore have a higher tensile strength.

Fillers always increase the stiffness or modulus of elasticity of the matrix polymer. Figure 3.5 shows a plot of the relative increase in modulus by additives above and below the glass transition temperature. The shape of the filler particle affects the modulus and other mechanical properties significantly. Spherical particles are less likely to initiate cracking than more slender particles. However, flat platelet-type particles tend to produce a stronger bond between the filler particle and the matrix, due to the increased surface area available for bonding. A quantitative measure of the shape of platelet-like particles is the aspect ratio, the ratio of the average platelet diameter to its thickness. The flatter the particle for a given mass (the greater the aspect ratio), the greater the increase in bonding area available. Special grinding and handling techniques are employed to maintain high aspect ratios in some fillers (such as mica) to develop special grades which produce unusually improved mechanical properties.

The tensile strength of a filled plastic depends on several factors. The interfacial bond strength determines the distribution of stress between the filler and the matrix. For many low-cost fillers, the interfacial bonding strength is rather low, and debonding between the particle and the plastic can occur at low stress levels. Treatment with coupling agents to increase the bonding between the plastic and the filler increases the tensile strength and decreases elongation. In general, the smaller the average size of the filler particles, the greater their surface area and the greater the bonding between the plastic and the filler particles. The effect of fillers on the elongation

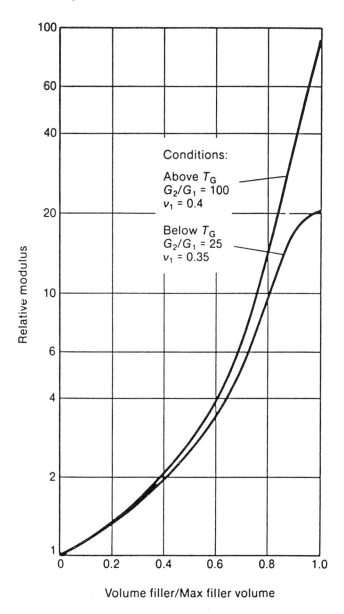

Relative modulus

Volume filler/Max filler volume

Figure 3.5. Effect of filler additions on the modulus above and below the glass transition temperature. [From T. B. Lewis and L. E. Nielsen, *J. Appl. Polym. Sci. 14*, 1449 (1970). Reprinted by permission of John Wiley and Sons, Inc.]

of plastics varies widely. Most systems exhibit decreased elongation. Some exceptions do occur, and the finer the filler particle size the more likely the filler is to increase the ductility of the plastic.

Plastics which partially crystallize such as PE show the largest improvement of tensile strength with filler addition. In these materials, crystallization is not impeded by the filler particles, as they are not strongly bound to the plastic, and flow and crystallization proceed in the same fashion as in the unfilled plastic. Short fiber fillers are more effective in increasing the strength in the direction of plastic flow during processing, as they also tend to align in the flow direction.

Coupling Agents

Many filler materials as mined or manufactured adhere very poorly to plastics. The strong bonding desired between the matrix and the filler is obtained with the aid of coupling agents. These are materials that react with the filler's surface and have good compatibility with the polymer matrix. Silanes are the principal coupling agents currently used. Silanes have the general formula

$$R—Si—(OR)_3$$

where R is a methane or ethane group. The silanes are converted into Si—OH groups, which are chemisorbed onto the filler's surface. The R group contains functional groups which can cause bonding of the wetted glass to thermosets. Attraction to thermoplastics is by secondary valence forces.

Filler Materials

Calcium Carbonate

Calcium carbonate (whiting) is one of the least expensive fillers and is accordingly utilized extensively. Calcium carbonate fillers are stable at temperatures up to the decomposition temperature of all plastics. They are completely nontoxic and odorless and can therefore be mixed in any proportion safely. The white color produced is pleasing and helps mask the natural yellowish color of many plastics. Calcium carbonate is extensively utilized in PVC compounds and is up to 50% by weight of PVC plastisols and up to 80% of the weight of floor tiles. It also acts as a processing stabilizer in these materials, as the hydrochloric acid and chlorine generated by partial decomposition of the PVC during processing react with the calcium carbonate and are removed from the melt, preventing further autocatalytic decomposition.

Glass Spheres

Glass spheres have excellent thermal stability. Solid particles are available with diameters between 0.004 and 5 mm (0.00016–0.2 inch). Hollow spheres are available with diameters of 0.01 to 0.25 mm (0.0004–0.01 inch). They are utilized where a low-density product is required and are usually added to foamed products during manufacture. Since they are spherical, there is little tendency for crack initiation to begin at the particle. Low molding forces are required, as the polymer can readily flow around the particles. The properties of the resultant parts are isotropic, no directionality occurring with spherical filler particles. This is in contrast to moldings containing fibrous fillers, which always orient to some extent during molding, resulting in highest strength in the flow direction.

Graphite

Carbon is one of the most widely used filler materials, as it affects many properties favorably. Graphite powder filler materials produce a product with self-lubricating properties and are most beneficial when the sliding interface is lubricated with water.

Kaolin

Kaolin is china clay and is a complex mixture of aluminum, silicon, magnesium, potassium, sodium, and iron oxides. Kaolin as an additive is excellent in improving the flow characteristics of the resultant compounded plastic. In reinforced thermoset plastics kaolin additives permit better flow in the molding operation so that the reinforcement is more uniformly dispersed in the plastic. Fibers are retained in the bulk of the sample and are not exposed on the surface of the molded part. Kaolin is therefore used as a filler in thermosetting bulk molding and sheet molding compounds and is an excellent filler in rubber compounds. It improves chemical resistance and electrical resistivity and decreases water absorption. It has an appreciable effect on increasing the hardness of the compound.

Mica

Mica is a complex potassium aluminum silicate. Grinding splits the mineral into extremely thin, flexible platelets. The highest aspect ratio causes mica to be an excellent reinforcing additive. However, the flakes are so thin that they easily break into small particles with lower aspect ratios during normal thermoplastic molding procedures, lowering the resultant properties of the molded part. Better properties are obtained by special powder and latex blending techniques which prevent degradation of the mica platelets. The flakes orient in the direction of flow during molding, producing anisotropic products. Mica fillers produce increased stiffness and dimensional stability. Mica compounds have outstanding dielectric properties, electrical resistance, and arcing resistance. As a result, mica is utilized in thermosetting epoxy, phenolic, polyester, and polyurethane resins for electronics and electrical appliances.

Talc

Talc is a natural mineral with the formula $3MgO \cdot 4SiO_2 \cdot H_2O$. Some impurities of iron and calcium oxide always occur. Although as a result of its low cost talc is classified as a filler, it has a significant effect in enhancing many mechanical properties and is often considered a reinforcement. This reinforcement is primarily a result of the shape of the particles. Talc is a naturally occurring fibrous mineral, and the platelets are generally long and thin with large aspect ratios and surface areas and capable of bonding strongly to the matrix plastic. This platelet type of structure aligns in the direction of flow during molding, tending to produce some anisotropy in the resultant part. Talc additions increase the modulus, stiffness, and hardness; decrease creep; and raise the heat deflection temperature. Dimensional stability is improved. The large increase in stiffness and measurable increase in tensile strength are to be noted, along with only a slight decrease in strain to failure and impact energy. As in the case of most fillers, coupling agents greatly increase the strength.

Colorants

Colorants are subdivided into dyes and pigments. Dyes are soluble in the medium they are coloring, whereas pigments are insoluble. Most of the colorants used in the plastics industry can be classified as pigments. The appearance of an item colored by pigments is affected by the particle size of the pigment and the accumulation of these particles into larger particles called aggregates. The crystal structure of inorganic pigments also affects the appearance of the finished product.

Light of a particular wavelength is absorbed by the pigment. The appearance of the product will correspond to the color of the remaining light reflected to the eye. If the refractive index of the pigment is significantly different from that of the plastic and if the size of the pigment particle is of the same order of magnitude as the wavelength of light, most of the light striking the pigment particle will be reflected at the interface. The plastic appears translucent or opaque, depending on the concentration of pigment.

Pigments are generally added in the concentration range 0.1 to 0.5%. Plastics which have strong intrinsic yellowish colors, such as impact-modified

PS and ABS, require higher pigment concentrations of up to 5%. Thin parts, which have little depth to develop a rich color, also require higher concentrations of pigment. For the lower concentration range, the effect of pigments on the mechanical properties is usually quite small and may often be considered negligible. The colorants may affect the viscosity of the melt when present at higher concentrations.

Stability of Colorants

Colorants are the most thermally sensitive of the the additives. The pigments change color as a result of heating during processing and over a period of time during use. Color changes are a function of the time of exposure to high temperatures during processing; reprocessing of a polymer can increase the color changes. The inorganic pigments are less thermally sensitive than the organic pigments and dyes. Thermal stability depends on the environment of the pigment, the thermal stability of a particular colorant varying with the polymer with which it is being used.

Light degradation of the colorant can cause color changes within the polymer. This degradation is also a function of the plastic with which the pigment is compounded. Mixing pigments affects the lightfastness. Dark colors may generate surface chalking, and large concentrations of heat and light stabilizers are usually required for dark-colored parts for outdoor applications. Pigments are expensive additives. To obtain proper color levels economically in many applications coextrusion is utilized, with an outer layer with high levels of titanium dioxide and stabilizers for weather resistance and increased concentrations of pigments. The inner layers have lower concentrations of the same additives.

Blooming and Bleeding of Pigments

Inorganic pigments are essentially completely insoluble in the plastics with which they are used. As a result, there is no diffusion of the pigment particles, and bleeding into liquids or other plastics in contact with the colored plastic part does not occur. Organic pigments can exhibit diffusion, and bleeding into contacting solvents or plastics can occur to some extent. Blooming of organic pigments dissolved to their saturation concentration at elevated temperatures and then cooled may cause precipitation on the specimens' surface producing surface discoloration or haze. This precipitation on the surface is controlled by the kinetics of nucleus formation of the precipitate and the growth of nuclei of the new phase. The nucleation is more likely to occur at some dirt particle, fingerprint, or oil spot on the surface or some surface imperfection such as a scratch or molding imperfection, resulting in random surface spotting by the precipitated pigment.

Pigment Materials

Many materials are available as pigments. Among the most common are the following.

Carbon Black

Carbon black pigments are not interchangeable with carbon black fillers. They contain different organic groups which affect the deepness of the color and the wettability of the particles by the polymer melt. The particles are significantly smaller than those used as fillers. The finer the particle the darker the shade, as more internal reflections of the incident light permit greater absorption and a lower overall reflectivity of the part. Although most pigments have little effect on other properties, carbon is such an effective UV absorber that, even at the low concentrations used for pigments, it increases weathering resistance and increases thermal degradation resistance.

Dyes

Highly transparent plastics such as polystyrene (PS) and styrene-acrylonitrile copolymer (SAN) appear glass clear when molded. Pigment additions would tend to cloud these transparent plastics due to light scattering at the pigment-matrix interface. For such articles, soluble dyes are utilized which dissolve in the plastic, producing a single-phase structure with no light scattering. Much lower concentrations are required (as low as 0.02%), as the intensity of coloration for dyes is much higher than for the pigments.

Fluorescent Pigments

Fluorescence occurs when an atom interacts with high-energy light in the UV range, generating an activated electronic energy state. The activated molecule then reradiates the energy as light in the visible range, the difference in energy between the incident and reradiated light dissipating as heat. Very strong attractive colors are produced by this radiation. Although the molecules which fluoresce are dyes, they are processed and sold mixed with plastic in an insoluble form and are handled in the same manner as pigments. However, they have a somewhat greater tendency to bleed and bloom as compared with the pigments and are more sensitive to degradation by light.

Organic Pigments

The organic pigments are somewhat more intense and varied in color than the inorganic pigments. Their use is somewhat less widespread than that of the inorganic pigments, as they tend to be more light and heat sensitive. In mixtures with titanium dioxide, their degradation rate is futher increased.

Oxide Colors

Inorganic oxides exhibit a variety of colors. The inorganic oxides in their highest oxidation state are very resistant to thermal or UV-accelerated oxidation. These heavy metal compounds are in general toxic, and there is increasing concern about the desirability of replacing these chemicals with others of lower toxicity.

Titanium dioxide (rutile) is the most widely used white pigment. However, it tends to accelerate photochemical oxidative degradation of plastics and the surface of most plastics tends to disintegrate rapidly in sunlight, producing a powdery surface layer (chalking).

Antistatic Agents

Most plastics are excellent insulators. The surface conductivity of plastics is of the order of 10^{20} times greater than that of metals. As a result, any surface charge developed on a plastic part will dissipate slowly. Surface charges tend to attract dust particles, producing contamination of the surface. Although the buildup of some surface charge would not affect the behavior of a part in many applications, such charge can cause serious problems in specialized applications. In electronic recording, this static charge and attracted dust distort sound reproduction. Surface charges built up on plastic flooring and floor coverings can produce unpleasant shocks for unwary individuals. Surface charges on sheet materials cause them to cling to each other and make the opening of plastic bags difficult.

Antistatic agents are divided into two groups, internal and external. Internal agents are additives which are mixed into the bulk of the plastic and are similar to other additives. External antistatic agents are hygroscopic substances such as glycerine, polyols, and polyglycols which are applied directly to the surface of the finished part after molding is completed. They are effective immediately but can be removed easily if the surface is washed or scrubbed. Since they are externally administered, they are not classified as plastic additives.

The additive antistatic agents are surfactants, and when added to the bulk plastic during compounding or molding, they diffuse to the surface of the molded part. The molecule has two components. One part is compatible with and remains dissolved and reacted with the plastic. The other part extends to the surface of the plastic and has bonding groups with an attraction for water molecules; hence a layer of water of molecular thickness is bound to the surface of the plastic part, decreasing the electrical resistance of the surface and permitting the surface charge to dissipate. The thickness of the water film generated depends on the humidity of the environment. Some antistatic agents are also hygroscopic, attracting and increasing the thickness of the water layer on the surface. A good antistatic agent decreases the surface resistivity by 10^6 on the average. The water layer developed on the surface is obtained from the humidity in the air; the relative humidity of the environment therefore affects the efficiency of antistatic agent operation.

For a surfactant antistatic agent to be effective, the additive must diffuse to and concentrate at the surface. This can best be accomplished by additives which are not completely compatible with the polymer, and the antistatic agent can easily be washed from the surface of the polymer. Environmental considerations are particularly severe for these additives, as they can be present in wash waters and can present greater hazards to workers than other additives due to their higher surface concentrations. Antistatic agents used in food and drug packaging must be approved by the FDA. The most common surfactant antistatic additives are amines, quaternary ammonium compounds, sulfonates, and phosphates.

Another technique for decreasing the static electricity on a material is to decrease the electrical conductivity of the bulk material. The bulk electrical resistivity must be 10^9 ohms or less. Carbon black is the most commonly used material in this regard. Metal powders are also utilized, as well as tin oxide, titanium dioxide, and mixed oxides. Among the various new additives are nickel-coated carbon flakes, stainless steel fibers, metal-coated ceramic microspheres, and aluminum flakes. Large fractions of these additives must be used. Concentrations of 50–60% of the total compound may be required for effective antistatic behavior. The effect of additions of conducting powders of Sb/SbO_2 on PVC and PP is shown in Fig. 3.6. This additive produces white conducting plastic.

Nucleating Agents

Above the melting point, the molecules of thermoplastics have sufficient kinetic energy to move independently. As the temperature is lowered, the kinetic energy decreases, and intermolecular forces tend to align the molecules to produce a uniform array. If the intermolecular forces are sufficiently strong, the polymer will crystallize. The main chemical factors tending to produce an easy crystallizable polymer are:

1. Simple, relatively branch-free chains, so that chain alignment is not hindered.
2. The presence on the molecules of polar groups which increase intermolecular forces.

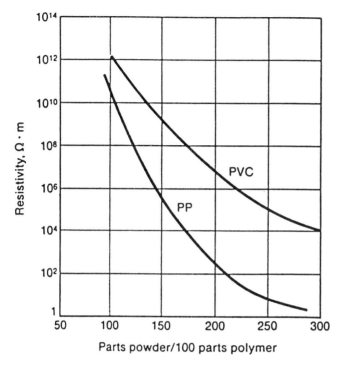

Figure 3.6. Effect of additives on surface conductivity. [From M. Yoshizumi and K. Wakabayashi, *Plast. Eng. 43*, 61 (1987). Reprinted by permission of the publishers.]

Crystallization requires the formation of a small nucleus of the required ordered atomic arrangement by random motion of the polymer molecules and subsequent diffusion of additional molecules to align themselves with the nucleus. Inorganic materials and organic molecules with low molecular weight generally crystallize completely. However, at the crystallization temperature polymer liquids are very viscous, the molecules are mechanically entangled, and branching and irregularities in molecular structure hinder alignment of the molecules. As a result, if crystallization does occur, it occurs slowly and is never 100% complete. Crystallization occurs in very small regions as long molecules fold on themselves to produce small orderly packets or lamella. Some molecules end within the lamella; others extend out to amorphous regions or into other lamella. When crystallization begins, a small region or nucleus develops into an interconnected mass of lamella and amorphous material. This growth outward is in a roughly spherical shape and the entire growing structure is the spherulite. Growth of the spherulite continues until it impinges on another solidifying spherulite. The rate of nucleation of these spherulites can be increased by several orders of magnitude by the addition of nucleating agents.

Nucleating agents accelerate the formation of crystallized regions. The nucleating agent remains as fine solid particles approximately 1 to 10 μm in size in the molten polymer, thereby serving as a nucleus for further crystallization. The most common nucleating agents can be divided into three general classes:

1. Mineral additives talc, silica, kaolin
2. Organic salts
3. Polymers

The concentration of nucleating agents is of the order of 0.5%. They are added primarily to polyethylene terephthalate (PET), polyamides, and polypropylene (PP). Figure 3.7 shows the effect of nucleation on the crystallization time and temperature of PET.

Increasing the percentage of crystallinity of a polymer in general improves almost all mechanical properties related to strength. The finer the size of the spherulites, the greater the increase in properties. The modulus of elasticity, tensile strength, yield point, hardness, elongation to failure, heat deflection temperature, and impact strength are all increased. Water absorption is decreased. Dimensional stability is improved, as the distortion of molded parts due to slow partial crystallization over long periods of time during storage or use is minimized if the part has already largely crystallized. Increased crystallinity also improves production rates, as once crystallization is complete the part is stable dimensionally and can be removed immediately from the mold.

Blowing Agents

Foamed products are prepared by generating gas bubbles during molding of the part. The primary objective of generating a cellular foam is to obtain a low-density material with a dense skin. The overall weight reduction is itself desirable in many applications. The savings in weight of material used is an additional benefit. Foamed material has improved heat, sound, and dielectric insulating properties. Packaging materials capable of withstanding large impact and shock loadings are usually foamed products.

Physical Blowing Agents

Physical blowing agents are gases which are pumped into the mold to produce the expanded finished product. The principal gases used are chloro-fluorocarbons (CFCs), methylene chloride, pentane, and butane. The chlorinated materials are under scrutiny because of their toxicity and deleterious

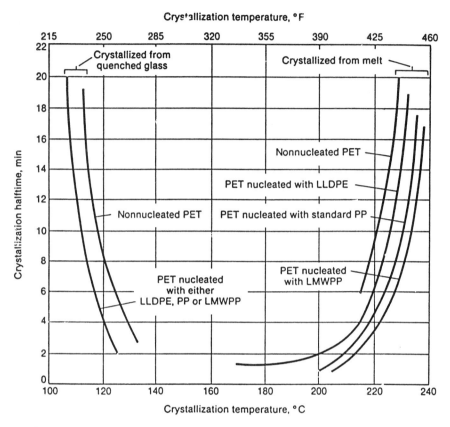

Figure 3.7. Changes in crystallization time with addition of nucleating agent to PET. LLDPE, linear low density polyethylene; LMWPP, low molecular weight polypropylene. [From L. Bourland, *Plast. Eng. 43*, 39 (1987). Reprinted by permission of the publishers.]

effects of CFCs on the ozone layer. As a result, their usage is decreasing. Although pentane and butane are relatively low in toxicity, their flammability is troublesome, and they are included in the listing of volatile organic compounds whose emission to the atmosphere is controlled. These are not additives to the plastic compound and will not be considered further.

Chemical Blowing Agents

Blowing agents are compounds which decompose at the processing temperature, producing a gas which generates a foamed, lighter structure. Blowing agents are usually fine powders, although there are a few liquid blowing agents. The blowing agent must decompose within the narrow temperature range of normal processing. The decomposition product must be a nontoxic and chemically inert gas; nitrogen is the most common product. Both the blowing agent and the decomposition product should be compatible with the plastic. The foaming occurs during processing, producing a material with a dense skin and a porous internal structure for structural members and whenever light weight is an important consideration. Processing of the plastic containing the blowing agent is similar to that of dense plastics. However, provision must be made to permit expansion of the gas to produce the cellular product.

In low-pressure processing, a fixed volume of compounded resin containing the blowing agent is fed into a mold cavity which is sufficiently large to permit the polymer to expand to fill the chamber when the blowing agent decomposes. The plastic in contact with the mold surface forms a dense outer skin, the foam generation being restricted to the central portion of the product. In high-pressure processing, the compounded plastic is injected into the mold cavity so that it completely fills the cavity. By means of movable slides, the mold cavity volume is increased, permitting expansion of the plastic and generation of the foam. This produces very good surface finish and somewhat less expanded central material.

Materials

Most common chemical blowing agents are azo compounds and hydrazine derivatives. An example of an azo compound is azodicarbonamide (H_2N—CO—N=N—CO—NH_2), which decomposes to form primarily nitrogen, but CO, ammonia, and HNCO are also produced. Azodicarbonamide is also produced as a complex additive containing plasticizer and surfactants to aid in compounding, activators to accelerate the decomposition, and nucleating agents to produce a finer cell or bubble volume. An example of a hydrazine derivative is 4,4'-oxybis(benzenesulfohydrazide), which decomposes to yield nitrogen and water vapor.

Flame-Retarding Agents

Prevention of combustion and smoldering of plastic components requires the addition of flame retardants. However, smoke and toxic combustion products are more dangerous than the heat generated during burning of the plastic. Although additives are available which can reduce the possibility of the plastic actually burning, some of these additives may contribute the major portion of the toxic combustion products, and low-smoke and low-toxicity formulations of flame retardants are becoming increasingly important. Flame-retardant additives containing antimony oxide, bromide, and chloride are extensively utilized. Newer formulations with reduced concentrations of these compounds are of great interest, since they tend to be among the principal generators of toxic fumes during combustion. Flame retardants in general degrade mechanical properties to some extent. Table 3.5 shows the effect on mechanical properties of typical flame retardant–polymer compounds.

Several mechanisms exist for flame retardation:

1. During combustion, a surface char develops which acts as an impermeable layer blocking further oxygen penetration to the surface.
2. During combustion, free radicals form. These combine with oxygen to produce other free radicals. Combustion-inhibiting agents react with the free radicals, breaking the chain reaction. Halogen compounds act by this mechanism.
3. Chemicals can be introduced which decompose at the temperatures generated during combustion, producing water vapor or other gases which quench the flame and displace oxygen.

Flame-retarding materials include the following.

Antimony Trioxide

Antimony trioxide is the basis of virtually all flame-retardant plastics. The chlorine and bromine compounds act synergistically with antimony oxide and are always used in coordination with this compound. It is believed that SbOBr and HBr are formed when this compound is heated during combustion of the polymer. HBr reacts readily with free radicals to terminate the chain reaction. At 250–290°C SbOBr decomposes into the more highly oxidized $Sb_4O_5 \cdot 3Br_2$ and $SbBr_3$:

$$5SbOBr = SbBr_3 + Sb_4O_5Br_2$$

and at about 500°C the $Sb_4O_5Br_2$ decomposes to Sb_2O_3 again, reinitiating the quenching reactions.

Antimony oxide is one of the major contributors to smoke generation in nylon and polybutadiene terephthalate (PBT). As a result, other inorganic compounds, ferric oxide, zinc oxide, and zinc borate, are being considered as replacements, but they have not found wide usage as yet.

Phosphorus Compounds

Phosphorus compounds are believed to form phosphoric acid, which reacts with the polymer producing an incombustible protective layer on the surface.

Table 3.5. Effect of Flame Retardant on Mechanical Properties

| Plastic | Flame-retardant type[a] | Amount of additive %[b] | Change in tensile strength | | Elongation at break, % | Flexural yield strength, % | Flexural modulus, % | Heat deflection temperature at 455 kPa, % | Impact strength, Izod notched % |
			At yield, %	At break, %					
Polyethylene	NP	30	10–15	—	—	—	—	50 to 75	−15 to −20
	SP	25–35	10–15	−20	—	—	—	50 to 60	−15 to −20
Polypropylene	NP	40	−25 to −35	−5 to −10	−60 to −70	7 to −12	50–70	—	−15 to −75
	SP	20	—	−9 to −11	—	—	—	—	—
Polystyrene	NP	20–30	—	−15 to −55	−25 to −40	−25 to −45	0	0	−10 to −70
	SP	17–20	—	−2 to −8	9 to 12	—	—	—	—
Impact polystyrene	NP	20	0	0	−5 to −75	—	2–15	−3 to −10	−20 to −40
	SP	15–20	—	—	—	—	—	—	—
ABS	NP	15–20	0	0 to −10	—	0	0	0 to −10	−3 to −40
	SP	15–20	—	—	—	—	—	—	—

[a] NP, nonplasticizing; SP, semiplasticizing.
[b] Blend of two to three parts per part antimony trioxide.
From: G. F. Chadwick, Flame retardants, in *Plastic Products Design Handbook*, Part A, E. Miller, ed. New York: Marcel Dekker, 1981, pp. 67–82.

Water vapor and other flame-extinguishing noncombustible gases are also produced and smother the flame.

Chlorine-Containing Compounds

These flame retardants are relatively inexpensive, making them commercially advantageous over the more expensive Br compounds. The principal compounds used are highly chlorinated paraffins. Unfortunately, they are easily decomposed by heat and can be molded only below 200°C. During processing of many plastics temperatures exceed this value, and other chlorinated compounds such as hexachloroendomethylene-tetrahydrophthalic acid are used.

Bromine-Containing Compounds

On an equal weight basis, bromine compounds are more than twice as effective as chlorine compounds. Their light stability is somewhat less than that of chlorine compounds. Because lower percentages can be used, they degrade the mechanical properties of the plastic to a smaller extent. Their compatibility is better in general than that of the chlorine-containing compounds, and there is little problem with bleeding or blooming.

Aluminum Trihydrate

Minerals which contain water of hydration, which can be released at the flame temperature, are currently of great interest, as this method of flame control does not produce any toxic by-product. Aluminum trihydrate is the most widely used mineral. It is not soluble in the plastic but acts as a filler material, increasing the heat transfer through the material, and it decomposes to generate water vapor when heated. This process absorbs heat, and the water vapor produced dilutes and cools the hot combustion gases. The particle size of the additive is important, as decreasing particle size improves the dispersion of the flame retardant throughout the polymer and increases the mechanical properties and appearance of the product. Figure 3.8 shows the loss of water of hydration between 220 and 400°C, with resultant heat absorption.

Fiber Reinforcements

Materials that have fibrous reinforcement are also called composite materials. Materials in this class have mechanical properties which are dependent on the properties of the fibrous reinforcing material, the properties of the matrix polymer, the volume fraction of each material, and the orientation of the

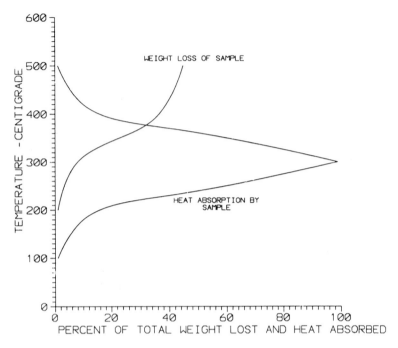

Figure 3.8. Heat absorption and weight loss due to water loss from aluminum trihydrate.

fibers with respect to the load. These materials are discussed in Chapter 10, which pertains to such composite materials exclusively.

3.5. RECYCLING OF PLASTICS

Thermosets having cross-links cannot be reprocessed. Two main problems exist with respect to recycling of thermoplastic waste. Such waste is likely to contain foreign materials, such as adsorbed grease or other fluids on its surface or solid particles mixed throughout the scrap. Cleaning costs may exceed the cost of the original material. The other difficulty is sorting the plastic into known types. This sorting involves an adequate means of quickly identifying the type of plastic. No such quick identification exists. On the other hand, recycling of scrap produced during processing is widely practiced. Such scrap is relatively clean, having been maintained within the plant, and its identity is known. However, because high temperatures and shearing forces exist during processing, the plastic may undergo some degradation on reprocessing.

4

Industrial Plastics

4.1. INTRODUCTION

A discussion of the various types of plastics and a detailed description of their mechanical properties are beyond the scope of this book. This is particularly true because each type of polymer is manufactured in various grades with different average molecular weights, different molecular weight distributions, different additives, and different manufacturing processes, all of which affect the properties. The following is therefore a brief introduction to the various classes of polymers. Due to the vast array of plastics available, only the most commonly used are described below. The mers of the plastics discussed are shown in Fig. 4.1.

4.2. THERMOPLASTICS

Polyethylene

Polyethylene (PE) is prepared by direct polymerization of ethylene (C_2H_2) in the presence of an appropriate catalyst:

$$
\begin{array}{cccccc}
\text{H} & \text{H} & & \text{H} & \text{H} & \text{H} & \text{H} \\
| & | & & | & | & | & | \\
\text{C} & = & \text{C} & \rightarrow & \text{C} - \text{C} - \text{C} - \text{C} - \\
| & | & & | & | & | & | \\
\text{H} & \text{H} & & \text{H} & \text{H} & \text{H} & \text{H}
\end{array}
$$

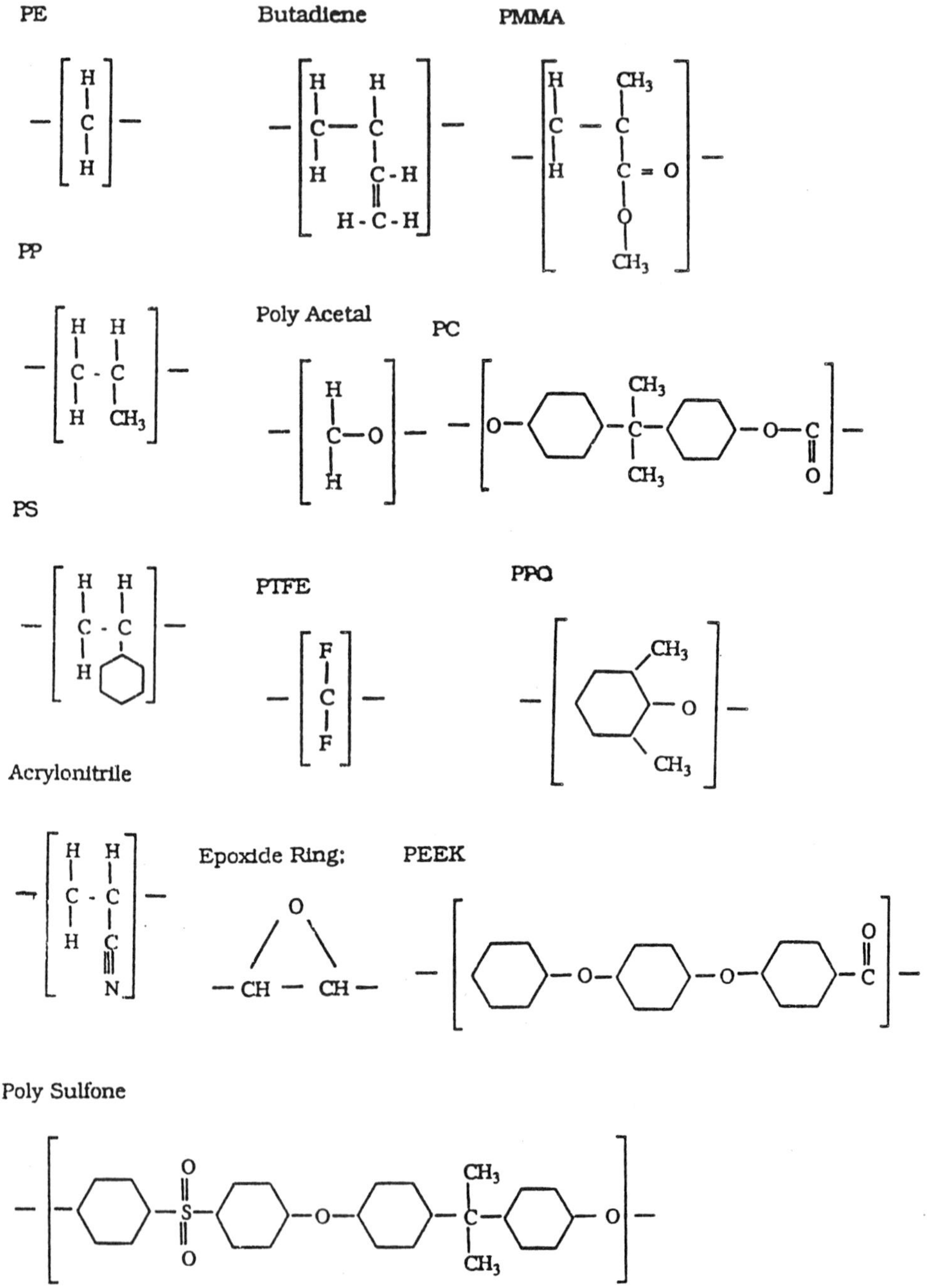

Figure 4.1. Molecular configuration for polymers.

Although this simple formula suggests that all PE molecules should be linear, under actual polymerization conditions some side chains are formed.

Polyethylene is the largest-volume plastic used in the United States. One of the reasons for its wide usage is the existence of many different grades, with a wide range of properties. The two factors which strongly control the properties are the molecular weight and the number of side chains existing on the mainly linear molecules. Grades with molecular weights greater than 200,000 are generally considered as the high-performance grades. The high molecular weight grades show improved toughness, improved tensile strength, a higher softening temperature, and other mechanical properties related to strength. Ductility and resistance to stress corrosion cracking tend to decrease with increasing molecular weight. The longer chain length and increased intermolecular attractive forces increase the visocsty in the liquid state and decrease the processibility of the grade.

The molecular weight and the number of side chains which exist on the mainly linear polyethylene depend on the techniques employed during the polymerization process. In the high-pressure method of polymerization, ethylene (C_2H_2) is contained in a reactor at pressures approaching 20,000 psi at a temperature of 400°F. Peroxides are the common catalysts. The reaction proceeds rapidly at these pressures and temperatures but, as described in Chapter 1, it is difficult to obtain high yields; the reaction product is therefore transferred through separators and the unreacted monomer returned to the reaction chamber. The polymerization is highly exothermic, and good temperature control is necessary to avoid overheating and the possibility of an explosion occurring in the reaction chamber. The high-pressure reaction results in many side chains and low molecular weight resulting in the production of low-density PE (LDPE). Typically, side chains are 1 to 65 carbon atoms long, and about 3% of the carbon atoms in the main chain may contain these side chains, with most side chains 1–6 carbon atoms long.

Low-pressure polymerization is performed at 1500 psi and 212°F, with the monomer either as the pure gas or in a suspension or a solution in an organic liquid. These low-pressure methods utilize catalysts which tend to produce more linear PE molecules, resulting in grades of higher density and higher crystallinity. The grading based on molecular weight, density, and number of side chains separates PE into the following types:

UHDPE, ultrahigh-density polyethylene
HDPE, high-density polyethylene
LLDPE, linear low-density polyethylene
LDPE, low-density polyethylene

Processing ease decreases for these grades roughly in the ratio of 0.1 : 0.4: 0.8 : 1.

PE type	M_n	M_w/M_n	Specific gravity	Ultimate strength (psi)	Elastic modulus (psi)
UHDPE	$3\text{--}6 * 10^6$	5	0.96	4000	150,000
HDPE	$2\text{--}5 * 10^5$	10	0.94	3000	100,000
LDPE	$1\text{--}3 * 10^5$	50	0.92	1800	30,000

The lower the M_w/M_n ratio, the smaller the distribution of chain molecular lengths. As expected, this reduction in molecular variability results in increases in impact strength, toughness, and softening point and increased resistance to stress cracking.

Although most PE is thermoplastic, cross-linked PE is manufactured by electron radiation and heating with peroxides to produce a material which is insoluble in most organic solvents. Its tendency toward stress corrosion cracking is decreased, its strength increased, and its fatigue resistance increased, as usual with a resultant decrease in strain at failure. It is most commonly employed for cable insulation. The cross-linking material is also sold with fillers and modifiers, further improving the mechanical strength and lowering the cost.

Polyethylene is copolymerized with many other polymers. Copolymerization with polypropylene (PP) increases the strength and impact resistance. Copolymerization with butene, hexene, and octene monomer introduces short side chains, producing very low density material for easy processing. Copolymerization with ethylene vinyl acetate (EVA) produces a wide range of properties depending on the amount of vinyl acetate (VA) added.

Blends of various grades of PE are also commonly used. For example, blends of LDPE and LLDPE (low density–linear low density) combine the processing ease of LDPE and the improved mechanical properties and resistance to stress cracking of LLDPE.

Desirable Properties

    Low cost
    Good toughness
    Near-zero moisture absorption
    Excellent chemical resistance
    Excellent electrical resistance
    Low coefficient of friction

Good moisture barrier
Good corrosion cracking resistance
Highest impact strength of any polymer
Good abrasion resistance—used for bulk handling equipment

UHDPE is especially applicable to bulk material handling, for ores, dump truck beds, chutes, conveyors, etc.

Typical Uses

Piping
Automobile interior parts
Containers
Cable insulation
Film and bagging material
Material handling equipment
Wear-resistant flat bearing material

Cross-linked polyethylene is most commonly used insulator in electrical distribution systems.

Polypropylene

The relatively tight small CH_3 group provides sufficient steric hindrance that good molecular alignment can occur. As a result, PP is highly crystalline, approaching 90% crystallinity. Although PP is mainly isotactic (as it must be to be so largely crystalline), about 5% of the atactic form is present in the commercially available materials. The fraction of atactic form varies between manufacturers, producing different mechanical properties. As with most crystalline polymers, increasing the rate of cooling during solidification reduces the crystallite size and improves the toughness of the product.

An unusual characteristic of PP is that very high molecular weights lead to poorer properties. The optimum properties are observed at a degree of polymerization (DP) of about 1000. The specific gravity of PP is 0.9, and, like PE, it is useful for solid items which need to float. In general, the properties are similar to those of PE, but the increased steric hindrance improves mechanical properties and increases the melting point (melting point = 170°C). This is advantageous for items which must be sterilized, and PP is therefore used for hospital trays, food items, etc.

The fatigue characteristics of PP are excellent, and it is extensively used for solid hinges. It has excellent moldability and has less shrinkage on solidification than PE.

Desirable Properties

Can be sterilized
Excellent fatigue resistance
Excellent moldability

Typical Uses

Hinges
Computer cabinets
Piping
Impellers

Polystyrene

Polystyrene (PS) is a rigid polymer due to the steric hindrance produced by the benzene ring. It is partly syndiotactic, partly atactic. The fraction of atactic material present is sufficient to prevent crystallization, and despite the stiffening of the molecule, PS is amorphous. PS comes in several grades with varying ease of moldability. The stiff molecule produces a viscous melt, and increased moldability is at the expense of mechanical properties. The following grades are available:

1. *General purpose*: This grade provides a balance of good flow properties for ease of molding and good mechanical properties.
2. *Heat-resistant grades*: All polymers contain small amounts of un-reacted monomer. Since the monomers have low molecular weights, they are more volatile than the polymerized material, and careful reduction of the amount of monomer remaining in the plastic reduces the overall volatility of the material.
3. *Improved moldability*: Lubricants are added to improve the flow characteristics. The lubricants tend to degrade mechanical properties to some extent.

General Properties

Low in cost
Good moldability
Low moisture absorption
Good dimensional stability
High transmission in visible light range
High refractive index gives brilliant appearance
Good electrical properties:
  High dielectric strength
  High electrical conductivity

Typical Uses

Piping
Automotive interior parts
Packaging
Cassettes, reels
Vending cups, food trays

HIPS (High-Impact Polystyrene)

HIPS is a very important modification of polystyrene. It is produced by dissolving polybutadiene rubber in the styrene monomer and then polymerizing. PS forms the continuous phase, and the rubber forms small separate particles (with smaller PS particles within the rubber particles). This HIPS mixture can also be blended after polymerization (with polyphenylene oxide primarily) to improve toughness and heat and chemical resistance.

For comparison, some properties of PS and HIPS are shown below:

	Tensile strength (psi)	Elongation (%)	Izod impact (ft-lb/in)
PS	4000	35	1.2
HIPS	3500	45	2.5

Note that the strength is decreased slightly, but the impact resistance is doubled for HIPS compared with conventional PS.

Severe environment-assisted cracking (EAC) can occur with HIPS in the presence of foods with high fat contents such as butter and with similar chemicals. All plastics must be carefully tested for the possibility of EAC with any organic materials they may touch.

Typical Uses of HIPS

The most common use is in food packaging, since HIPS is easily thermo-formed or extruded to form cups, plates, etc. The low monomer concentration (less than 500 ppm) is very desirable.

Acrylonitrile

The triply bound nitrogen produces very strong dipole bonding between neighboring molecules. This results in a very high melting point and a resultant liquid which is very viscous. As a result, acrylonitrile is very difficult to mold. To improve the moldability, it is usually copolymerized with 20% PS. The PS-acrylonitrile copolymer has improved properties due to increased intermolecular forces. As the percentage of acrylonitrile increases, the strength increases and moldability decreases. The improved properties include:

1. Improved resistance to hydrocarbons, oils, greases
2. Increased softening point
3. Increased impact strength
4. Transparency retained

Typical Uses

Barrier layer for transmission of gases or flavor in packaging
Used primarily in bottling carbonated beverages

Butadiene

The side chains tend to prevent the main chains from approaching closely, lowering the intermolecular forces. The double bond on the side chain reduces the number of hydrogen molecules on the chain, also reducing the intermolecular forces. These two factors increase the flexibility of the material, and butadiene is classified as a rubbery substance. Cross-linking with sulfur produces a thermoset material with widely spaced cross-links. Cross-linking produces thermosetting rubber, and the relatively few cross-links permit rubbery kinking of the molecule between the cross-links and give wide interchain spacing. Blends of butadiene and acrylonitrile yield a rubbery material with improved oil resistance.

Typical Uses

All applications of rubbery materials:
Coal, ore, and equipment moving belts
Tires
Shock absorbers
Bumpers

ABS

ABS is a blend of PS and acrylonitrile (styrene-acrylonitrile, SAN) and acrylonitrile-butadiene acrylonitrile (AB) (about 65% SAN, 35% AB). It is widely used because of the possibility of varying the amounts of the components, yielding a wide variety of grades with different combinations of strength and rubberyness. The addition of different additives also increases the total range of properties attainable.

Morphology

The material consists of two phases, with a matrix of SAN and small rubbery particles of AB. However, these two materials are essentially insoluble in each other, and such a material would have little interfacial strength across the boundaries of the two phases. To increase the strength of bonding between the two phases, SAN side chains are grafted onto the main chains of butadiene. These SAN side chains at the interface bond strongly to the matrix phase (also SAN).

The rubbery phase is interspersed with particles of SAN, producing a complex microstructure. The size, concentration, and distribution of the rubbery phase can vary widely between grades. Typically, the rubbery particles are 50 μm in diameter.

This variation of morphology and the ability of the ABS to accept many additives because of the wide range of intermolecular forces within the polymer result in a very wide range of possible properties. Blends with other polymers are also used: ABS-PVC (polyvinyl chloride), ABS-PC (polycarbonate), ABS-PA (polyamide).

General Properties

> Combined toughness due to rubbery phase
> Chemical resistance due to acrylonitrile
> Excellent moldability due to PS
> Good strength (SAN)
> Maximum service temperature about 200°F—this is low compared with other commercial polymers

Typical Uses

> Piping and fittings—drain, waste, water distribution
> Instrument and appliance housings
> Tool housings—hand drills, electric screw drivers
> Automotive instrument panels
> Home appliances—low environment-assisted cracking

High-Strength Polymers

Polyacetals

Polyacetal is manufactured by polymerization of formaldehyde (CH_2O). The polymerization reaction is

$$n \; \overset{\displaystyle H}{\underset{\displaystyle H}{C}}{=}O \quad \rightarrow \quad -[\overset{\displaystyle H}{\underset{\displaystyle H}{C}}{-}O]_n-$$

This reaction can be initiated by a number of materials, including acids, bases, and metallic compounds of cobalt and nickel. The nomenclature "acetal" refers to the carbon-oxygen bond. These materials are also referred to in the literature as polymethylene, POM, or simply acetal.

The strong attractive forces between the carbon atoms on one chain and the oxygen atoms on the neighboring chain produce strong intermolecular attractions. This combination therefore generates a largely crystalline polymer. Typical crystallinity is in the range 75–80%. While the large crystallinity increases strength and tends to improve dimensional stability, the transformation from liquid to crystalline solid on cooling produces large shrinkage changes. Mold design must accommodate this shrinkage and requires more careful evaluation than for amorphous polymers.

The two common trade names for acetals are Delrin and Celcon. Delrin is homopolar, and Celcon is a copolymer. These polymers are highly regular in structure, permitting close packing of adjacent molecular chains.

General Properties

> High strength. Typical ultimate tensile strengths are in the neighborhood of 65 MPa
> High stiffness. Typical elastic modulus is 3 GPa. Ductility, while being good, is not as great as for other high-strength materials, with deformation being principally by dilatation, rather than by shear (Flexman et al., 1988). Typical strain to failure values are in the neighborhood of 50%. Polyacetal typically fails shortly after yielding and does not exhibit necking or cold drawing.
> Low friction
> Excellent dimensional stability
> Service temperature to 110°C
> No solvents at room temperature

As expected, the mechanical properties depend strongly on temperature. Hashimi and Williams (1985) reported that acetal is in general brittle below approximately $-80°C$, exhibits some ductility between -80 and $20°C$, and is ductile above $20°C$. The strength increases with rate of testing. The variation of yield stress with cross-head speed fits a logarithmic equation well. The K_{ic} value reported by Bandyopadhyay et al. (1993) depended on testing speed, varying from close to 5 MPa m$^{0.5}$ at a test speed of 0.5 mm/min to about 4.0 at a test speed of 1000 mm/min.

Typical Uses

Applications are for parts requiring the combination of high strength and low friction such as bearings, bushings, gears.

Fuel supply parts, pump parts—no solvent interaction
Water system parts, plumbing parts—low water absorption
Food processing—resistance to hot water absorption, decomposition
Gears—low friction, wear, dimensional stability

Cellulosics

Cellulosics are made from naturally occurring materials, such as cotton. The starting product is steam heated in a dough mixer and mixed with a plasticizing agent (most commonly camphor), and reaction with an appropriate acid generates the resultant polymer. Natural cellulose has the structure shown in Fig. 4.1. Cellulosics have complex structures related to the original cellulose which vary with reaction chemicals added. The cellulosics have very stiff backbones, with added polar groups, resulting in highly crystalline polymers. They have very good optical properties, good resistance to outdoor weathering, and good strength. However, because oxygen is present on the main chain, water absorption rates are high. On excessive heating, the backbone decomposes, producing flammable gases.
Among the cellulose products most commonly used are:

Ethyl cellulose (EC)
Cellulose acetate butyrate (CAB)
Cellulose acetate propionate (CAP)

Typical Uses

Knobs, tool handles
Steering wheels
Light fixtures

Polytetrafluoroethylene

The most well-known trade name is Teflon. The polymer owes its strength to the tightly bound, stiff molecular chain, in which the fluorine atoms produce steric hindrance and prevent molecular flexibility. No branching occurs during manufacture, and the nearest-neighbor molecules can readily align closely to produce a highly crystalline substance. The crystallinity observed in polytetrafluoroethylene (PTFE) is between 94 and 99%, yielding one of the most crystalline of polymers. The chain lengths are quite high, with molecular weights on the order of 10^7. Although individual fluorine atoms would tend to yield strong polar forces, the complete replacement of all the hydrogens produces a molecule with very low van der Waals forces. This results in little attraction with other molecules and aids in yielding the very low frictional values of PTFE. The combination of PTFE against stainless steel has the lowest known friction coefficient. PTFE-PTFE contacts produce frictional coefficients that are significantly higher. Teflon has a high melting point of 342°C, but the liquid has such high viscosity that on melting there is so little change that the resultant material still retains its shape, like a gel. The polymer does not decompose until well above its melting point.

Processing

Although the melts of all polymers are very viscous, PTFE is so viscous that the liquid cannot be extruded to fill a mold under normal conditions. PTFE parts are therefore mainly produced by coining under high pressure and elevated temperature in a process similar to powder metallurgy. A preform is molded by compression in a suitably shaped mold using powdered starting materials. These fine powder particles can be formed as a water-based dispersion by suspension polymerization in water. The applied pressure compacts the powder so that it maintains its shape on removal from the mold. This preform is then heated, and sintering of the powdered particles occurs. The fine powder starting material may also be used to produce a thin coating on metals, which is then sintered to produce nonstick surfaces, largely for cooking pans.

Properties

The ultimate strength is of the order of 3000 psi (not particularly high), and the ductility is low, of the order of 3%. The chemical stability is excellent. Teflon is virtually unaffected by any solvents, acids, etc. Therefore EAC (environment-assisted cracking) is not usually of concern.

The temperature range over which PTFE can be used is exceptionally high. Due to its high melting point and temperature stability, it can be used continuously up to 260°C. Its low-temperature behavior is almost unchanged down to absolute zero, and it can therefore be used for any cryogenic application.

Typical Uses

Nonlubricated bearings
Chemically resistant piping and Chambers
Gasketing and sealant material
Body part replacement

Poly(Ether Ether Ketone)

Poly(ether ether ketone) (PEEK) has excellent mechanical properties and retention of these properties to high temperatures. It has a high melting point and is extremely resistant to solvents. The only known solvents are diphenyl sulphone, benzophenone, α-chloronaphthalene, and a mixture of 1,2,4-trichlorobenzene and phenol. (Gupta and Salovey (1990)). It is one of the promising candidates for use as a thermoplastic matrix material for composite materials to replace thermoset reinforced plastics. It can be prepared in the amorphous form by rapid cooling from above the melting point. It crystallizes on reheating to just above the glass transition temperature. Such low-temperature crystallization produces spherulites and lamellae an order of magnitude smaller than those produced by direct cooling from the melt. This results in improved properties for the same extent of crystallization.

Polymethyl Methacrylate

Polymethyl methacrylate (PMMA) as manufactured is a mixture of the various forms. A typical analysis would show that the molecular morphology is partly syndiotactic, atactic, and isotactic. As a result of this irregular arrangement of the molecules, PMMA is mainly amorphous, generally exhibiting less than 10% crystallinity and many grades with no detectable crystallinity.

Processing for Sheet Material

Cast sheet of the highest quality is made by pouring a syrup between two spring-loaded glass sheets and permitting the syrup to polymerize. The syrup is a slightly polymerized material, which has been prepared by adding the initiator and heating for several minutes to produce some polymerization. The syrup can also be prepared by dissolving the polymer in the monomer and adding catalyst. The filled cells are then heated for about 16 hours at a temperature slowly increasing from 40°C to just below that at which the monomer will boil off (100°C). After the sample is removed from the casting mold, some unreacted polymer is still present. To attempt to improve the completeness of the polymerization and to remove any residual stresses, the sheet is afterbaked at 140°C.

Stress Orientation

Although the acrylic is amorphous and below the glass transition temperature, the molecules can be oriented by an applied stress. The toughness and strength in the direction of stretching are greatly increased. Biaxial stretching of the sheets (heated so as to increase the molecular orientation speed) greatly increases the toughness and strength in the plane of the sheet.

Typical Properties

Molecular weight $= 10^6$
Glass transition temperature $= 104°C$
Tensile strength $= 10,500$ psi, among the strongest plastics
Strain to failure $= 3\%$
E (modulus of elasticity) $= 430,000$ psi, among the stiffest plastics
Excellent light transmission
Outstanding resistance to weathering
High-impact grades by adding copolymers of other acrylics

Typical Uses

Outdoor signs, lighting
Knobs, covers
Packaging
Not dishwasher safe: crazing with repeated cycling, soaps

Polycarbonate

Due to the large number of polar molecules and the stiff backbone, it would be expected that PC would be crystalline, but with normal cooling rates crystallization is not observed, and PC is classified as an amorphous polymer. Since the glass transition temperature is 145°C, it is glassy at room temperature. However, despite the fact that it is used well below its glass transition temperature, it is a very ductile, tough plastic. Many attempts to explain this apparent anomaly have been advanced, and these will be discussed later.

Properties

Ductile

Excellent toughness (12–15 ft-lb/in)—vandal-resistant window glass, can resist a baseball thrown at full force. However, the toughness exists only at thickness below 1/4 inch.

Low dimensional changes on molding

High strength

Very low creep

Very susceptible to stress corrosion from ozone in air, which is a detriment for windows. Also susceptible to bases, acetone crazes in hot water, and hydrocarbons.

Typical Uses

Light-transmitting housings, windows (tougher but more scratch susceptible than PMMA)

Impact-resistant housings

Polyamides

Polyamides are the products of the reaction of an amine group $(-NH_2)$ and an acid. The characteristic reaction is

$$-NH_2 + HO-\underset{\underset{O}{|}}{C}- \quad \rightarrow \quad -N-\underset{\underset{O}{|}}{C}- \qquad \text{(the amide group)}$$

Nylon is the household name for the polyamides. Nylons are named for the number of carbon atoms in the two parts of the monomer on either side of the amide group. Nylon 6/6 has six carbon atoms on each side and is manufactured from hexamethylenediamine, containing a total of six carbon atoms $[H_2N(CH_2)_6NH_2]$, and adipic acid, also containing six carbon atoms $[HOOC(CH_2)_4COOH]$. Nylon 6/12 is also manufactured from hexamethylenediamine and from an acid containing 12 carbon atoms in the chain, $HOOC(CH_2)_{10}COOH$. Nylon 6 is manufactured by a different route, involving the hydrolysis of a ring molecule, and has a smaller mer than Nylon 6/6. A comparison of the two is shown below:

Nylon 6/6:

$$-[NH(CH_2)_6NHCO(CH_2)_4CO]_{\overline{n}}-$$

Nylon 6:

$$-[NH(CH_2)_5CO]_{\overline{n}}-$$

Nylon 11 is manufactured from aminoundecanoic acid, which contains the amine and acid groups in one molecule containing 11 carbon atoms:

$$NH_2—(CH_2)_{10}—COOH$$

Note that the two functional groups which result in the condensation polymerization do not need to be on different molecules. Other nylons with different carbon atom chains are also available, as are copolymers of various nylons. The larger the number of carbon atoms, the smaller the polar bonding per atom and the more flexible and lower in strength the nylon. Nylons with even numbers of carbon atoms are stronger than those with odd numbers. The molecules are linear and contain polar atoms. As a result, nylons are highly crystalline. The crystallization tends to continue after molding, and for many parts the dimensions change slowly over several years as a result of very slow continued increases in crystallinity. This is coupled with variations in dimensions due to moisture absorption, making complex dimensional changes occur. The nonpolar sections of the molecule impact flexibility, the flexibility increasing with increased length of the CH_2 regions. The melts have low viscosities compared with other polymers, also due to the flexible CH_2 components of the molecule. This low viscosity makes fiber drawing possible, leading to nylon clothing and ropes.

Properties

> Good abrasion resistance
> Good tensile strength, 8000–11,000 psi
> e_f (strain at failure) = 100–200% high modulus
> Good fatigue resistance
> Good impact strength, 2–4 ft-lb/in
> Resists organic and nonpolar solvents
> High moisture content absorption, accompanied by some dimensional changes, and large decreases in strength with the water absorption. When saturated with water, the yield strength decreases by 30% as compared to the dry state, while the modulus decreases by 60%

Cross-Linked Nylon

Cross-linking is possible using trifunctional amines as monomers.

Typical Uses

> Gears, cams, machine parts: toughness, abrasion resistance, impact resistance

Nylon can easily be produced as thin fibers, which are then wound into cord or rope. These are used extensively in tire cords and marine applications where resistance to fatigue loading is one of the more important mechanical properties. Although the various fibers tend to rub against each other during fatigue loading, data for single fibers and ropes made by various interlacing methods all exhibit the same fatigue resistance, indicating that such rubbing is unimportant. Under tensile fatigue, during which tensile strain is accumulated during each cycle, failure is observed when a critical strain is reached. That strain is the same strain at which the sample fails in a direct stress-strain test, indicating that the cyclic fatigue lifetime is controlled solely by creep rupture, and failure is affected only by the time under load, rather than the number of cycles of loading.

Polyphenylene Oxide (PPO)

Due to this rigid backbone structure, it would be expected that this material would be crystalline. However, the glass transition temperature is close to the melting point, and it is not possible to have sufficient undercooling to nucleate the crystallites before the material becomes glassy. As a result, PPO is a glassy polymer with a high T_g of 210°C, making it one of the high-temperature use polymers. This high T_g results in excellent dimensional stability and low mold shrinkage. However, the viscosity of the plastic is very high and molding is difficult. To improve molding characteristics, it can be blended with PS. These two polymers can be blended in any proportions, being completely soluble in each other. It is also blended with Nylon 6/6 to improve the high-temperature performance and moisture resistance of the nylon. The copolymer which has blocks of mers containing 2 and 3 CH_3 groups attached to the benzene ring is called PPEC (poly phenylene ether).

Properties

> Outstanding moisture resistance
> Low coefficient of friction
> Good impact resistance
> Excellent dimensional stability
> Low warpage or shrinkage on molding—glassy
> Excellent resistance to degradation or dimensional changes in water
> But—very susceptible to attack by halogenated and aromatic hydro-carbons

Typical Uses

> Housings which are intricate, require dimensional accuracy
> Electrical switches, cable connectors, electronic parts supports
> High-temperature parts in contact with water in chemical industry

Polysulfone

Despite its name, the amount of sulfur in the plastic is rather small. Despite the stiff chain and linear arrangement of the backbone, polysulfone is amorphous, probably because the glass transition temperature is so high that crystallization does not have time to occur. Since it is amorphous, it is transparent and can be used for light-transmitting applications. The $T_g =$ 185–230°C, depending on manufacturer, with a heat deflection temperature of 175°C and a continuous use temperature of 160°C. Annealing for about 10–20 minutes at 160°C improves the mechanical properties and resistance to crazing.

Properties

  Excellent strength: tensile strength = 10,200 psi
  Excellent thermal stability
  Excellent chemical stability
  Resistance to burning
  Excellent hydrolytic stability: Resistance to high temperatures and
      steam, hot water. Can be repeatedly sterilized

Uses

  Medical and food requirements involving repeated sterilization
  Microwave cookware (resistance to burning, washing)
  Printed circuit boards (resistance to burning)
  Appliance parts

4.3. RESINS FOR COMPOSITES

Epoxy Resins

Epoxy resins include a very wide range of plastics manufactured from different monomers. The common connection is that they all contain the epoxide ring (C_2O) shown in Fig. 4.1.

  Epoxy resins in industrial applications are generally thick, viscous fluids which are used for encapsulation of electrical parts or impregnation of fiberglass or carbon fiber components. Other uses involve epoxy resins in suitable solvents for coating or painting applications. In all these applications, the resin consists of a partly polymerized material with M_n of the order of 3000–4000. Many epoxy chemical formulations exist, but the most common is DGEBA, named for the two starting materials which generate the monomer. The abbreviation stands for the diglycidyl ether of bisphenol A. Since the mer is complex, with a molecular weight of 338, the resin contains molecules with an average DP of about 10. These relatively short chains produce a liquid resin which can be worked into the desired shape and then polymerized further to produce a strong final product. Since the epoxide group is the functional group involved in the polymerization process, the useful resins start from monomers which contain several epoxide groups. The monomer of DGEBA is:

The polymerization proceeds through breaking of the bonds in the epoxide ring by means of a cross-linking additive. The opening of the rings results in covalent bonding of the monomers, and either thermoplastic or thermosetting material can develop depending on the additive, although thermosetting epoxies are the rule. Reaction by use of the epoxide group can be performed by adding a variety of reactive agents (called hardeners) by a condensation reaction. The advantage of this type of condensation reaction involving the epoxide group is that no small molecule is produced during the cross-linking, so that opening of molds for breathing and the formations of voids and blowholes are avoided. An additional advantage of the lack of production of any small molecules arises during the application for en- capsulating electronic components. Such encapsulation is frequently carried out under vacuum, and the vaporization of such reaction products would contaminate the system.

The wide choice of hardeners contributes to the wide range of properties for different types of epoxies. Some hardeners react at room temperature; others (those utilized in the aerospace industry) require long-term (16–24 hours) elevated temperatures for complete reaction. The two most common curing agents are amines and acids (or acid anhydrides). The curing reaction employing a tertiary amine as a catalyst generates a reaction similar to the addition type of polymerization. Amines can also act to produce a condensa- tion reaction. Polyamines can also be added to produce cross-linking due to the presence of several functional groups. The other common hardening agent type is an acid. As with amines, polymerization occurs by reaction with the epoxide ring and then continues through the resultant —OH groups generated.

Additives

Usually, the resin that is utilized as the matrix binder for epoxy-glass or epoxy-carbon composites is made by polymerization to produce materials with sufficiently low molecular weights that they still flow. As mentioned above, typical resins for infiltrating composite fibers have molecular weights of about 3000, corresponding to a DP of the order of 10. This low molecular weight resin still, however, has very high viscosity, and proper flow is difficult. Solid epoxide resins which are somewhat easier to handle in bulk can be prepared by increasing the molecular weight somewhat, still keeping the M_n low enough so that further polymerization will occur on the addition of a hardener.

Unmodified epoxies have excessively high viscosities, and the resultant polymers have very low elongations to failure. Other disadvantages are the large shrinkage and large heat generation that occur on polymerization. Since the cost of epoxy resins is very high, the addition of lower-cost components is very desirable. Fillers, including sand, metal fibers, and metal powders, improve mechanical properties and lower cost.

Conventional plasticizers are added to produce more flexible epoxies. Lower molecular weight polysulfides (HS—R—SH) with M_n of about 1000 are also added. These will react during polymerization to become incorporated into the polymer, so that bleeding will not occur. Besides improving the flexibility, they also lower the molding shrinkage. Addition of rubbery polymers to the brittle epoxy improves the fracture toughness by distributing rubber particles throughout the epoxy. The most important effect is to produce cavitation within the sample, which relieves the triaxial tension at the base of the crack. This results in the formation of shear bands and crack multiplication, all of which are energy-absorbing mechanisms. Diluents are added to reduce the viscosity of the resin. Their addition also lowers the heat generation and shrinkage during polymerization. The most common are lower-DP epoxy resins.

Properties

Typical values of the mechanical properties of epoxies are shown in Table 4.1. The properties of epoxies can be summarized as:

    Good strength
    High toughness (1–8 ft-lb/in depending on hardener)
    Low shrinkage during curing

Table 4.1. Mechanical Properties of Epoxies

Property	Value
Tensile strength	6–12 ksi
Compressive strength	11,000–15,000 ksi
Elastic modulus	400–500 ksi
Flexural modulus	350–450 ksi
Elongation to failure	1–5%
Heat deflection temperature	90–170°C
Impact strength (Izod)	0.5–1.5 ft-lb/in

Adhesion to many materials
Alkali resistance
Excellent electrical properties
Low moisture absorption

Typical Uses

Encapsulation of electronic materials
Matrix material for use with glass, carbon fiber
Powder coating of metal substrates

Polyesters

Polyesters are the general reaction products of an acid and a glycol (organic chemicals with more than one OH group). Although long-chain high molecular weight polyesters can be prepared directly by reacting the acid and the glycol, laminating resins are prepared by first producing a product which has been polymerized to a low degree, so that the average DP is only 8–10. This produces a thick syrupy liquid, suitable for use as the binder for a fibrous filler material.

To permit reaction which will form a thermoset and harden to yield the final composite, the polyester in this form also contains some reactive double bonds. An excessive number of such bonds will produce a hard, brittle polyester, so the number of such reactive sites is controlled by preparing the initial resin by reacting several components:

1. Saturated acid (no double bonds). The greater the concentration of this acid, the fewer cross-links and the more flexible the polyester. The most common acids used are isophthalic acid and adipic acid:

Isophthalic Acid

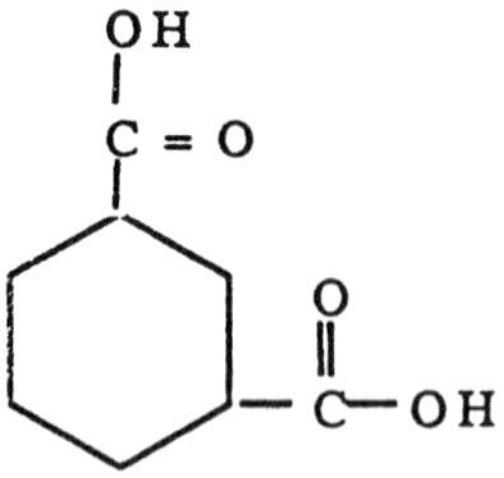

Adipic acid

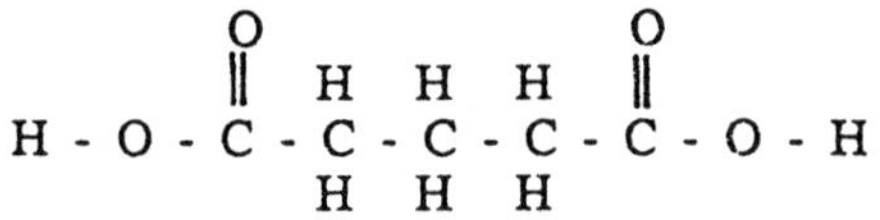

Note that isophthalic acid produces a molecule with a very stiff backbone, yielding a strong, low-elongation product with a high distortion temperature. Adipic acid yields a much more flexible resin due to its more flexible backbone.

2. To produce cross-links, an unsaturated (containing double bonds) acid is also used. The most common acid is maleic acid:

Maleic acid

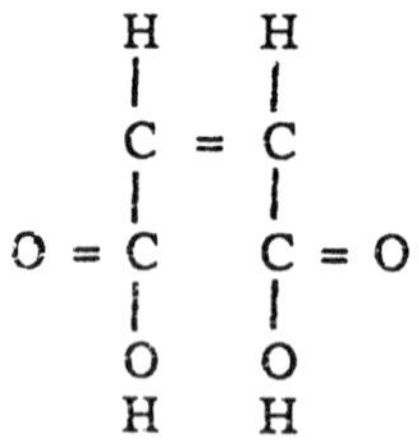

3. The glycol most commonly used is 1,2-propylene glycol:

1,2 - Propylene Glycol

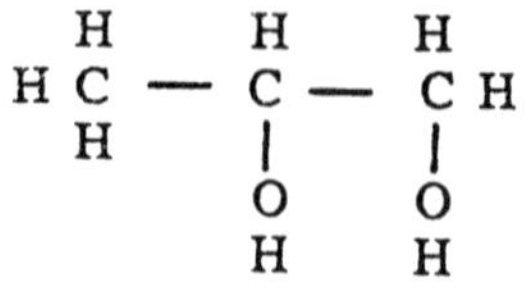

A typical formulation would be:

Propylene glycol	50% (by number of molecules)
Maleic acid	33%
Isophthalic acid	17%

4. To reduce the cost of the product, the prepolymerized material is dissolved in styrene. Since styrene contains double bonds, during the hardening process the styrene will become copolymerized with the polyester.

Polyester is most commonly used as the laminating resin for fiberglass. The cross-linking occurs either at room temperature using methyl ethyl ketone peroxide (MEKP) or at elevated temperatures in press moldings using benzoyl peroxide. Since polymerization continues slowly after the viscosity of the mixture becomes high due to the increase in chain length, maximum strength is not obtained until more than a week has passed. Oxygen inhibits the polymerization. To prevent air contact with the surface of the product, some wax can be added to the resin. When polymerization progresses, the wax becomes insoluble, and is forced to the surface, preventing the contact with air.

Uses

Laminating resins
Dough molding compound
Compression molding: resin, lubricant, reinforcements with short fibers, etc. are compounded.

5
Manufacture of Plastic Parts

5.1. COMPARISON OF THERMOSET AND THERMOPLASTIC MOLDING

Whereas die casting of metals consists of forcing the molten metal into sealed molds under high pressure, most metal castings are performed by permitting the molten metal to flow into the mold under gravity. The viscosity of molten plastics is orders of magnitude greater than that of liquid metals, and high pressure is required for the molding process. Thermosets initially are unreacted small molecules, and the molding process not only forms the material into the desired final shape but also provides the time and temperature conditions required for the thermosetting reaction. Consequently, the time required for thermoset molding is generally longer than for thermoplastics. This longer molding cycle time results in an economic penalty, and consequently thermoplastics are usually preferred. If wall thicknesses are smaller than 1/8 inch, the heat transfer during the molding process is sufficiently rapid that thermoset molding cycle times can be reduced sufficiently to make them attractive materials. During molding, the plastic must be transferred through various runner systems to the mold. The material left in these systems can be reground and reused for thermoplastics but must be discarded for thermosets. Typically, about 15% of the weight of the part is lost in the runner system, also tending to make thermoplastics economically more appealing, despite their somewhat higher initial cost per pound.

Many of the newer thermoplastics have excellent mechanical properties which are retained at high temperatures. Improvement of these properties by the addition of fiber reinforcements with fiberglass and graphite is also possible and puts them on a par with thermosetting materials. Sheet composites of thermoplastics have many promising advantages. The main two are:

1. The addition of these reinforcements would permit the production of sheet material which could be molded by the rapid methods available for thermoplastics.
2. The impact resistance of these thermoplastics is decidedly superior to that of the cross-linked thermosets. This increased toughness is of critical value in many applications. Of course, the impact resistance is affected by the type of reinforcement and the strength of the interfacial bond between the plastic and the fiber.

5.2. EFFECT OF MOLDING ON MECHANICAL PROPERTIES

The mechanical properties of polymers depend more strongly on the manufacturing process than those of most other materials. Consequently, the data available in the literature and from the manufacture cannot be taken as to be expected from a specific molding process. This is to be expected for both reinforced and unfilled polymers. The part thickness of the molded section will affect the properties, since the smaller the cross sectional area, the greater the shear developed within the polymer during flow and the greater the alignment of molecules and reinforcing fibers in the flow direction. The shear stress is greatest close to the mold walls, and the molecular and fiber orientation will be correspondingly greatest. The mechanical properties therefore become increasingly anisotropic near the part surface in thin sections, with greatest orientation and mechanical properties parallel to the flow direction and lowest properties perpendicular to that direction. Lower shear stresses are developed in thicker sections, and less orientation can be expected. In all types of molding, variations in properties occur as a result in changes in temperature, pressure, and raw materials. It should be noted that, in general, molding of low molecular weight thermoplastics proceeds more easily, since the melt viscosity is lower than for high molecular weight material. Thus processing ease tends to be opposed to maximizing the mechanical properties of the product, which are improved with increasing molecular weight.

5.3. MOLECULAR ORIENTATION

Molecular orientation occurs during any manufacturing process. There are no methods available which can accurately predict the orientation that occurs. However, X-ray analysis and other techniques can be used to obtain the orientation of finished parts. Such evaluations are time consuming and costly, and the analysis is complex.

One technique employed is the study of *birefringence*. This is the difference in refractive index measured in two perpendicular directions. It is a difference between the values in the molecular chain direction and in the direction transverse to the chain direction and is caused by the difference in electronic density along the chain axis and between adjacent molecules. This difference results in a difference in the polarizability, so that the velocity of light is different in the two directions. Since a random polymer will have molecules oriented in all directions, the greater the birefrigence, the greater the orientation. For semicrystalline polymers, birefringence will exist as a result of the crystalline regions, which are oriented, and will be additive to the orientation in the amorphous regions.

Thermal retraction is also a useful tool for evaluating orientation. The amount of permanent deformation is very small, and the elongated molecules can return to their original coiled configuration on heating. The extent of length change on such heating accurately measures the extension and orientation during deformation or manufacture.

5.4. RESIDUAL STRESSES

Residual stresses are developed within molded parts, as the high-temperature polymer melt flows into a mold at a much lower temperature. If the molding process will yield an amorphous product, the mold temperature is usually well below the glass transition temperature; if the process will yield a crystalline product, the mold is well below the melting point of the material. The plastic cools rapidly, and residual stresses are generated by the following sequence:

1. The material in contact with the mold walls cools rapidly and shrinks, either as a result of solidification shrinkage or solely due to thermal contraction.
2. This surface contraction causes the interior to be compressed. Since the interior is still hot, it has little mechanical strength, and the applied stress produces permanent deformation of the interior material.

3. The center then cools and contracts, tending to generate compressive stresses on the exterior. However, the exterior is frozen, and the resultant stresses are resisted by elastic contraction of the exterior.

In general, it can be expected that the magnitude of the residual stresses will decrease with increasing melt temperature and mold temperature, because the higher temperatures will permit accelerated stress relaxation. If the mold temperature approaches or exceeds T_g for an amorphous polymer, relaxation of residual stresses can be expected to be much faster than if the mold temperature is well below T_g.

Since stress relaxation requires time, during mold filling, relaxation of the residual stress can be expected to be greater for the portions of the product farther from the gate and the residual stresses to be highest at the gate. Figure 5.1 shows the residual stress profile in an amorphous polyamide sample at various distances from the gate (Siegmann et al., 1987). The sample was an edge gated sheet. The data shown is for samples cut 72 mm away from the gate. The figure shows the resultant stress distribution from the center toward the surface of the sample at various molding temperatures. Higher melt temperatures and greater distances from the gate lower the residual stresses.

Figure 5.2 shows the stress distribution in PC produced by quenching. The resultant stress distribution in both cases involves compressive stresses on the surface and balancing tensile stresses in the interior of the part. This stress distribution can be beneficial or harmful. The surface compressive stresses tend to improve fatigue resistance, as fatigue failures are caused by tensile surface stresses. Toughness is improved, and the compressive stress reduces the harmful effect of surface notches and grooves, which act as stress concentrators. As a result, it is often desirable to introduce residual stresses intentionally. This can be accomplished by heating the part and then quenching. However, the residual stresses add to applied exterior loads, and the interior tensile stresses can cause failure at unexpectedly low applied loads. Furthermore, if part of the surface is abraded or machined away, the resultant residual stresses become unbalanced and the part may warp away from its designed shape. It is important to recognize the difference between residual stresses and molecular orientation. If the part is reheated above T_g and molecular orientation exists, the part will distort and change shape as the molecules rekink toward the equilibrium random configuration. This distortion is not a result of residual stresses within the material.

If uneven cooling occurs in the mold, the residual stresses may be unbalanced initially, with compressive stresses on one side only balanced by

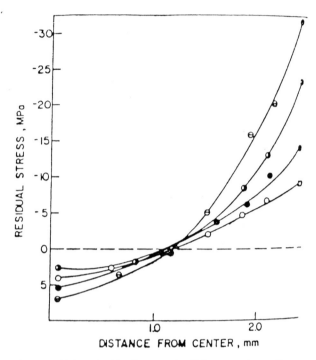

Figure 5.1. Residual stress in PA at various distances from the center of an injection molded sample. Higher injection temperatures give more time for relaxation and lower residual stresses. The stresses are compressive in the outer layers and tensile toward the center. The residual stresses are more pronounced closer to the gate. Molding temperatures of 295°C, (circles with horizontal line), 310°C (half-dark circles), 330°C (dark circles), 350°C (open circles). [From A. Siegmann, S. Kenig, and A. Buchman, *PES 27*, 1069 (1987). Reprinted by permission of the publishers.]

residual tensile stresses on the other, causing warping on removal from the mold. Molding procedures which produce slower, more uniform cooling should reduce the magnitude of residual stresses but are contrary to rapid production practice.

The most common rapid determination of residual stresses is done by placing the part in a liquid which induces cracking where the tensile stresses are high (stress-corrosion cracking). Fine cracklike crazes will form on the surfaces at the locations of highest tensile stress. However, compressive residual stresses will not produce a noticeable effect. Other methods include removal of material by machining or drilling holes in the sample to eliminate part of the material and unbalance the remaining residual stresses. The extent of distortion of the part can be used to calculate the residual stress

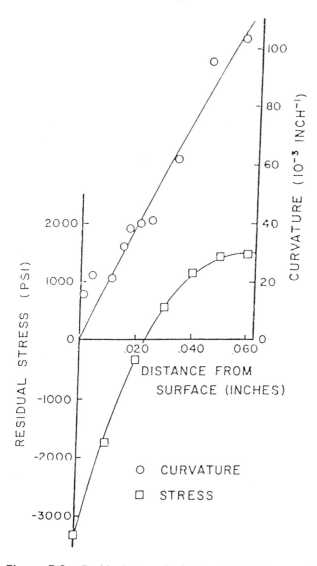

Figure 5.2. Residual stress distribution in PC after quenching. The residual stress and resultant curvature for 1/8-inch-thick PC sheets quenched from 150°C. [From P. So and L. J. Broutman, *PES 16*, 785 (1976). Reprinted by permission of the publishers.]

distribution. Although this technique is effectively employed with metals, machining of some plastics is difficult and may introduce additional surface stresses.

Residual stresses affect the fatigue life and impact strength of the molded part strongly, the major effect being due to the residual stresses at the surface, where fatigue cracks initiate and propagation of these cracks begins. Compressive residual stresses are beneficial, but tensile stresses are generally harmful. For example, So and Broutman (1976) reported that quenching polycarbonate samples in an ice bath produced compressive residual stresses at the surface of 3000 psi, sufficient to increase the Izod impact strength by a factor of 2.

5.5. MELT ELASTICITY

On heating above the melting point, an amorphous polymer does not exhibit any sudden discontinuous change. Rather, the material can be evaluated by the same analysis as for the rubbery state, with a relaxation time that decreases as the temperature increases. Therefore, it is permissible to consider the melt as a viscoelastic material. During processing, such as injection molding or extrusion, hydrostatic stresses develop within the melt. When these are removed as the extruded material leaves the die, the liquid swells considerably, as any elastic material would. This is called *die swell*. It is described by the *swell ratio function*:

$$B = \frac{\text{diameter of extrudate}}{\text{diameter of the die}}$$

B is a function of several production parameters. The general dependence is a result of extended times under the hydrostatic stress to affect the molecular orientation of the melt. The production parameters which increase B are:

1. Increased extrusion rate
2. Increased die length/diameter ratio
3. Decreased ratio of die diameter to liquid pool diameter

5.6. MELT FRACTURE

If the flow rate through a die is increased excessively, the resultant extrudate develops several surface imperfections, which increase in severity with increasing flow rate. These, in order of increasing severity and flow rate, are:

1. Bamboo surface: a slight variation in diameter
2. Sharkskin: surface roughness

3. Melt fracture: large-scale variation in diameter and surface appearance

The explanation for these defects is still somewhat unclear. Due to the very high viscosity of polymer melts, the Reynolds number for flow in processing is generally close to 1, and laminar flow is always assured. During this flow, the viscoelastic behavior results in internal fracture.

5.7. WELD LINES

In most molding techniques, the polymer enters the mold chamber through a runner system, and the velocity and shape of the entering fluid are controlled by the design of the entrance or gate into the mold. The high viscosity of molten polymers prevents rapid flow, and several gates may be designed for a mold. Separate polymer streams entering through these gates will eventually impinge. Because molecules tend to align in the flow direction, the line of such interaction (the weld line) does not have the same molecular configuration as the remainder of the part. Weld lines are also generated whenever the polymer liquid is forced to diverge and reunite in flow around an insert. The degradation in properties depends on the ability of the molecules within the impinging streams to diffuse together to form a homogeneous material. Cloud et al. (1976) studied several polymers and found that most polymers had losses in tensile strength at the weld line. Polysulfone appeared unaffected, however. An additional problem generated by the presence of flow lines is the appearance of a slight surface groove along the weld line due to air trapped in the mold between the two melt fronts. This groove not only generates a displeasing appearance but also provides a stress concentration at a location of inherent weakness. As an example of the weakness which can be produced by weld lines, Crawford and Benham (1974) observed that the fatigue life of acetal copolymer parts was reduced to 0.02 to 0.04 times that of samples without weld lines.

Proper design requires sufficient gates that the plastic will join while it is hot enough for proper welding to occur. If the weld lines exist in regions of high stress or where appearance is important, it may be more appropriate to drill the holes afterward. Another way to avoid weld lines is to mold holes only partway through the product, so that plastic flow will be unimpeded by a core running entirely across the plastic flow pattern. The holes are then drilled out after molding. Reinforcing fibers tend to align strongly in the flow direction. At the weld line, where the two flowing liquids intersect, the flow of each is forced parallel to the weld line. As a result, throughout the part the fibers tend to run parallel to the initial flow direction, but at the weld line the fibers turn and tend to align perpendicular to the initial flow direction, decreasing the strength of the fiber-reinforced part greatly.

The importance of weld lines is demonstrated in a study by Takemori (1988) in which he prepared tensile specimens by double-gated injection molding, so that a weld line was generated in the center of the specimen. He observed that a notch was formed at the weld line 6 to 10 μm deep with a crack opening displacement of 3–4 μm. The weld line had a sharp notch tip, with molecular orientation or residual strain around the notch, extending about 20 μm around the weld line. Notches for initiation of fatigue failures are commonly produced in such studies by pressing a razor against the surface. It has been found that the fatigue lifetime of samples with such mechanically induced notches is the same as that of samples with weld lines (Matsumoto and Gifford 1985).

5.8. OBTAINING ACCURATE DIMENSIONS

Tolerances

It is not possible to obtain the same tolerances with molded plastic parts as it is with machined metal parts. Although the mold can be machined very accurately and filled completely with molten plastic, shrinkage will occur due to thermal contraction and crystallization. These will depend on the temperature at which the gates freeze shut, the rate of cooling of the product, and the molding pressure. The shrinkage will vary with direction in the finished part, since the extent of molecular orientation will depend on flow directions within the mold. Shrinkage allowances are always added to mold dimensions and vary for different plastics and grades. Typical values for polystyrene are 0.002–0.004 inches per inch of material, while for acetals they range from 0.02 to 0.025 inch/inch. Using these two materials as examples, since shrinkage factors can vary depending on slight variations in processing, tolerances for polystyrene molded parts cannot be expected to be greater than 0.002 inch and those for acetal parts 0.005 inch. When accurate dimensions are needed, prototyping is required. One technique is to manufacture the mold with the best estimate of shrinkage, mold a part, and then modify the mold dimensions to approach the desired part dimensions. The mold may be machined further if the finished product is too small; the mold may be chromium plated if the product is too large.

Best designs are obtained by maintaining thin walls with as uniform a thickness as possible. Heat transfer determines the rate of cooling, and thicker walls have less orientation and more crystallization and spherulite structure than thinner sections. These differences, coupled with residual stresses due to the differences in cooling rate, affect postmolding distortion. In addition, thicker walls decrease production rates due to the longer time required for the part to remain in the mold until it is sufficiently strong to be ejected. Because thick sections shrink to a greater extent than thin sections on cooling, *sink marks*, regions where shrinkage produces a surface depression, are likely at such sections. These are most likely to occur where a thick section changes thickness. The best way to avoid such sink marks is to maintain uniform thickness. Strength can be maintained by incorporating ribs perpendicular to the walls, rather than having thicker walls. However, excessively thin walls make mold filling difficult, and the minimum wall thickness is about 0.025 inch.

Reduction of Shrinkage During Molding

Shrinkage can be reduced by adding materials with low shrinkage. Such added materials should not change dimensions during molding (e.g., fillers which remain solid) and do not melt during molding. Typical filler materials utilized for this purpose are calcium carbonate, mica, glass fibers, flakes, and balls.

Reduction of Postmolding Shrinkage

Glassy plastics may continue to shrink after molding as a result of slow reduction of free volume toward the equilibrium value. Crystalline plastics may shrink, as crystallization may proceed slowly even at room temperature. Nylon and urea-formaldehyde parts continue to shrink for several years after molding. These postmolding shrinkages may be reduced by annealing at slightly elevated temperatures to speed the process. Annealing temperatures should be kept low, however, as annealing may affect other mechanical properties unfavorably.

Utilizing Amorphous Materials

Since amorphous materials do not change phase during molding, there is no large-scale volume change, and their use is recommended in applications where dimensional tolerance is critical. Second, amorphous materials are desirable in that whatever shrinkage occurs is isotropic, as compared to the anisotropy of the solidification of crystalline polymers. However, if fillers are added, they produce some anisotropy in amorphous materials. Since long-fiber fillers usually increase the mechanical properties and the anisotropy to a greater extent, a combination of both long fibers and flakes may be used.

Because all thermosets are amorphous, compression molding shrinkage of these materials tends to be controlled primarily by the chemical reaction involved in the thermosetting process.

However, as discussed in Chapter 2, glassy polymers which have been rapidly cooled are not at their equilibrium free volume and slowly shrink toward that volume. Furthermore, the cooling rate will affect how close the sample is to equilibrium: the slower the cooling, the closer to equilibrium. The density of a glassy polymer will therefore depend on the cooling rate. For injection-molded samples of polyetherimide, the density increases from the surface to the center. This variation in density is attributed to the variations in cooling rate within the sample. The surface is closer to the cold mold wall and is cooled more rapidly, freezing in a larger free volume. However, other effects also occur. The residual stresses generated by the unequal cooling rate also affect the density, and contradictory results have been found for other materials more sensitive to surface compressive residual stresses.

Changing the Cooling Rate

The cooling rate of the part must be uniform to avoid distortion and uneven shrinkage. This, of course, is in conflict with the desire to remove parts as rapidly as possible from the molding machine to increase productivity, and usually high production rate is the controlling factor. The extent of crystallinity increases with lower cooling rates. The cooling rate is a function of the coolant supply to the mold but is also dependent on heat transfer through the molten polymer. Since the thermal conductivity of polymers is very low, the cooling rate in the interior of the part is much lower than at the walls, even for relatively thin sections. Due to the increased fraction of crystallization, the density often increases toward the center. An increase in the density of a polyethylene molding of 0.1% may be observed over distances as small as 0.05 inch. Similarly, increasing the mold wall temperature increases the amount of crystallinity. Steinbuch (1964) reported that increasing the mold temperature of nylon from 20 to 100°C resulted in an increase in specific gravity from 1.084 to 1.122.

If the polymer does not crystallize readily, the following morphologic changes throughout the part may be observed:

1. The region near the mold walls cools rapidly, resulting in an outer amorphous skin.
2. Some crystallization occurs toward the center.

If the polymer is rubbery, three general regions may form:

1. An outer amorphous region
2. An intermediate oriented region due to flow orientation and subsequent freezing
3. An inner amorphous region, since the flow shear stress is small and sufficient time exists for recoiling of the molecules

5.9. COMPRESSION MOLDING

Compression molding is utilized primarily for thermosets. The process has two mold halves (usually manufactured from metal) which are heated and supported within a press mechanism. The starting material is thermosetting molding compound, a mixture consisting of the resin system with fillers and other additives, which can include pigments, accelerators, lubricants, etc. The molding compound is usually purchased by the molder and used as received. The molding compound is available in granular, nodular, flaked, diced, or pelletized forms. The molds for compression molding require that during the compression of the initial charge to the final product, the change in volume (called the bulk factor) be of the order of three or less. Typically, about 5% excess materials is added. The excess causes a small amount of flash around the parting line, where the two mold halves join. The parting line should preferably be designed to be in one plane for simplicity of mold design and is most commonly placed where the perimeter of the part is large. Several general types of molds exist. Flash, material which flows between the two mold halves, produces an unpleasing appearance and can be minimized by redesign of the mold. The existence of a flash line on the product defines the type of mold as being of the flash type. In other types, such as the positive mold type, flash is minimized by having the flash run vertically instead of horizontally. The amount of flash can also be reduced by using carefully weighed charges and double contacts in the vertical direction. Figure 5.3 shows several mold configurations.

To assure that the charge that is placed in the mold cavity is sufficiently compact to fit properly, the initial charge may be premolded to the approximate shape of the mold cavity. Use of such tablets helps determine the exact weight of the charge, reducing both the amount of flash and the shearing forces, which tend to break or deform the inserts. In addition, the depth of the mold can be reduced because the volume of the charge is smaller. This results in lower mold machining costs. The tablet is compressed and in better contact with the mold walls, improving heat transfer during molding. Ease of material handling is also improved. In other uses, the charge may be powder that is simply poured into the mold cavity. Inserts and cores are placed in the mold while open. The surfaces of the inserts are usually knurled

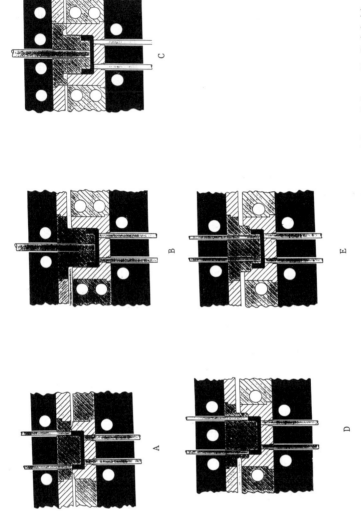

Figure 5.3. Various types of molds for compression molding. (Adapted from R. Bainbridge, Compression molding; in *Plastics Products Design Handbook*, Part B, E. Miller, ed., Marcel Dekker, New York, 1983.) (A) Flash mold; (B) direct positive mold; (C) landed positive mold; (D) semipositive horizontal flash mold; (E) semipositive vertical flash mold.

or undercut and flanged to produce a better bond to the plastic. The pressures during mold closing are very high, and large shearing pressures are developed on the inserts. Their location and construction are critical in assuring that they do not deform or move during this step in the molding cycle.

The upper part of the press (called the force) is mounted on a heated platen and contains knockout mechanisms, just as the lower mold cavity. The process, which can be automated or hand operated, consists of the following steps:

1. With the mold open (the force raised so that the operator or supplying robot can add material to the cavity), the cavity is cleaned of any plastic which may have adhered to any surface after the previous cycle.

2. If any metal inserts are part of the product, they are placed within the cavity.

3. The molding compound or premolded part is then placed in the cavity.

4. The force comes down, closing the mold, and a low pressure is applied. Heat from the platen heaters causes the resin's viscosity to be reduced, and the resin flows easily, filling the cavity. Various heating techniques are employed. Steam, hot oil, and electric heating are all used, although electric heaters are most common. Typical molding temperatures are in the range 300–400°F, with pressures ranging from 2000 to 4000 psi. As the cavity remains sealed, the pressure is increased to assure good surface replication of the mold. The heat causes cross-linking of the polymer, and the viscosity starts to increase very significantly.

5. If a low molecular weight by-product is produced during the polymerization (such as water or ammonia), the mold is briefly opened (breathing) to permit its escape and then reclosed until polymerization is complete.

6. The mold is opened, the knockouts remove the part, and the process is repeated.

7. After baking in a separate furnace to assure complete polymerization may be performed. This speeds production and frees the molding machine for further use. Afterbaking temperatures are the molding temperatures or slightly higher, the specimen being held in the afterbaking furnace for 5–15 hours. The increase in extent of cross-linking during this process increases the rigidity of the material, and the glass transition temperature (all thermosets are amorphous) is raised, often as much as 15%.

8. Polymers have poor thermal conductivities, and heat transfer to the polymerizing material is difficult. Undercure resulting from too low a temperature or insufficient time in the mold results in lowered mechanical strength and a lowered resistance to solvents and environmental chemicals. If sufficient heat transfer is attempted by using excessively high temperatures or excessively long holding times, the mechanical properties will also tend to be degraded due to oxidation and pyrolysis, but generally to a lesser degree than with undercuring. Heat discoloration will also result, and the financial burden of longer cycle times is always present.

Uses of Compression Molding

Compression molding is best used when thermoset parts which require close tolerances are needed. Since the mold closes when the resin is of low viscosity, the surface replication is good and no parting lines are usually visible. No scrap or wasted material results, as the charge into the cavity should be the weight of the finished product. The system itself is not overly complicated, and the cost of a compression molding machine is low compared with that of other methods of molding. If the part is very complex and has a large number of undercuts, requiring slides in the mold, the cost can become much larger.

When compression molding of thermoplastics is performed, in addition to heater assemblies to melt the material, water cooling coils must be included to harden the material sufficiently rapidly that reasonable production rates can be obtained.

Materials for Compression Molding

The materials that are most commonly compression molded are phenolics, urea-formaldehyde resins, melamine, diallyl phthalate (DAP), polyester molding compounds, and epoxies.

Sheet Molding Compound

The molding material for compression molding is compounded prior to the molding operation. Sheet molding compound (SMC) is most commonly a polyester-based resin reinforced with 25–30 wt% glass fibers about 1 inch long and is prepared as shown in Fig. 5.4. The molding compound consists of a paste of the resin and catalyst, fillers, and desired additives. It is fed onto two carrier films and compacted on either side of the fiber reinforcement. Random fibers are always used for finish coatings for appearance. The designations for SMC are:

R	Random chopped short fibers	SMC-R
C	Long continuous fibers	SMC-C
C/R	Combination of random and long	SMC-C/R
25	wt % of fibers specified	SMC-R25

Bulk Molding Compound

Molding thicker sections requires better flow properties, and the high weight percent and length of the fibers in SMC would make flow to all corners of many parts difficult. Bulk molding compound (BMC) therefore has only 10 to 20 wt % fibers of reduced length (typically 1/4 to 1/2 inch long).

Mechanical Properties of Compression-Molded Parts

For filled compression-molded objects, some orientation of the fibers usually occurs. Depending on the direction of flow, the properties may be enhanced by this alignment. Since most of the flow occurs in the temperature range in which the viscosity is very low, the residual stresses are generally low. The shrinkage that occurs after the thermosetting process during cooling is fairly uniform; thus low residual stresses are maintained.

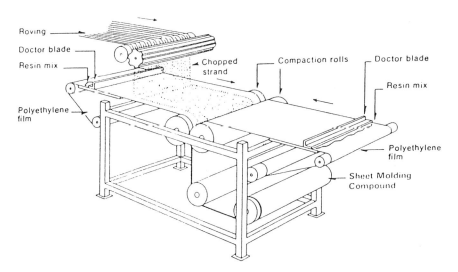

Figure 5.4. Preparation of sheet molding compound. (From: S. S. Schwartz and S. H. Goodman, *Plastics Materials and Processes*, Van Nostrand Reinhold, New York, 1982. Reprinted by permission Van Nostrand Reinhold Co.)

5.10. TRANSFER MOLDING

Compression-molded parts which contain long, thin, or delicate inserts and regions with very thin mold walls often become distorted due to the large shearing stresses generated during the closing of the mold and the compacting of the material into the desired configuration. Often, the line on the finished product caused by the flash that forms at the parting line of the mold is undesirable, even if the flash is carefully removed. For such parts it is necessary to have the mold closed before the material enters the mold. In transfer molding, the charge to be molded is placed in a separate chamber and heated until it can flow readily, either by convection in an oven or by high-frequency dielectric heating. The charge is usually a tablet-shaped preform. The low-viscosity charge is then forced by a plunger into the closed mold, usually situated directly below the heating pot. In many respects this technique is similar to injection molding, except that the mechanism is similar in structure to that for compression molding, in that the two half-molds and the heating pot are mounted vertically above each other in a press. Injection molding has tended to replace transfer molding except for applications in which inserts are required.

To permit the material to flow readily, the compounds used are generally of lower viscosity than for direct compression molding. Due to the greater flow required in transfer molding, the amount of orientation of filler material is greater, the properties tend to be more anisotropic, and greater shrinkage is observed than in compression molding. However, in general the mechanical properties tend to be somewhat better than for a compression-molded part, since heat transfer and mixing of the preform are improved.

Transfer molding has other advantages compared with compression molding. Since the mold is closed, the dimensional tolerances of the finished product tend to be better. Fewer blowholes and less porosity in the product result, and the residual stresses are even lower than in compression molding. However, some material must be left in the heating pot, and scrap loss is high compared with compression molding. The mechanical properties of transfer-molded products tend to be slightly improved over those of compression-molded items due to the improved heat transfer and mixing of the preform.

5.11. STAMPING

Flat sheets of thermoplastics or filled unpolymerized thermosetting material are cut into desired shapes with a sharp-edged cutter under a large applied force. The similarity to cutting cookies from thin dough is unmistakable.

Sheet molding compound is commonly used for this purpose. The process is often run continuously. For thermoplastics, the sheet can be preheated by passing it through a furnace to soften it, simultaneously thermoformed between mated upper and lower dies, and then cut using presses. Thermosetting cut parts are cured in furnaces after such forming and cutting.

5.12. PULTRUSION

In the pultrusion process, continuous strands of fiber reinforcing material are passed through a bath of the impregnating resin (Fig. 5.5). The bath is contained in a tank sufficiently long that the immersion time is about 1 minute. The reinforcement can be continuous strands, mats, woven cloths, rovings, etc. The fibers are then pulled through a stripper to remove the excess resin. The coated reinforcement then enter a forming guide, which shapes the polymer-coated fibers to the desired shape. A predie shaper forces the impregnated reinforcement to the general shape and pulls the reinforcement over any required core material. The final die is preheated so that some polymerization occurs, in the tool. The tools are substantial because they must not distort as the resin expands during the polymerization. The fibers then pass through a furnace, typically 3 to 5 feet long, which causes complete polymerization to occur. Two general devices used to pull the fibers through the bath and furnace are the opposed caterpillar and the hand-over-hand gripper. The caterpillar has two opposed clamps attached to caterpillar pullers. A large number of gripper pads on both caterpillars fit the part. The large number of pads and continued contact may apply excessive crushing pressure to the part. Hand-over-hand devices have only two grippers, which repeatedly interchange contact with the product. This tends to generate less

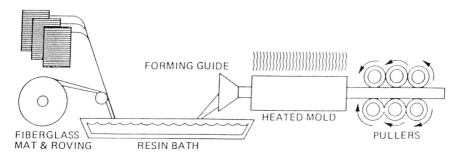

Figure 5.5. The pultrusion process. (From J. Martin, Pultrusion, in *Plastics Products Design Handbook*, Part B, E. Miller, ed., Marcel Dekker, New York, 1983.)

crushing force as the pulling force is separated from the gripping motion. The disadvantage is the initial higher expense of the system. The process is continuous, and after being pulled through the heater, parts of the desired length are cut by a flying cutter. Virtually any shape with constant cross section can be manufactured, and the length can be any value. Among the common items are flagpoles, fishing rods, I-beams, and aircraft flooring. Bonding of various pultrusion-manufactured parts can produce complex shapes such as built-up beams or construction walls.

The start-up costs of the process are small, and the cost of tooling and equipment is low compared with other thermosetting processes, such as compression molding. The process is not labor intensive, and automatic flying cutters which move with the finished product can generate a continuous process. Pultrusion can also be used with polymer baths which contain fillers. If the fillers settle to the bottom of the polymer tank, variations in properties will result, and care must be taken to keep the bath well stirred.

Mechanical Properties

Pultrusions are composite materials, and the mechanical strength is largely that of the reinforcement. Longitudinal reinforcements are most common, but woven mats or other two-dimensional or three-dimensional reinforcements can also be utilized. The mechanical properties can be computed by the standard techniques for composites, described in Chapter 10. Very high percentages of reinforcements (as high as 80% glass) can be used.

5.13. ROTATIONAL MOLDING

This process is most commonly applied to thermoplastics, but it can be readily applied to thermosets. It is utilized to produce hollow parts by placing a limited amount of the polymer in a mold and heating and rotating simultaneously so that the walls of the mold become coated with a thin layer of the polymer. The mold is rotated about two perpendicular axes, but the rotational angular velocity is not the same on both axes. Figure 5.6 shows the various stages in a rotational molding. The ratio of rotation in the two axes can vary from 2:1 to 8:1 depending on the shape of the part. The mold material is most commonly aluminum, 1/4 to 1/2 inch thick, because of its high thermal conductivity and ease of machining. The molds need not be very strong, as no stress is developed during the molding operation.

After the entire polymer mass becomes fused into one continuous part, the mold is cooled and the product removed. Thermoplastic particles (by far most common) or thermoplastic plastisols (thermoplastic material partly

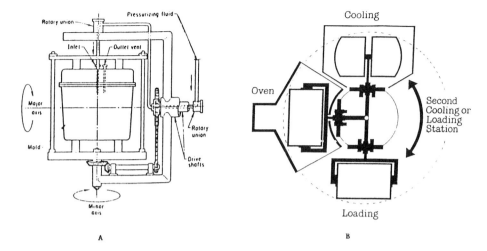

Figure 5.6.   Rotational molding. (A) Mold rotation occurs about two independent axes. (B) Separate molding arms with separate molds permit several molds to be in different stages of production simultaneously. Machines may have many arms to speed production. (From D. J. Ramazzoti, Rotational molding, in *Plastics Engineering Handbook*, J. Frados, ed., Van Nostrand Reinhold, New York, 1976. Reprinted by permission Van Nostrand Reinhold Co.)

dissolved in a plasticizer with some small particles remaining) are most commonly used. The method is also employed with liquid thermosetting resins. For the thermosets, the heating process causes polymerization.

During cooling, in a well-sealed mold, the entire surface (including the parting line) is coated with plastic. As material cools, this produces a vacuum within the mold, which can pull product away from the walls and cause distortion. A vent (small tube through the mold) prevents a vacuum from forming; however, moisture may be drawn in, causing blowholes.

The particle size of the charge should be small enough to permit the particles to move about, so that when the mold wall reaches the melting (or flow) temperature the particles will stick to the walls, and not to each other, to prevent balling. The best polymer material would ideally have particles identical in size. The smaller the size distribution, the better the product properties. A mesh size of 80 is most commonly employed.

The mold is a closed volume, and if only part of the mold is desired in the finished product, then the part which is not to be part of the final product is insulated so that the plastic does not stick to it. For example, household garbage cans are easily rotation molded and the top opening is generated by such blanketing. Since the polymer is free to flow on the mold surface, accurate control of the wall thickness is difficult, and the process is most suitable for objects in which dimensional accuracy in the thickness direction is not critical.

Advantages of the Rotational Molding Technique

1. Virtually any shape which is hollow can be made. If the part is less than 0.5 inch thick, the product need not be hollow.
2. Composite parts are easily manufactured. Combinations of compatible materials can be molded in layers. For example, a tank can be made with an outer skin of one plastic for impact resistance and strength and an inner skin for acceptable contact with food or beverage. The composite part can be manufactured either by opening the mold after the first part is melted and molded or by placing the second polymer in an insulated box within the mold, which is opened after the outer skin is melted and covers the mold.
3. The molds can be manufactured inexpensively, as the applied stresses are very small and rotational speeds low.
4. No scrap is produced in the process, as the finished product is used as molded.
5. Since the polymer is not injected or compressed into the mold, the finished part is stress free and problems of warping, crazing, and failure due to internal stresses are avoided.
6. Inserts can easily be placed in the mold without concern about distortion due to stress.

Products

Garbage cans, light fixtures, and motorcycle fairings are made by rotational molding.

5.14. INJECTION MOLDING

Injection molding is the most common process for part production in the plastics industry. It is ideally suited to large production runs, 50,000 parts being considered typical for a full-scale multishift operation. Smaller runs, of course, are common, utilizing less automated operation and molds of less costly materials. Virtually any thermoplastic can be injection molded. The starting material is normally thermoplastic pellets which must be heated until the melt viscosity is sufficiently low to permit injection into the closed mold. The temperature range of operation must be closely controlled. Too low a temperature will result in material of such high viscosity that the plastic will not flow sufficiently to fill the mold completely. Excessively high temperature will result in decomposition of the plastic, resulting in gas-generated pores and a lowering of the mechanical properties of the product. The standard industrial machines currently employ screws rotating within closely fitting barrels to melt and mix the polymer granules by the shear work performed as the material is carried forward in the threads into the mold cavity. Most plastics are hygroscopic and absorb moisture from the atmosphere on standing. This moisture must be removed, or it will generate blowholes in the final product. Moisture removal can be accomplished by preheating the pellets, and pellet delivery systems which prevent excessive contact with moist air are necessary in automated systems. Several types of injection molding machines are employed.

Plunger-Type Injectors

These are the oldest machines and are being replaced rapidly by the more efficient screw-type machines. In the plunger injection molder, a hydraulically driven plunger in a barrel initially is behind a hopper which contains granular particles of the plastic to be molded. A specifed amount of the plastic drops into the barrel in front of the plunger. The plunger then moves forward, forcing the granules around a torpedo-shaped spreader, which spreads the granules against the inner sides of the heated barrel, melting the plastic. Continued forward motion of the plunger forces the plastic through the nozzle of the injection molder into the mold cavity. Poor mixing of the plastic and melting of the granules result from this simple straight line motion of the plastic granules, and modifications include the use of two separate plungers. One plunger melts and fuses the granules. The output of this plunger is then fed to another, which injects the plastic into the mold.

Screw-Type Injectors

The difficulty in completely homogenizing and melting the plastic when a plunger is used led to the development of the screw-type injection molding machine. The mechanical shearing of the plastic melt as a result of the rotation of the screw produces sufficient heat to melt the plastic, and cooling must often be provided to the barrel to prevent overheating of the plastic. Heating bands are also provided to melt the plastic during start-up. The general configuration of an injction molding machine is shown in Fig. 5.7, and a typical plasticizing screw is shown in Fig. 5.8. Thermoplastic polymer melts have very high viscosities due to their high molecular weights, and high pressures must be generated to force the material into a closed mold. This pressure is generated in a reciprocating screw machine by driving the screw forward after plasticization of the polymer is complete. During

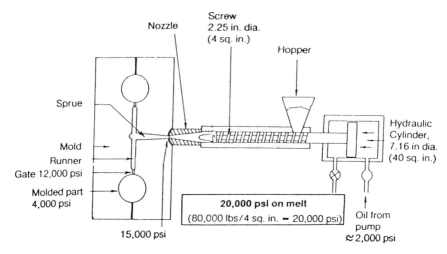

Figure 5.7. Reciprocating screw injection molding. The high pressures developed during injection molding are shown. (From *Plastics Processing Data Handbook* by Donald V. and Dominick V. Rosato, Van Nostrand Reinhold, New York, 1990. Reprinted by permission Van Nostrand Reinhold Co.)

injection molding the high pressure generates large shear forces, which produce significant orientation of the molecules during flow.

The screw therefore must be designed to produce the following:

1. Entrance into the barrel of pellets of the cold plastic dropping from the hopper. Sufficient space must exist between the flights (screw threads) so that enough material can be transferred ultimately to the mold. The portion of the screw designed for this purpose is called the *metering* section. This section occupies approximately 50% of the length of the screw.

2. Rotation of the screw must plasticize (heat, homogenize, compress, and melt) the plastic. During this process any entrapped and adsorbed gases are removed. This is the *transition* section. For shearing action to occur, the coefficients of friction of the barrel and the screw must be different with respect to contact with the plastic. If the two were identical, the plastic would rotate within the flights as a stable plug, and no shearing and heating would occur. The sticking of the plastic to the barrel results in work being performed on the plastic, resulting in eventual melting.

3. Generation of sufficient pressure to force the molten plastic into the mold through the sprue and runners. Pressures as high as 20,000 psi may be required. This last section is the metering or pumping

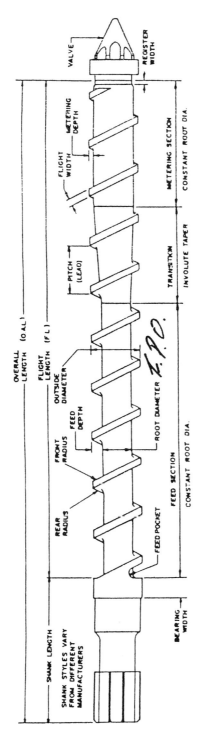

Figure 5.8. Injection molding screw construction. The various sections with screw nomenclature are shown. (From *Plastics Processing Data Handbook* by Donald V. and Dominick V. Rosato, Van Nostrand Reinhold, New York, 1990. Reprinted by permission Van Nostrand Reinhold Co.)

167

section of the screw. This pressure is generated by reducing the depth of the flights in the metering section compared with the feed section, and in the transition section the depth is slowly varied between these two values.

Nozzles

Transfer of the molten plastic to the mold from the barrel occurs in the nozzle, which may contain a spring-actuated check valve to prevent the material from dripping out prior to the injection. When the injection pressure rises sufficiently to overcome the resisting force of the spring, the valve is forced open and injection starts. When the pressure drops sufficiently, the spring recloses the valve. This permits continuous rotation of the screw without loss of material. If the material within the nozzle freezes, no injection is possible, and separate heating bands are usually required. Thermocouples inserted directly into the path of the molten plastics are used for the most accurate temperature control.

Molds

The plastic enters the mold cavity through a *sprue*, which is the cylindrical cavity entrance. If the part to be molded is large, mold filling through one opening is likely to cause freezing of the plastic before the mold is completely filled. Mold filling from several positions simultaneously is therefore necessary, and multiple runners supply molten plastic to the mold. If the parts being molded are small, it is common for the mold to contain more than one cavity, and smaller passages from the sprue, the runners, carry the molten plastic to the various cavities. The direct entrance from the runners into the mold cavity is reduced in cross-sectional area and is called the *gate*. Often, several gates are designed to permit rapid mold filling. For appearance, the gates should be small, so that the marks on the finished product produced when the excess material is removed are unnoticed. The reduction in area at the gate permits easy separation of the runners from the product and permits freezing of the gate to prevent any loss of hot plastic from within the mold if the part is ejected before complete freezing occurs. This results in considerable time saving, as an excessively long time would be required to wait until the center of the molded part was completely cool. The location of the gate may affect the mechanical properties, as the plastic must flow rapidly through the gate to fill the mold, and large shear stresses are introduced in this region, resulting in oriented material with large residual stresses. The net effect is to produce weaker material in this region, and if possible the gates should be located in relatively low-stressed regions.

The part is ejected from the mold by knockout pins located on the ejector side of mold (or *tool*), the side away from the injection barrel. Ejector pins should also be placed to eject runners.

Proper design of a mold encourages easy mold filling and avoids stress concentrators in the finished part.

1. Sharp corners should be avoided to prevent stress concentrations in the finished part. Equally important, they produce stress concentration within the mold itself, which may result in breakage of the mold during molding. Good practice is to avoid any radius of less than 0.025 inch at any corner.
2. *Draft*, a slight angle of 1° to 2°, on any depression walls should be designed into the mold to permit easy removal of the molded part. Even if removal is possible, the force required to pull the molding from the mold will result in rapid wear of the mold walls.
3. Wall thicknesses should be as uniform as possible to prevent uneven shrinkage and uneven cooling, which can change the amount of crystallinity from one part of the molded product to another.
4. Coring may be required for holes and depressions in the product. The generally accepted standard for such cores is to design height to thickness ratios of less than four. Cores which exceed this ratio are too slender to resist bending moments caused by the plastic entering the mold and can deflect or break.
5. Venting is required to permit air to escape as the plastic flows into the mold. Insufficient venting may cause air bubbles or surface voids in the finished product, and the back pressure generated will slow down the overall production rate. Most venting is provided along the parting lines of the mold, with thin (of the order of 0.003 × 1/4 inch) grooves extending to the surface of the mold.
6. Ejector pins. After the part is cooled sufficiently so that it will not deform while being ejected from the mold (about 10–15 seconds), the mold is opened, and ejector pins push forward to remove the part. While the ejector pins are commonly on the mold half away from the plasticating barrel, the part normally shrinks away from the mold walls onto any insert, and the pins must be located so as not to separate the part from these inserts.

Foamed Products

These products usually contain individual small bubbles of gas. The bubbles are produced either by injecting a gas directly into the molten plastic during the molding process or by having an additive incorporated into the plastic which decomposes, generating a gas at the temperature of molding. Under the high pressures within the barrel of the injection molding machine the gases are so compressed that they cannot produce foaming. However, within the mold, where the pressure is reduced, the foaming action occurs.

Injection Molding of Thermosets

The thermoset injection molding process utilizes the screw mixer to raise the temperature of an uncured thermosetting mixture sufficiently to reduce its viscosity so that it will flow readily into the mold. However, the temperature and residence time in the barrel must be such that the thermosetting reaction does not occur to a significant extent. A typical temperature in the barrel is 90°C. Thermoset injection molding must therefore be controlled more carefully, since curing must be permitted to occur only in the mold, not in the runner system or the barrel. Polymerization within the barrel is very serious, as no easy method of removal exists. Freezing of a thermoplastic within the barrel presents no problem, as reheating will melt the material. To prevent runaway temperatures in the barrel and to permit shutdown when desired, the barrel has a large water-cooled jacket, which can lower the barrel temperature rapidly if required. Since the starting material is not a polymer, it has a much lower viscosity than the polymerized material in a thermoplastic injection molding system. This permits easier mold filling, smaller runner systems, and good reproducibility of mold features. Thermoplastics may be injection molded with variations in temperature and cycle times which may vary widely without any disastrous results, since the material remains a liquid until cooled sufficiently. The unpolymerized material is injected into the mold, which is electrically heated sufficiently that the prolymerization process occurs rapidly (between 170 and 200°C). This is in contrast to the more common thermoplastic injection molding, in which the mold is cooled to induce freezing of the plastic. The mold must remain closed and heated until polymerization is complete. This results in a longer molding cycle time for thermoset injection molding. In addition, the scrap from the runners and sprues cannot be reground and reused. Since thermosets have relatively low viscosities before curing, they tend to ooze from the mold at the parting lines, and the flash must be removed afterward by tumbling, blasting, or cutting. Among the thermosets molded are diallyl phthalate (DAP) urea-melamine foams, epoxy, phenolics, and polyester BMC.

Mold Filling Studies

Studies of mold filling have been used to determine the flow of the polymer into molds, so as to permit some prediction of the resultant properties. The most common procedures are:

1. *Short shots*: Injection of insufficient material to fill the mold completely. Varying amounts are injected, and the resultant series of products shows how the mold fills.
2. *Direct observation*: Special molds can be constructed with transparent sides, so that the flow of polymer can be directly observed. This is possible only for simple mold shapes, and it is difficult to obtain three-dimensional visualization of the mold filling.
3. *Pigmented flow lines*: Injection of two colored materials into the mold through an appropriate sprue generates colored lines which lie parallel to the flow direction. Cutting through the molded part permits three-dimensional analysis of the flow.
4. *Computer simulation*: Finite element programs have been developed to predict the flow within molds. At present, these are used principally to determine the injection pressure, the size and number of runners, and where gates should be placed in the mold cavity. They also aid in predicting where weld lines will occur and the molecular and fiber orientation to be expected. Obviously, weld lines must be located in regions where the expected stress will not be critical. The accuracy of these programs is increasing rapidly, and their use is becoming widespread rapidly.

Heat transfer programs aid in determining and minimizing the total cycle time. Since warping will occur if there is a substantial difference in cooling rates throughout the part, these programs are also used to assure uniform cooling.

Results of these studies indicate that several stages of flow can occur:

1. *Jetting*: The initial high pressure in the injection produces a jet of molten material, which spews across the mold, hitting the side opposite to the gate. After contact with the mold wall the plastic folds over onto itself, producing weld lines in which the polymer cools before sufficient molecular entanglements between the various parts of the injected plastic can occur. After the mold has partially filled, the flow becomes more laminar, but further weld lines may occur between the jetted material and the material that enters the mold afterward.
2. *Normal mold filling (fountain flow)*: After the initial jetting, the remainder of the mold is filled by laminar flow. Additional weld lines may occur between the jetted material and the plastic that fills the rest of the mold. During this second stage the plastic flows in contact with the walls of the mold. Due to friction at the walls, the velocity is zero at the walls and reaches a maximum at the center of the mold. Therefore the material at the center must spill toward the

walls, and velocity components exist both parallel and perpendicular to the flow direction from the gate.

3. *Molecular orientation*: These forces produce aligned molecules along the flow pattern direction, which is therefore not necessarily parallel to the mold walls. Even in amorphous polymers, these shear forces tend to produce some alignment of the molecules. The orientation is large at the walls because the shear is greatest, and the polymer molecules are rapidly frozen into their aligned position. In semi-crystalline polymers, the size of the crystallites depends on the temperature gradient at the solid-liquid interface. A thin skin tends to form at the mold walls and is amorphous, since the melt cools so rapidly that crystallization cannot occur. As with completely amorphous material, this layer has some molecular alignment parallel to the mold walls. Outward from this skin zone, a trans-crystalline zone is often found, with crystalline material growing toward the center in the direction of the temperature gradient. Closer to the center of the molding, the shear stresses generated by flow are small, and the liquid can cool to produce a spherulitic-type structure. Thin moldings are less likely to exhibit this region, as it requires absence of flow-induced shear. Generally, there is some variation in percent crystallinity along the mold, with highest crystallinity nearest the gate, since that is the last part to fill and therefore cools more slowly.

Imperfections in Molded Parts

Distortion

Large-scale distortion evident on removal of the part from the mold is usually due to removal of the part when it is still soft. Such problems are usually a result of an attempt to increase production rates excessively.

Warping

Changes in shape over a period of time after the part is removed from the mold are called warping. These slow changes are a result of residual stresses which vary in magnitude over the dimensions of the part. The warping is a result of creep generated by these stresses. To correct these effects, the flow patterns are checked, usually by short shots, and the gate locations or cooling of the mold varied to produce a more uniformly filled mold.

Flow into the mold is not uniform, leading to warping. The molten plastic, on flowing through the constriction of the gates, experiences a higher pressure and greater flow rates, leading to greater orientation in these regions. In addition, the pressure is greater near the gate, leading to higher density and greater internal stresses; relief of the stresses when the pressure is removed on opening the mold leads to warping.

Thermal degradation of the polymer may occur in the barrel of the injection molding unit. Excessively long residence times result in overheating of the charge. This is particularly severe in polyvinyl chloride (PVC), in which the chlorine generated during thermal decomposition can attack the injection barrel and also catalyzes further decomposition of the polymer. The mold size should therefore be matched to the size of the injection unit, since use of a larger-capacity machine with a small mold results in only a small portion of the charge being injected at each shot.

Thinner sections have lower fractions of crystallinity and will therefore warp more due to later crystallization. Postmolding shrinkage can be reduced by annealing to induce crystallization and shrinkage. This is an added step that raises the cost of the product, and it is therefore not a common operation. When performed, annealing should be just below the heat distortion temperature. Constrictions at the gate which restrict flow into the mold result in greater molecular orientation and crystallization. Excessive restriction can also result in additional heating of the material as it passes through the gate, resulting in degradation of the polymer and generation of volatile components, which can cause bubbles in the molded part.

Feeding the polymer from several different gates reduces shrinkage and property variation, since there are shorter flow paths, resulting in less orientation. More random flow directions within the mold also produce more uniform properties.

Reaction Injection Molding

Reaction injection molding (RIM) is a process for thermosetting materials. The two monomers are kept in separate tanks and are injected into the mold through separate metering systems and a mixing system. Polymerization occurs within the closed mold. Since monomers are used directly, the viscosity of the materials is very low, and the pressures and temperatures are much lower than for injection molding. Typical conditions are 50 psi and 75°C. However, since the polymerization occurs rapidly, the viscosity of the mixed fluids in the mold increases very rapidly, so that complete mixing of the two monomers must be accomplished prior to the introduction into the mold. Usually this mixing is accomplished by impinging the two separate streams to produce turbulence or by rapid mechanical agitation. The mechanical mixing occurs in flighted rotary mixers operating at speeds up to 20,000 rpm. To ensure good shear and mixing, the gap between the rotor and housing is made quite small, and the mixing system requires considerable energy and heat generation. This heat must be removed, or else polymerization will occur within the mixer rather than in the mold.

Automobile bumpers, panels, fenders, furniture parts, and other large objects are readily produced by this process. Reinforced reaction injection molding (RRIM) utilizes the same procedures, but reinforcements are added to the components.

Compared with compression molding, the advantages include the following:

1. Since the mold is closed when the monomers are injected, higher accuracy of mold reproducibility is possible. The monomers have much lower viscosity and can easily flow and completely reproduce the mold surface.
2. Reinforcements can be placed in the mold before injection of the monomers. The low pressures required will not displace the reinforcements or distort the fibers. Similarly, cores can be used which are manufactured of inexpensive foam, because their strength need not be high.
3. Since flow occurs in the monomers and not in the long polymer molecules, orientation and residual stresses are lowered.
4. Larger parts can be manufactured, since the low pressures require smaller pressuring units.

Injection Compression Molding

When molding large parts, normal injection molding presents difficulty in mold filling, because a large amount of the plastic (with high viscosity) must flow through narrow runners and gates. The requirement for short reinforcing fibers to avoid jamming the gate area has already been mentioned. To overcome these disadvantages for large parts, injection compression molding has been developed.

In this process the mold is not completely closed during the injection process. The plastic is injected through larger gates, permitting rapid mold filling. When injection is almost complete the mold is closed, compressing the molten plastic and causing it to fill the mold without the necessity of the last flow through the gates, which may tend to freeze shut at the end of the cycle. If the mold starts to close after completion of the injection step, weld lines may be formed where the plastic folds onto itself. The advantage of injection into an incompletely closed mold is that less pressure is required, so higher injection rates and smaller injection molding units are possible. Because of the lower pressures and larger flow passages, the product is likely to have less orientation and less warping. As the final packing of the mold is produced by pressure exerted over the entire mold face, rather than flow through a narrow gate, properties are likely to be more uniform. An additional benefit of the larger gates is that reinforcing fibers can be longer, since they are less likely to jam and stick in the gate or other narrow passages.

5.15. EXTRUSION

The extrusion process utilizes a screw plasticizer similar to that described for injection molding. The screw melts the plastic granules introduced through a hopper and forces the molten plastic through to the end of the barrel under high pressure, through a series of meshes to remove dirt, and out through a shaped die to produce a continuous product of constant cross-sectional dimensions. Since thermoplastic melts have very high viscosities, the material can leave the extruder above the melting point. After extrusion, the material must be cooled below its melting point for crystalline polymers, or below the glass transition temperature for amorphous materials, before it can be roughly handled. This is performed by passing the extrudate into a water tank or by water or cool air sprays. After extrusion, the strength can be increased in the extrusion direction by applying an external tensile stress with a pulling device. The most commonly used technique is to apply the stretching force with opposed rubber rolls, tires, or caterpillar belts. The extruded product may be drawn down considerably, and the extrudate may be reduced as much as 30% in the stretching process. Due to this orientation, properties vary considerably between the extrusion and transverse directions. Pipes, sheets, rods, films, and structural members can be produced and cut into desired lengths with a flying cutter attached to the draw table if desired.

Although the extrudate is fairly uniform in dimensions on leaving the die, it can be distorted during the cooling and stretching processes, and for dimensional accuracy it can be resized after sufficient cooling. For accurately dimensioning of piping, the pipe passes through the inside of a tube which has vacuum ports so that trapped air can be withdrawn. The extruded pipe is expanded with internal pressure and/or external vacuum until it accurately fits the external pipe. The entire extrusion system must in general be considered as a whole. The size and shape of the extrusion die and the postextrusion devices depend on the characteristics of the extrusion screw and barrel, and they cannot in general be moved to a different machine without adversely affecting the shape of the product.

Film Production

Film extrusion can be direct, with a thick extruded film passed directly over a chill roller and then tensioning rollers which stretch it to the desired thickness. For the high draw ratios required for thin films, the extension produces extreme orientation in the draw direction and poor mechanical properties in the transverse direction. Biaxial orientation is obtained by reheating the sheet and grabbing the sides with a hooking mechanism which exerts both a direct tensile stress along the extrusion direction and a simultaneous transverse widening force. Stretch ratios of up to 4:1 in both directions are employed. The result is a material with the molecules elongated in the plane of the sheet but randomly arranged within the sheet, with little anisotropy. Takeoff devices apply tension to the sheet to increase molecular orientation and strength.

Film may also be produced by utilizing a die which contains an air blowing port. The die is shaped so that a thin tube of plastic emerges around the air supply port. The air pressure expands the tube to a film, also producing biaxial hoop stresses. After cooling, the film is slit and wound.

5.16. BLOW MOLDING

Blow molding (Fig. 5.9) is similar to injection molding or extrusion but produces thin hollow parts. It is principally used for bottle manufacture; no other process compares with it for manufacture of such hollow containers. However, many parts which are not usually considered as being hollow are manufactured by blow molding. The mold shape can be quite varied, and, in fact, the two halves of the mold may generate entirely different shapes. Among the various parts which are blow molded are instrument panels and computer enclosures. The main design considerations are that the product should not have rapid changes in shape and that a region of weakness will exist where the parison is pinched together to seal it. The only portion of the product which has accurate external and internal dimensions is the neck region, where the parison is held. The remaining blown region has wall thicknesses which depend on the extent of expansion to make contact with the mold walls. Programming the rate of extrusion to vary the thickness of the parison helps control the wall thickness.

The steps in blow molding are:

1. An extruder ejects a fixed amount of material around a blow head to produce a tube sealed at one end. The extruder is similar to the ones used in injection molding. The objectives of extruder operation are to:

   a. Melt the polymer by work performed.
   b. Move the polymer forward.

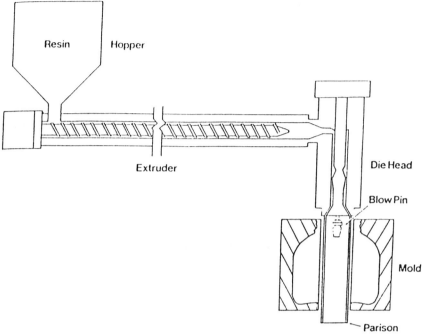

Figure 5.9. Blow molding. The extruder screw produces the parison, which is expanded in the mold by air from the blow pin. (From S. F. Raber, Blow molding, in *Plastics Product Design Handbook*, Part B, E. Miller, ed., Marcel Dekker, New York, 1983.)

    c. Acting as a plunger, push melted material out of the barrel through the die to form a preform, or parison. This is a relatively simple shape.

    d. Continuous extrusion can occur by having the molds move.

    e. If different parts of the parison are expanded by different amounts, the final product will have different wall thicknesses. To avoid this, the parison can be programmed to have different thicknesses by controlling the rate of extrusion.

2. When a sufficient tube has been ejected, the mold is closed, and air is emitted through blow head, expanding the part against the walls of the mold.

Injection Blow Molding

The parison is injection molded into a mold that shapes the bottle top. The rest of the bottle is then blown in a separate step.

Biaxial or Stretch Blow Molding

As well as radial stretching by air, longitudinal stretching by a stretch rod improves mechanical properties by aligning molecules in two directions. In the extrusion of the parison, the molecules tend to align along the direction of stretching and flow (the extrusion direction), and the material is anisotropic, with high strength in the extrusion direction but lower strength in the transverse direction. The part may therefore fail under low loads if applied in that direction. Flow in the two dimensions of the thin part will produce an isotropic material, better able to withstand off-axial loads. In general, since higher alignment is obtained, better mechanical properties in general are obtained:

1. Better clarity for transparent bottles
2. Increased strength, impact strength, reduced creep
3. Better gas and water vapor penetration resistance

5.17. THERMOFORMING

Thermoforming involves heating a sheet of a thermoplastic until it is sufficiently soft to conform easily to the shape of a tool. Although most thermoforming is performed on amorphous plastics, it has also been applied to crystalline ones. For amorphous materials, the sample is heated to above T_g; for crystalline plastics the material must be heated to just below the melting point to soften it sufficiently. It is fascinating to note that much of the thermoforming deformation in amorphous plastics is a result of removal of kinks within the molecules at the elevated temperatures, and the elongated molecules are then frozen into this configuration on cooling back to room temperature below T_g. If the sample is heated above T_g, the rekinking of the molecules can return the sample to its original shape. A sample with a large deformation, such as a flat sheet bulged into a hemisphere, can be reversibly returned to the original flat sheet of uniform thickness by reheating. The severity of the deformation is customarily indicated by the *draw ratio*, which can be measured in several rather ill-defined ways. The most common way is to identify the ratio of the depth of the product to its diameter. Another less common ratio is that of the area of the product to the area of the undeformed sheet.

The greater the amount of deformation and the sharper the curvature, the thinner the resultant part at that location. Thermoformed parts can vary by as much as a factor of 10 in thickness at various sections. Uniformity of strain is also affected by:

1. Molding conditions, such as molding speed, pressure, and temperature.
2. The thickness of the sheet.
3. The ease of creep at the molding temperature (called the elongational viscosity). Molecular orientation in the sheet material will cause uneven draw around the mold and uneven thickness.

The sides of the sheet are clamped and the center pressed against an appropriately shaped form to develop the final product. The mold can either be a cavity (negative) mold or a positive tool. Since the mold is generally colder than the sample, rapid cooling and resultant reduction in ease of deformation occur when the sheet makes contact with the mold. As a result, most of the deformation occurs in the portion of the sheet which has not touched the mold. Typical product shapes resemble cups or domes of some sort, and thinning of the sheet in the regions which have the greatest deformation occurs. This thinning is generally greatest along the walls, but all parts can exhibit thinning.

Many variations of the fundamental thermoforming process have been developed to reduce the variation in thickness in the product. In even the simplest techniques vacuum is applied to the negative mold through small holes drilled through the mold walls. Air pressure above atmospheric can also be applied on the other side, either with or without vacuum. If no vacuum is applied, venting of the air between the sheet and the mold must be available. Other variations use plug assists, in which the sheet is first pushed into the mold with a formed plug and then vacuum applied to complete the molding. Free forming is commonly performed to generate light transmitting domes and skylights. In this technique the heated sheet, clamped around its edges, is inflated by air pressure until it expands freely to a smooth curve until a specified height is reached. The air pressure is then controlled to maintain that height until the sample has cooled sufficiently to be removed from the fixture.

5.18. TECHNIQUES FOR COMPOSITES

Thermosetting polymer products are economically manufactured because the monomers have relatively low viscosities and can rapidly spread into and coat the fiber mats. As a result of the low viscosity, this impregnation can be accomplished using low pressures. This therefore permits the use of molds manufactured from easily machined or produced materials, lowering the overall cost of part production. For small production runs, reinforced thermoset products can be manufactured using simple manual processing. Automated processing is available for larger production runs.

Composites are usually not manufactured directly from separate filament and matrix materials. A prepreg is prepared initially from the desired proportions of the fibers and the resin, with the fibers in the desired array and volume fraction. For aerospace and other applications where maximum properties are required, the resultant properties are very dependent on the quality of the prepreg. Uniformity of the fiber spacing is of vital concern, and the number of broken fibers must be minimal. Use of a prepreg has the advantage over hand layup of fibers impregnated with resin in that the fiber-resin ratio is much better controlled and more uniform. The formulation of the resin component ratios can be metered more accurately than in the small-batch operation of hand layup.

Manufacture of Prepregs

Prepregs are primarily manufactured of glass or carbon fibers and polyester or epoxy resin. The glass is obtained as rolls of the dry fibers, which are passed through the appropriate resin solution and partly polymerized in tower furnaces. To prevent adherence of the layers of prepreg, a backing film is rolled on the take-up roll along with the prepreg. For carbon fiber–epoxy prepregs, Mylar backing is most common. The completed prepregs are in the partially cured or B stage. Since complete polymerization occurs at room temperature, refrigeration is required unless quick use in production is expected. Even at refrigeration temperatures, polymerization proceeds, and the typical working life for the prepreg after production is from 3 days to 1 month at room temperature and about 6 months at 0°C. Careful control of the prepreg inventory is therefore necessary. Many unsatisfactory products can be traced to material which had polymerized excessively before use.

Hand Layup

Composite laminates are constructed from the prepregs, which usually have unidirectional fibers. The complete assembly is obtained by laying the prepreg layers upon each other in the appropriate orientation (Fig. 5.10). When hand layup is performed, special plastic film templates of the required shape for each layer are frequently cut first. Mylar is the most common template material. This procedure produces the most accurate alignment of the layers and is employed for complex layups where maximum properties are required. Each prepreg layer can also be laid directly on the preceding

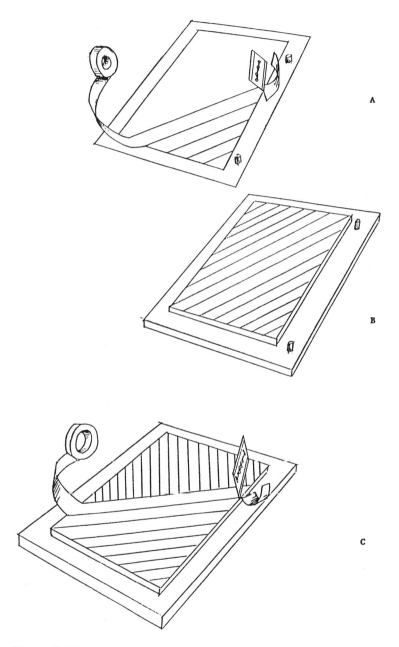

Figure 5.10. Hand layup technique. (A) Preparing one lamina. (B) Completed lamina. (C) Next lamina with different orientation.

layer with no template and cut to the appropriate shape ("black on black" for carbon fibers).

Tape Laying Machines

Various tape laying machines improve the production rate. In addition, they help reduce the variation in orientation and reduce the void content in the final product. The tape laying machine can include rotation tables so that the correct tape orientation is automatically obtained.

Filament Winding

A mandrel of the desired shape must first be manufactured. Fibers from a continuous roll are then woven over the mandrel either by having the mandrel rotate or by having the automatic winding machine rotate around the mandrel. Axial motion of the winding machine simultaneous with rotation of the mandrel produces fibers crossed on the mandrel, so that fibers can be oriented in the direction of required strength. Various resultant winding configurations are shown in Fig. 5.11.

The winding machine maintains uniform tension in the fiber as the winding proceeds. The dispensing head positions the tape and simultaneously applies enough pressure that the succeeding layers remain firmly in contact. Since the total stress developed on the mandrel increases with the number of turns, the mandrel must be able to withstand the accumulated stress of all the windings. The resin can be added by passing the fibers through a resin bath prior to winding, or the completed wound mandrel can be coated with resin in a painting operation or by dipping the completed mandrel into the resin bath.

Mandrels may of necessity be fairly complex to permit their removal after winding. Solid metal mandrels may be used if one end is open or the finished product is cuplike in shape, since the mandrel must be removed without collapse. Collapsing mandrels must be designed for more complex parts. The collapsing mechanism must be within the mandrel, so that the smallest-diameter mandrel of this type which can be constructed is of the order of 3 inches.

Consolidation of the plies is achieved by the application of tension to the impregnated fabric tow during the rotation of the mandrel. During the polymerization of the thermoset, which takes place in an autoclave or furnace in the temperature range 250–350°F, the thermal expansion of the mandrel applies internal pressure which further compresses and consolidates the part. The part cross-links at the elevated temperature. On cooling to room temperature, as with other tooling, the mandrel contracts, whereas the

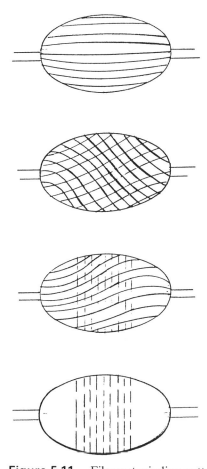

Figure 5.11. Filament winding patterns.

reinforced thermoset part has a much smaller coefficient of expansion and contracts only slightly, permitting easy retraction of the mandrel.

There are significant advantages to filament winding compared with hand layup:

1. For large-scale production, the cost of the prepreg preparation is avoided, lowering total costs.
2. Fiber placement is exact to produce the maximum properties in the required directions.
3. The fibers are truly continuous. The lack of joints increases the strength of the product.

4. Higher fiber volume percentages are possible than with other methods of fiber placement.

However, there are compensating difficulties:

1. Special shapes are required, since the mandrel must be removed. Complex shapes can still be constructed, but special collapsing mandrels are required, greatly increasing the overall cost.
2. The internal surface is pressed firmly against the mandrel and develops a smooth finish. Conversely, the external surface is formed freely and does not have the same high finish.
3. The mandrel cannot have excessively sharp curvature, as the fibers may not bend sufficiently to reproduce its shape.

Curing of the Prepreg

The resin will cure at room temperature, but the process is speeded by raising the temperature in the range of 250 to 350°F for epoxy resins. Pressure is simultaneously applied to eliminate air pockets between the layers, to assure intimate contact between the layers, and to obtain the maximum volume fraction of fibers by squeezing out excess resin. Pressures are commonly between 50 and 100 psi. The combination of heat and pressure to effect the cure is obtained by placing the sample in an autoclave.

Vacuum Bag Molding

This is the most common technique used in the aerospace industry. Advantages of vacuum bagging include:

1. The cost of tooling is low. The usual competition is matched metal molds. The molds do not need to be machined, and the tool can be made of inexpensive materials such as wood or plaster if only a few parts are required, since the forces used are low in vacuum bagging. However, the dimensional accuracy is improved by using steel or aluminum molds, and the number of parts produced from the mold is greatly increased.
2. Parts of virtually any size can be made. Most parts in the aerospace industry manufactured by vacuum bagging are much too large to be economically possible with matched metal molds.
3. Complex shapes can be molded easily, sharp corners being possible by building them up stepwise to generate the final tool.

Although many shapes can be manufactured by this process, it is particularly applicable to parts which do not have sharp or rapid changes in radius. After the layup process is complete, the part is hand rolled to

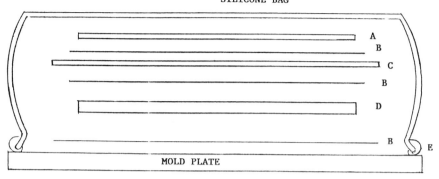

Figure 5.12. Bagging assembly for curing composite laminates. (A) Caul plate. (B) Peel plies. (C) Bleeder fabric. (D) Completed laminate ready for curing. (E) Rubber vacuum seal.

remove excess resin and air bubbles. A peel ply of Mylar or other nonsticking polymer film is placed over the laminate, and a bleeder fabric is placed over that (Fig. 5.12). The bleeder fabric absorbs additional excess resin when the pressure is applied. A final aluminum caul plate may be included to distribute the pressure uniformly. A silicone rubber bag is then placed over the assembly and sealed to the mold plate, either simply with silicone sealer or by forcing its thickened end into a machined groove in the mold plate.

A vacuum is then applied to the layup through holes in the mold and mold plate, forcing the silicone bag against the assembly and forcing some of the excess resin into the bleeder fabric. The entire assembly is then placed in an autoclave. Air is circulated through the autoclave to maintain a uniform temperature while the autoclave is heated at 3 to 5°F/min to a temperature at which polymerization is still relatively slow (250°F). The temperature is maintained for 30 minutes. The elevated temperature reduces the viscosity of the resin, and more resin is forced from the layup, increasing its density. Although polymerization also begins, since the process is a condensation reaction, the average molecular weight does not increase rapidly initially, and the reduction in viscosity is the more important effect. Pressure is then applied (50–100 psi) and held for 30 minutes to complete the densification. The assembly is then heated to the final polymerization temperature (350°F typically) and held for several hours to complete polymerization. The assembly is then cooled under pressure. For some assemblies, additional heating outside the autoclave is performed to assure complete polymerization.

6

Mechanical Properties of Plastics

6.1. INTRODUCTION

Viscoelastic Behavior

The molecular structure of polymers produces mechanical responses which are quite different from those of elastic materials. In the rubbery region below the melting point and above T_g, molecular motion is extensive and the molecular shape continually changes, kinks developing and disappearing rapidly. The average distance from tip to tip of the molecule is far less than the length of the extended molecule. Applied stress will straighten the molecule, but the rate of extension depends on the rapidity of atomic motion, and a slow retarded elongation may occur under constant load. Upon releasing the load, random molecular motion will return the molecule to its kinked position, also resulting in a slowed response. Such responses are *viscoelastic*, containing aspects of the viscous responses of liquids and the elastic responses of solids. This time-dependent viscoelastic response therefore requires emphasis on time-dependent mechanical properties, and such data are of principal importance for polymers.

Internal Energy and Entropic Deformation

Two competing effects control the elastic behavior of plastics. The first law of thermodynamics can be applied to the elongation of a sample.

Recognizing that the changes in potential and kinetic energy of the sample are negligible:

$$dQ - dU = dW$$

where dU = change in internal energy of the sample
dQ = heat absorbed by the sample during the process (tensile test)
dW = work done by the system

The loading can be considered a reversible process, and the change in entropy S during the extension of the sample at temperature T is

$$dQ = T \, dS$$

The work done on the sample by a force F during elongation dl is

$$dW = -F \, dl$$

Substituting these terms into the first law above and solving for the applied force at constant temperature, we obtain:

$$T \, dS - dU = -F \, dl$$

$$F = \frac{dU}{dl} - T\left(\frac{dS}{dl}\right)$$

The force necessary to elongate a sample therefore results in an increase in internal energy (commonly called strain energy) and a decrease in the entropy of the sample. For metals and ordered materials, whose internal structure does not change significantly during deformation, the entropic term is negligible and the increase in strain energy is the only term that needs to be considered. For rubbery plastics, the molecules become ordered during the deformation, and the resultant change in entropy controls the properties. The applied force straightens the molecules, and the random thermal vibrations of the individual atoms within the molecular chains tend to randomize and twist the chains, reducing their length. This entropic restoring force must be opposed by the external force. Note that for such materials, the entropic force required for a specified deformation increases as the temperature is increased.

The Tensile Test

The tensile test is applied universally to metals to characterize their resistance to static loading, their resilience, and their plasticity. Design stresses for metals are frequently specified in terms of the yield strength or ultimate strength as determined by the tensile test. Due to the strong time dependence of the viscoelastic properties of polymers, the data obtained from a tensile test cannot be employed in design of polymer parts in as confident a manner as for metals but must be augmented or completely replaced by creep data, which give the changes in properties under load with time. However, tensile test data are readily obtained and available for most polymers and provide useful information about resistance to short-term loading and large-scale deformation characteristics.

The stress-strain diagram for plastics is determined in equipment similar to that for metals. The size of the tensile samples permissible is specified in ASTM D638 and can vary over a wide range. The specimen can be a fiber, sheet, plate, tube, or rod. Typically, the samples are specified to be injection molded or machined from compression-molded plates, but experimenters frequently choose any sample size convenient for their product or molding operation. The specimen is mounted between a fixed and a movable grip, and a steadily increasing load is applied to the specimen until failure occurs. For rubbery plastics, where elongations of several hundred percent are common, it is common to report changes in length as the *extension ratio*, the ratio of the final to the initial length. For glassy and semicrystalline plastics, with much lower elongation limits, it is usual to report the engineering stress and engineering strain (dependent on the initial dimensions of the sample), despite the fact that even these forms of many polymers elongate extensively, making the true stress and strain significantly different from the engineering values.

Effect of Moisture

The moisture content of plastics affects their mechanical properties quite appreciably, and most mechanical property tests specify that the sample be conditioned prior to the test. Conditioning corresponds to maintaining the sample in a specified relative humidity for a specified time. ASTM D618 specifies the exact conditions depending on the size of the sample. Typically, the conditioning is performed at 50% relative humidity (RH) for either 40 or 88 hours at 73.4°F and should be at the same temperature as further testing. Although the conditioning time is standardized, it is not sufficient to achieve uniform moisture content through all samples, as the diffusivity and solubility of water vary widely from polymer to polymer. Despite this conditioning, therefore, samples can have different moisture contents depending on their prior history.

Effect of Temperature

As with most metals, the lower the temperature, the lower the ductility and the higher the yield stress. For glassy polymers, the yield stress decreases almost linearly and slowly with increasing temperature until the glass transition temperature is reached and then decreases much more rapidly in the rubbery region.

Elastic Modulus

The variation of elastic modulus of a polymer containing amorphous material changes rapidly around T_g. Figure 6.1 shows the change of the elastic modulus of polymethyl methacrylate (PMMA) as the material passes through the various transitions. Most polymers exhibit a ductility transition temperature below which little ductility is observed. Figure 6.2 shows the transition from a ductile to a brittle type of failure for cellulose acetate. The ductility transition temperature is close to the glass transition temperature, but this is not always the case: polycarbonate becomes brittle at temperatures close to $-60°C$ and has a glass transition temperature of $+140°C$.

Blends

For blends, if there were no interaction between the two phases, the rule of mixtures would give mechanical properties as the weighted average of the

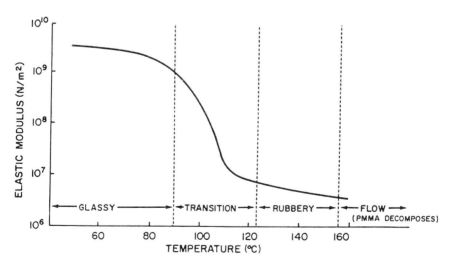

Figure 6.1. Elastic modulus of PMMA as a function of temperature. (From *Plastics Products Design Handbook*, Part A, E. Miller, ed., Marcel Dekker, New York, 1981.)

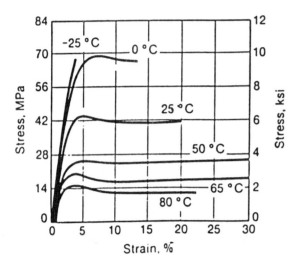

Figure 6.2. Effect of temperature on cellulose acetate. [From T. S. Carswell and H. K. Nason, *Mod. Plast. 21*, 121 (1944). Reprinted by special permission from *Modern Plastics*, McGraw-Hill, Inc., New York, NY 10020.]

properties of the two components. However, the interactions must be recognized, and Nielsen (1978) proposed the modification:

$$E = w_1 E_1 + w_2 E_2 + \beta w_1 w_2$$

where E, E_1, E_2 = the moduli of the blend, and components 1 and 2, respectively

        w_1, w_2 = the weight fractions of the components

        $\beta = 4E_{12} - 2E_1 - 2E_2$, where E_{12} is the modulus for a 50–50% composition of the mixture

Modulus Determination

The elastic modulus of plastics is close to an order of magnitude less than that of steel and other structural metals. The resultant larger elongations in the classically defined elastic region exacerbate problems associated with buckling under compressive loading. Since most plastic parts tend to be manufactured in relatively thin cross sections, the possibility of such collapse must be always considered in any design. Ribs are often added in plastic design to increase the moment of inertia. The stress-strain diagrams of most plastics are usually somewhat curved in the region usually associated in metals with elastic behavior. This presents some difficulty in defining the elastic modulus of the plastic.

     The modulus determined from stress-strain diagrams is of far less importance in design than for metals, so the exact value is not as critical. Various sizes of samples are possible. The most common sample is dumbbell or dogbone shaped, with a typical sample being 2 inches long by 1/2 inch wide by 1/8 inch thick. The instantaneous slope at zero stress is defined as the tangent modulus in the ASTM standard (see Fig. 6.3). However, in practice it is virtually impossible to operate the test without errors in the readings at low stresses due to backlash in the mechanical parts of the loading train and in the extensometers. The initial slope is often inaccurate, and nonlinearity is neglected. This definition is difficult to use accurately.

A definition that avoids this difficulty is that of the British Standards. The samples are specified to be rectangular in shape, with a pulling rate of 1 mm/min, approximately five times slower than the ASTM loading rate. The secant modulus is determined by drawing a line from the origin to intersect the curve at 0.2% strain. This determination is also shown in Fig. 6.3. The slope of this line is the secant modulus. A secant to 1% strain is also utilized. A compromise is the suggestion that a modulus which is 85% of the ASTM slope be used. This approximates the British Standards modulus. Since the stress-strain diagram always demonstrates a decreased slope with

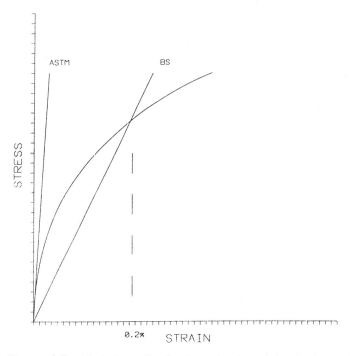

Figure 6.3. Techniques for the determination of the elastic modulus.

increasing deformation, the secant modulus, however defined, is always less than the instantaneous modulus defined by ASTM.

Since the stress-strain relation is not linear, but rather curved below the yield stress, the secant modulus decreases continuously with increasing strain. If the initial modulus is employed in the standard design equations developed from elasticity theory, the part may deflect a greater amount than calculated. Employing the secant modulus for the expected strain eliminates this error. If the strain is not known prior to the calculation, the modulus can be assumed and the calculation of strain performed. A new modulus can then be found and the calculation repeated.

6.2. YIELDING

Glassy Polymers

Polymers, like metals, are capable of exhibiting yielding, large deformation which may be limited to localized regions of the sample. Such yielding is a result of shear stresses resulting in shear deformations. In elastic-plastic materials (metals), the distinction between elastic behavior and plastic or permanent deformation is readily identified. In viscoelastic polymers, the difference is not as apparent, because even large deformations may eventually disappear due to the regeneration of molecular kinks if there is enough time for molecular relaxations to occur. In glassy materials, heating above T_g may result in recovery of extremely large deformations. Polymers which are glassy at low temperatures generally exhibit large increases in ductility on heating above the glass transition temperature. Some glassy polymers may, however, exhibit ductility in the glassy region. Polycarbonate is widely used due to its excellent impact resistance in the glassy region. The glassy polymers that exhibit ductility all have stress-strain diagrams such as that shown in Fig. 6.4, in which the stress reaches a maximum value and then decreases with increasing strain. This phenomenon is called strain softening. After some additional strain, hardening then begins. At low temperatures, below T_g, the material becomes brittle, with little total strain before failure.

Since T_g is considered the temperature at which molecular motions are greatly reduced, it is surprising that many glassy polymers are quite ductile and exhibit large deformations beyond the yield point. The yield phenomenon in these materials has been explained by considering that the glass transition temperature is reduced by the applied stress, and when the stress is sufficient to reduce the glass transition temperature to the test temperature, yielding occurs. Several mechanisms which produce yielding have been proposed, but none are generally applicable at this time.

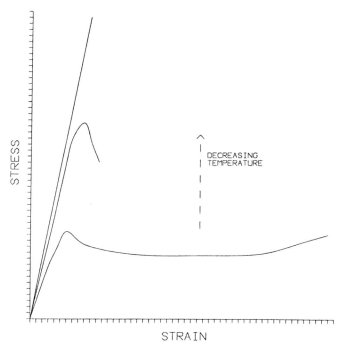

Figure 6.4. The stress-strain behavior of an amorphous polymer at different temperatures.

Yielding is a localized occurrence, and shear bands are observed in glassy polymers in which yielded material occurs in narrow bands surrounded by undeformed material. The deformation must involve stretching and alignment of randomly kinked molecules. The yielding may not involve motion of entire molecules relative to neighboring molecules, as much of this deformation can be recovered by heating above T_g. Such recovery can occur only if the molecules still retain most of their length within the confines of their nearest-neighbor molecules, and kinks or partial movement of the molecule occurs during deformation. Under extensive deformation, cracks may develop at the intersection of shear bands generated in different directions.

The stress-strain diagrams of amorphous and crystalline polymers which are fairly ductile bear similarity to those of plain carbon steel in that the stress reaches a maximum value, indicative of a yield point occurring. For steels, the strain at the yield point is approximately 0.1 to 0.5%. Polymers exhibit the maximum stress at strains of the order of 5 to 15%. This difference is not primarily due to the difference in energy to distort the atomic bonds and to separate the molecules slightly as a result of the applied stress. Rather, the mechanism of deformation is fundamentally different. Even below the glass transition temperature, some chain motion occurs, and in the region of strain below the yield point, the deformation in plastics is largely due to large-scale motion of the molecular chains. This produces a greatly decreased slope of the stress-strain curve of plastics before the maximum stress is reached. Although it is common practice to call this maximum stress value the yield stress of the plastic, it should not be taken as the clear separation point between regions of elastic recoverable strain and permanent deformation, as for metals. For polymers, it is more closely related to the transformation from uniform deformation to the beginning of necking. As a result, this value is a poor one for use in design calculations. The yield strength of crystalline polymers increases with decreasing temperature. At sufficiently low temperatures, a transition occurs to brittle behavior as the glass transition of the amorphous regions is approached. Brittle failure occurs at stresses below that at which yielding will occur and is dependent of surface smoothness and small nicks and scratches on the surface. As a result, the strength of the sample decreases once brittle behavior is encountered. Figure 6.2 shows the transition from ductile to brittle behavior in cellulose acetate. The data show the decrease in brittle fracture stress with decreasing temperature.

Semicrystalline Polymers

Crystalline polymers tend to deform by slip along crystallographic planes within the unit cell, and localized shear bands are less likely to be formed than in glassy polymers. Since all commercial crystalline polymers are partly amorphous, the resultant deformation is a combination of slip within the crystalline regions and shear band formation within the amorphous region. The net result is also localized regions where deformation occurs. The morphology is somewhat different from that in glassy polymers, and the bands are called *kink bands*. Within these bands the direction of the chain orientation changes sharply, and the separation between deformed and undeformed material is more clearly defined than in shear bands in glassy materials.

6.3. STRESS STATES

It is customary to resolve the stress at any location into convenient components by considering a small unit volume of the material at the point of interest. The stress applied is then resolved into normal (σ) and shear (τ) stresses on each face of the cube of material under consideration at that point. The actual stress at that point can be at any angle to the arbitrarily chosen coordinate axes, and the shear stress can be further resolved into two components directed along these axes. The usual convention is that σ_x is the normal component directed along the x-axis. For shear stress, τ_{xy} identifies the shear stress acting perpendicular to the x-axis and parallel to the y-axis (that is, within the $x - y$ plane).

Many problems can be approximated as having a two-dimensional state of stress. This approximation is particularly valuable when the sample is relatively thin, so that the stress perpendicular to the thickness is negligible. Such a condition is defined as the condition of plane stress. Under these conditions the stress state consists of the two normal stresses, σ_x and σ_y, and the shear stress τ_{xy}. Since the sample is not rotating, the shear stresses acting on all sides of the rectangle (the two-dimensional projection of the small cube) must be equal and have opposite directions of action along the x and y directions. Obtaining the components of σ_x and σ_y along a plane at an angle θ to the x-axis:

$$\sigma_\theta = \frac{\sigma_x + \sigma_y}{2} + \frac{\sigma_x - \sigma_y}{2} \cos 2\theta + \tau_{xy} \sin 2\theta \tag{6.1}$$

and

$$\tau_\theta = \frac{\sigma_x - \sigma_y}{2} \sin 2\theta + \tau_{xy} \cos 2\theta \tag{6.2}$$

At the θ value at which the shear stress is zero, the two normal stress values reach maximum and minimum values and are called the principal stresses. The planes along which these stresses operate are the principal planes. The Mohr circle representation of these two equations is performed by plotting the normal stresses along the x-axis and the shear stress along the y-axis. The principal stresses lie at the extremities of the circle along the x-axis (where the shear stress is zero). From Eq. 6.1, the angle on the Mohr circle diagram from the x-axis is twice the angle of a plane from the principal plane. Since the maximum shear stress is equal to the radius of the circle:

$$\tau_{max} = \frac{\sigma_1 - \sigma_2}{2} \tag{6.3}$$

where σ_1 = maximum principal stress
        σ_2 = minimum principal stress

and the maximum shear stress is on planes which are 45° to the principal planes. Note however, that the normal stress is *not* zero on the planes of maximum shear stress.

For a simple uniaxial tensile test, then, Eq. 6.3 becomes

$$\tau_{max} = \frac{\sigma_{yield}}{2} \tag{6.4}$$

In three dimensions, three principal planes exist on which no shear stress exists. If all three principal stresses are equal, the stress is defined as *hydrostatic*. As in the two-dimensional case, the shear stress on a plane 45° from a principal plane is given by

$$\tau = \frac{\sigma_a - \sigma_b}{2}$$

where the subscripts a and b refer to any principal stress. For each pair of principal stresses two planes of equal shear stress exist. However, not all these shear stresses are the maximum value.

The three-dimensional Mohr circle representation can also be used (Fig. 6.5). This representation produces three circles, with the endpoints of the

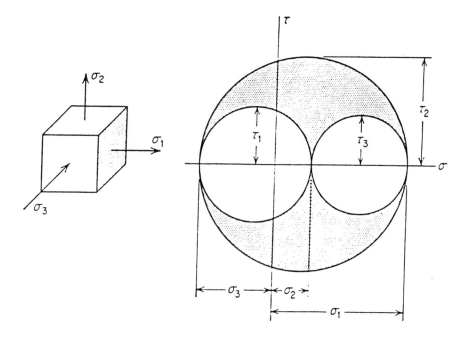

Figure 6.5. Three-dimensional Mohr circle. (From *Mechanical Metallurgy* by G. R. Dieter, McGraw-Hill, New York, 1961. Reprinted with permission of McGraw-Hill, Inc.)

circles along the x-axis (zero shear stress) being the three principal stresses. Necessarily, two circles with endpoints determined by the intermediate principal stress σ_2 must lie within the circle determined by the minimum and maximum principal stresses. Although the determination of the stresses acting on any plane at an arbitrary angle to the principal planes is not as easy in the three-dimensional representation, the diagram greatly aids in visualizing many common stress states. If we compare uniaxial tension (Fig. 6.6a) to biaxial tension (Fig. 6.6c), maintaining the maximum principal normal stress the same, we can see that the maximum shear stress on two of the three sets of planes which can have large shear stresses is reduced by biaxial tension, but the maximum shear stress is not affected. In the limit of uniform biaxial tension the maximum shear stress is still the same as in uniaxial tension, but the shear stress on other planes has been reduced to zero, reducing the ductility somewhat. However, for any amount of triaxial tension (Fig. 6.6d), the maximum shear stress is reduced, the reduction increasing as the three principal normal stresses approach each other. If a state of hydrostatic tension exists (all three normal principal stresses equal), no shear stress is developed on any plane within the sample. Since deformation is a result of shear stresses, this reduction results in a corresponding reduction in yielding and resultant brittle behavior. Conversely, if one of the principal stresses is made compressive, the maximum shear stress increases (Fig. 6.6e). In calendering and coining, the applied compression increases the shear stresses and permits greater deformation of the sample than would otherwise be possible.

Descriptions of Stress Components Which Produce Yielding

The two common methods of describing the decomposition of the applied stress into the portions which are responsible for yielding are given below.

Stress Deviator

In most applications, the stress is not uniaxial. If we recognize that yielding must be a result of shear stresses acting on the material, then the radii of the Mohr circles in three dimensions are the important criteria (since they measure the shear stresses). The position of the circles along the x-axis (the normal stress axis) is therefore unimportant, and different states of stress which just consist of shifting the circles along the x-axis should have no effect on yielding. This suggests that it is appropriate to divide the stresses into the sum of a hydrostatic (three equal principal stresses) stress and a deviator stress, the difference of the actual stress in each coordinate direction and the hydrostatic stress. The hydrostatic stress can produce only a volumetric

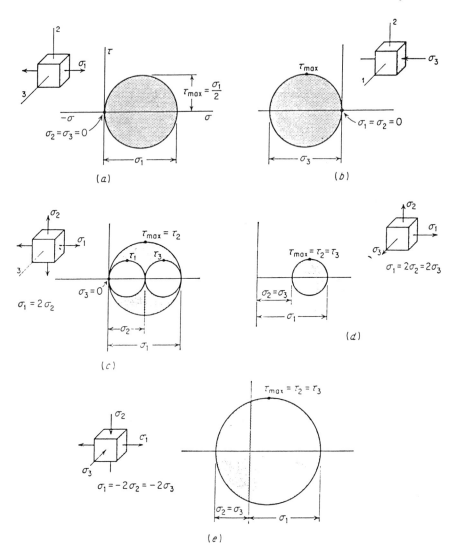

Figure 6.6. Different stress states producing different shear stresses. (From *Mechanical Metallurgy* by G. R. Dieter, McGraw-Hill, New York, 1961. Reprinted with permission of McGraw-Hill., Inc.)

change in dimension without any shear tending to cause deformation. The hydrostatic stress is defined as

$$\sigma_{hydro} = \frac{\sigma_1 + \sigma_2 + \sigma_3}{3} \tag{6.5}$$

The Mohr circle drawing for this hydrostatic stress would degenerate into a point lying on the normal stress axis. Therefore, the hydrostatic stress would be invariant regardless of which plane we consider cutting through the material. The hydrostatic stress is therefore also called the *spherical stress*, due to its uniform application regardless of orientation. It is also labeled the *dilatational stress*, since it extends the sample equally in all three directions.

The *deviator stress* is the effective stress which tends to produce shear. The components of the deviator stress would be

$$\sigma_{1d} = \sigma_1 - \sigma_{hydro} = \frac{2\sigma_1 - \sigma_2 - \sigma_3}{3}$$

$$\sigma_{2d} = \sigma_2 - \sigma_{hydro} = \frac{2\sigma_2 - \sigma_3 - \sigma_1}{3} \tag{6.6}$$

$$\sigma_{3d} = \sigma_3 - \sigma_{hydro} = \frac{2\sigma_3 - \sigma_2 - \sigma_1}{3}$$

and it is these components of the stress that are responsible for yielding and permanent deformation. The maximum deviator stresses occur on the same planes as the principal stresses.

Octahedral Stresses

Shear stresses are generated on planes other than the principal planes. In three dimensions, the planes which exhibit the hydrostatic normal stress make equal angles with the three principal planes. These planes produce an octahedron around the cube which would describe the principal planes (Fig. 6.7). In this diagram, the arrows indicate the directions of the principal stresses. On these planes, the normal stress is equal to the hydrostatic stress:

$$\sigma_{oct} = \sigma_{hydro}$$

and does not contribute to shearing or yielding. In three dimensions, the

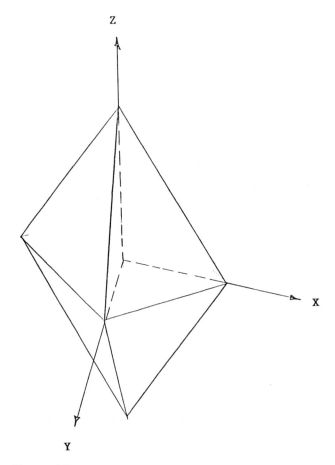

Figure 6.7. The octahedral stress planes.

shear stress on a plane at an arbitrary angle to the planes of principal stress with direction cosines l, m, n is given by (Nadai, 1950)

$$\tau^2 = l^2\sigma_1{}^2 + m^2\sigma_2{}^2 + n^2\sigma_3{}^2 - (l^2\sigma_1 + m^2\sigma_2 + n^2\sigma_3)^2$$

$$= (\sigma_2 - \sigma_1)^2 l^2 m^2 + (\sigma_2 - \sigma_3)^2 m^2 n^2 + (\sigma_3 - \sigma_1)^2 l^2 n^2$$

For octahedral planes, which lie at an angle of $54° 45'$ from the principal planes:

$$l = m = n = \frac{1}{\sqrt{3}}$$

The shear stress on the octahedral planes (τ_{oct}) is therefore given by

$$\tau_{oct} = \frac{[(\sigma_2 - \sigma_1)^2 + (\sigma_3 - \sigma_1)^2 + (\sigma_3 - \sigma_2)^2]^{1/2}}{3} \tag{6.7}$$

Since the octahedral planes develop only a hydrostatic normal stress, this octahedral shear stress is considered to be the best measure of the stress which generates shear deformation in the sample. Conversely, the hydrostatic or dilational stress tends to promote rupture. The greater the ratio of the octahedral to hydrostatic stress, the more likely the material is to deform by shear yielding rather than fail by rupture.

6.4. CRITERIA FOR YIELDING

The yield point can be directly determined in uniaxial tensile tests, but in most applications the applied stress field is either bi- or triaxial. For example, when an internal pressure is applied to a plastic tube, both hoop stresses and axial stresses are generated. It is therefore important for design to be able to predict whether yielding will occur under a complex, three-dimensional stress state. This is of critical importance for metallic and ceramic structures. For plastics, where failure often occurs as a result of excessive creep, the criterion is not of as great significance.

For metals and ceramics, the yield strengths in tension and compression are quite similar, and other experimental data concerning the effect of an applied hydrostatic stress indicate that yielding is relatively unaffected by such a stress. Therefore, for these materials the hydrostatic or dilational stress should not appear in the equations developed to predict yielding under multiaxial stress conditions. In contrast, all polymers exhibit higher yield strengths in compression than in tension. Furthermore, tests performed on samples under a hydrostatic stress indicate that the yield stress varies with the hydrostatic stress. Since the strain prior to yielding is so much larger in plastics than in metals, a significant volume change may occur, which would lead a change in free volume and a resultant dependence of yield on the hydrostatic stress. The criteria established for metals must therefore be modified to include the hydrostatic pressure for plastics.

Maximum Shear Stress Theory

The Tresca yield criterion states that yield occurs when the shear stress on any plane in the sample exceeds a critical value τ_{crit}:

$$\sigma_{max} - \sigma_{min} = 2\tau_{crit} \tag{6.8}$$

where σ_{max} = largest principal stress
σ_{min} = smallest principal stress

Note that in both biaxial and uniaxial tensile tests the minimum stress is zero and in both cases the principal stresses which will produce yield are equal. The second tensile load therefore does not affect the initiation of yield. However, since the shear stress on some planes is reduced, the ability of the sample to conform to some complex shape may be limited, and the ease of working or shaping the sample in biaxial tension will be reduced.

The τ_{crit} value is usually determined by the maximum shear stress obtained at the yield point in a uniaxial tensile test. For this test Eq. 6.4 holds, and

$$\tau_{crit} = \tau_{max} = \frac{\sigma_{yield}}{2} \tag{6.9}$$

Modification for the Effect of Pressure

This equation must be modified for plastics to take into account the fact that hydrostatic pressure affects the yielding process. The question arises whether the normal stress or the hydrostatic stress is the appropriate stress to be included in the correction. Since the differences in yield strength are most likely due to changes in the free volume within the plastic, the most reasonable conclusion is that the hydrostatic pressure is the important parameter, because this will affect the volume directly. Equation 6.9 can then be modified to include the fact that tensile and compressive stresses produce yielding at different values by including the mean hydrostatic component σ_{hydro}:

$$\tau_{crit} = \tau_{cr0} - \mu\sigma_{hydro} \tag{6.10}$$

where μ is a coefficient which is determined by the importance of the hydrostatic stress, that is, on the difference in yield strength between tension and compression, and τ_{cr0} is the critical stress in the absence of any hydrostatic component. The term containing the hydrostatic component has a negative sign because tensile stresses are defined as positive and compressive stresses are negative. This recognizes the fact that compressive hydrostatic stresses reduce the free volume and therefore raise the yield strength of the polymer. The critical shear stress in the absence of a hydrostatic stress, τ_{cr0}, can be found from tests performed in pure shear. One such test is a biaxial one in which equal tension in one direction and compression in the transverse direction are applied. In this case:

$$\sigma_1 = -\sigma_2, \qquad \sigma_3 = 0$$

and

$$\sigma_{\text{hydro}} = \frac{\sigma_1 + \sigma_2 + \sigma_3}{3} = 0$$

The yield stress as measured in a plane strain compression test is raised considerably above the value in pure shear, whereas the yield stress in uniaxial tension is lowered considerably. Using such biaxial tension-compression tests to develop pure shear, Bowden and Jukes (1972) obtained the $\tau_{\text{cr}0}$ and μ values for several plastics shown in Table 6.1. The experimental scatter of the data ranges from ± 0.02 to ± 0.05. It can be concluded that a value of 0.15 for μ might be chosen for a plastic if no experimentally determined value is available.

Maximum Distortion Energy Theory

The other commonly employed yield criterion considers that yielding will occur when the elastic strain energy stored in the material reaches a critical value. Since the hydrostatic stress cannot contribute to yield, only the deviator stress contribution to elastic strain energy need be considered. The elastic energy is the area under the stress-strain diagram. If we consider the material to have a stress-strain relationship which is linear in this region, then the strain energy per unit volume (U) generated by the application of any stress σ (in this context this symbol refers to any stress, normal or shearing) producing a strain ε in a linear elastic material with elastic modulus E is

$$U = \frac{\sigma \varepsilon}{2} = \frac{\sigma^2}{2E} \tag{6.11}$$

Table 6.1. Critical Shear Stress and Hydrostatic Correction Factors

Plastic	$\tau_{\text{crit}0}$ (MN m^{-2})	μ
PMMA	47.4	0.158
PS	40	0.25
PET	31	0.09
PVC	42	0.11
Epoxy I	49	0.133
Epoxy II	42	0.19
HDPE	17.4	0.138

Expanding this to include all stresses in a three-dimensional stress state:

$$U = \tfrac{1}{2}[(\sigma_x \varepsilon_x + \sigma_y \varepsilon_y + \sigma_z \varepsilon_z) + (\tau_{xy}\gamma_{xy} + \tau_{yz}\gamma_{yz} + \tau_{xz}\gamma_{xz})]$$

where ε = the normal strains
        γ = the shear strains

In a three-dimensional stress state, the strain in any direction is the sum of the contributions of the strains resulting from all three stresses. The strain in the x-direction due to a stress in the y-direction is, from the definition of Poisson's ratio (v).

$$v = -\frac{\varepsilon_x}{\varepsilon_y}$$

$$\varepsilon_x = -\varepsilon_y v$$

Since $\varepsilon_y = \sigma_y/E$, the strain in the x-direction due to a stress applied in the y-direction is

$$\varepsilon_x = -\frac{v\sigma_y}{E}$$

and expanding to include a similar term to the z-direction:

$$\varepsilon_x = \frac{\sigma_x - v(\sigma_y + \sigma_z)}{E} \qquad\qquad (6.12a)$$

Similarly:

$$\varepsilon_y = \frac{\sigma_y - v(\sigma_x + \sigma_z)}{E} \qquad\qquad (6.12b)$$

$$\varepsilon_z = \frac{\sigma_z - v(\sigma_x + \sigma_y)}{E} \qquad\qquad (6.12c)$$

The relation between shear stresses and strains for any direction is:

$$\tau = \frac{\gamma}{G}$$

where G is the shear modulus

and the shear elastic energy would correspondingly be

$$U_\tau = \frac{\tau^2}{2G}$$

for each component. However, if we restrict our attention to the principal stresses, the shear stresses are all zero. Substituting the equations for strains in Eq. 6.12 into Eq. 6.11 for the strain energy and applying it to the principal stresses yields:

$$U = \frac{\sigma_1^2 + \sigma_2^2 + \sigma_3^2}{2E} - \frac{v(\sigma_1\sigma_2 + \sigma_2\sigma_3 + \sigma_3\sigma_1)}{E} \tag{6.13}$$

Now we wish to remove the hydrostatic component of stress. Setting each of the components equal to σ_{hydro} and substituting into Eq. 6.13, the resultant strain energy is

$$U_{hydro} = 3\left[\frac{\sigma_{hydro}^2}{2E} - \frac{(v\sigma_{hydro}^2)}{E} \right]$$

$$= \frac{3(1 - 2v)\sigma_{hydro}^2}{2E} = \frac{(1 - 2v)(\sigma_1 + \sigma_2 + \sigma_3)^2}{6E} \tag{6.14}$$

To obtain the contribution of the deviator stress, we can substract Eq. 6.14 from Eq. 6.13 and substitute the relation between shear modulus G and elastic modulus E:

$$G = \frac{E}{2(1 + v)}$$

to yield:

$$U_{deviator} = \frac{(\sigma_1 - \sigma_2)^2 + (\sigma_3 - \sigma_1)^2 + (\sigma_3 - \sigma_2)^2}{12G} \tag{6.15}$$

The fundamental concept underlying this approach is that there is a maximum deviator strain energy the material can absorb without yielding. This will be true regardless of the applied stress field and will therefore also be true under simple tensile loading. Substituting that condition:

$$\sigma_1 = \sigma_{yield}. \qquad \sigma_2 = 0, \qquad \sigma_3 = 0$$

then

$$U_{dev\text{-}max} = \frac{\sigma_{yield}^2}{6G}$$

Substituting this condition into Eq. (6.15) gives the general relation for any stress state at which yield will occur:

$$(\sigma_1 - \sigma_2)^2 + (\sigma_3 - \sigma_1)^2 + (\sigma_3 - \sigma_2)^2 = 2\sigma_{yield}^2 \tag{6.16}$$

This is the form usually applied to metals. The three principal stresses are calculated, and the right-hand side of Eq. 6.16 must be less than the left-hand side for yield strength not to be exceeded.

This form does not include any differences produced by the hydrostatic stress, which is important for plastics. To include this term, it is convenient to recast this equation in terms of the octahedral shear stress by using Eq. 6.7:

$$\tau_{oct} = \frac{[(\sigma_1 - \sigma_2)^2 + (\sigma_3 - \sigma_1)^2 + (\sigma_3 - \sigma_2)^2]^{1/2}}{3} \tag{6.7}$$

and in terms of the critical value of the shear stress on the octahedral planes to produce yielding, this equation may be rewritten

$$(\sigma_1 - \sigma_2)^2 + (\sigma_3 - \sigma_1)^2 + (\sigma_3 - \sigma_2)^2 = (3\tau_{oct\text{-}crit})^2 \tag{6.17}$$

Comparing Eq. 6.16 and Eq. 6.17 yields

$$\tau_{oct\text{-}crit} = \frac{\sqrt{2}\sigma_{yield}}{3} \tag{6.18}$$

Since the hydrostatic stress has an effect in controlling yield in plastics, the modification to include the mean value of the hydrostatic stress is

$$\tau_{oct\text{-}crit} = \tau_{oct\text{-}0} - \mu\sigma_{hydro} \tag{6.19}$$

where μ is a parameter that represents the dependence of yield strength on the hydrostatic component for each particular plastic, and $\tau_{oct\text{-}0}$ is the critical octahedral shear stress to produce yielding in the absence of a hydrostatic stress (pure shear). Substituting Eq. 6.19 into Eq. 6.17 yields

$$(\sigma_1 - \sigma_2)^2 + (\sigma_3 - \sigma_1)^2 + (\sigma_3 - \sigma_2)^2 = 3^2(\tau_{oct\text{-}0} - \mu\sigma_{hydro})^2 \tag{6.20}$$

The values of μ and $\tau_{oct\text{-}0}$ can be found from tensile and compressive yield strength data. For such uniaxial testing, $\sigma_1 = \sigma_{yield}$, $\sigma_2 = \sigma_3 = 0$.

For a tensile test:

$$\sigma_{hydro} = \frac{\sigma_{yT}}{3} \tag{6.21a}$$

where σ_{yT} is the uniaxial yield strength in tension.

For a compression test:

$$\sigma_{hydro} = \frac{-\sigma_{yC}}{3} \tag{6.21b}$$

where σ_{yC} is the uniaxial yield strength in compression, or negative stress. From Eqs. 6.18, 6.19 for the tensile test,

$$\tau_{crit\text{-}oct} = \frac{\sqrt{2}\sigma_{yT}}{3} = \tau_{oct\text{-}0} - \frac{\mu\sigma_{yT}}{3}$$

and for the compression test,

$$\tau_{\text{crit-oct}} = \frac{\sqrt{2}\sigma_{yC}}{3} = \tau_{\text{oct-0}} + \frac{\mu\sigma_{yC}}{3}$$

Solving these two equations for μ and $\tau_{\text{oct-0}}$ yields

$$\mu = \frac{\sqrt{2}(\sigma_{yC} - \sigma_{yT})}{(\sigma_{yC} + \sigma_{yT})}$$

(6.22)

$$\tau_{\text{oct-0}} = \frac{(2\sqrt{2}/3)\sigma_{yC}\sigma_{yT}}{(\sigma_{yC} + \sigma_{yT})}$$

Thus by determining the yield strength in tension and compression, the effect of hydrostatic stress can be evaluated.

Example 1. A polymer is under triaxial compression, with all three stresses equal. What stress would be required for yielding?

Ans. Whether the stresses are tensile or compressive, substitution of three equal stresses into Eq. 6.17 indicates that no shear stresses will be produced and yielding will not occur.

Example 2. A polymer with a tensile yield stress of 12,000 psi and a compressive yield strength of 16,000 psi is under equal biaxial tensile stress of 5000 psi. What tensile stress in the third direction would produce yielding?

Ans. Since the tensile and compressive yield stresses are not the same, the effect of hydrostatic stress must be included. Substituting the values of the tensile yield strength and compressive yield strength into Eq. 6.22:

$$\mu = \frac{\sqrt{2}(\sigma_{yC} - \sigma_{yT})}{(\sigma_{yC} + \sigma_{yT})} = \frac{\sqrt{2}(16,000 - 12,000)}{(28,000)} = 0.202$$

$$\tau_{\text{oct-0}} = \frac{(2\sqrt{2}/3)\sigma_{yC}\sigma_{yT}}{(\sigma_{yC} + \sigma_{yT})} = \frac{(2\sqrt{2}/3)(16,000 \cdot 12,000)}{(28,000)} = 6464$$

Substituting these values and $\sigma_1 = 5000$, $\sigma_2 = 5000$ into Eq. 6.20:

$$(\sigma_1 - \sigma_2)^2 + (\sigma_3 - \sigma_1)^2 + (\sigma_3 - \sigma_2)^2 = 9(\tau_{\text{oct-0}} - \mu\sigma_{\text{hydro}})^2$$

$$2(\sigma_3 - 5000)^2 = 9\left[6464 - \frac{0.202(10,000 + \sigma_3)}{3}\right]^2$$

$$\sigma_3 = 15,123 \text{ psi}$$

This can be compared to a stress of 17,000 psi calculated from Eq. 6.16. The hydrostatic stress correction lowers the calculated stress value for yield.

Example 3. What is the maximum equal biaxial tension stress possible before yielding would occur in the material given in Example 2?

Ans. Substituting the three principal stresses, $\sigma_1 = \sigma_2, \sigma_3 = 0$ into Eq. 6.20:

$$2\sigma_1^2 = 9\left[6464 - 0.202\left(\frac{2\sigma_1}{3}\right)\right]^2$$

$$\sigma_1 = 10{,}665 \text{ psi}$$

Rubbery Polymers

It should be emphasized that the equations developed in the previous section apply only to crystalline and glassy polymers. In the rubbery region no accurate equation exists to predict failure under multiaxial loading.

Effect of Loading Rate

For metals the tensile data are relatively insensitive to pulling rate. Plastics are much more rate sensitive; the ultimate stress, yield stress, and modulus are much stronger functions of loading rate than for metals. The effect of strain rate on high density polyethylene is shown in Fig. 6.8. Furthermore, the dependence on loading rate is a function of temperature. It is small below T_g, since no kink movement can occur. The dependence is also small well above T_g, since the kinks move fast enough to completely respond to the applied stress. The dependence is greatest in the range T_g to $T_g + 50°C$. In this temperature range the temperature dependence is very large. ASTM D368 lists four possible standard testing speeds from 1/2 to 500 mm (0.05 to 20 inches) per minute. Rates of 5 and 50 mm/min are suggested for rigid and nonrigid polymers, respectively.

The higher the strain rate, the less the ductility at any given temperature, and the higher the temperature at which the ductile-brittle transition appears. The yield strength also increases with strain rate. Theoretical expressions have been derived relating these two quantities, but the number of experimentally required constants does not allow them to be useful for predictive purposes. A usable equation applicable to glassy polymers which has been experimentally verified is

$$\sigma_y = \sigma_0 + B \log \hat{e} \tag{6.23}$$

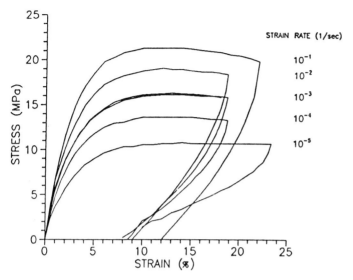

Figure 6.8.   Effect of strain rate on PE. [From C. F. Popelar, C. H. Popelar, and V. H. Kenner, *PES 30*, 577 (1990). Reprinted by permission of the publishers.]

where σ_y = yield stress at the specified strain rate
        σ_0, B = experimentally determined constants
        $\hat{e}$ = strain rate

Since in many applications the load is applied substantially faster than in testing, the materials' properties may be different in service than indicated in the test results.

The effect of temperature on the yield strength may be incorporated into Eq. 6.23 by using a shift factor a_t:

$$(\sigma_y)_T = (\sigma_0)_R + B \log(\hat{e} \cdot a_t)$$

where a_t = strain rate for a specified yield stress at temperature T/strain
             rate for the same yield stress at the reference temperature
        $\hat{e}$ = the strain rate
        $(\sigma_y)_T$ = yield stress at temperature T
        $(\sigma_0)_R$ = experimentally determined value of σ_0 at the reference temper-
             ature

The shift factor a_t is obtained by plotting curves of the yield stress vs. strain rate for various temperatures. A reference temperature is chosen, and

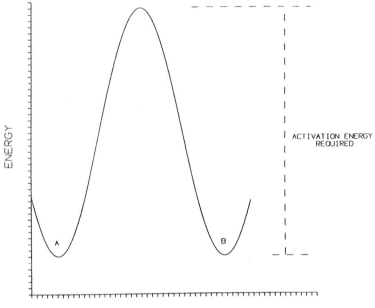

(a)   ATOMIC OR MOLECULAR POSITION

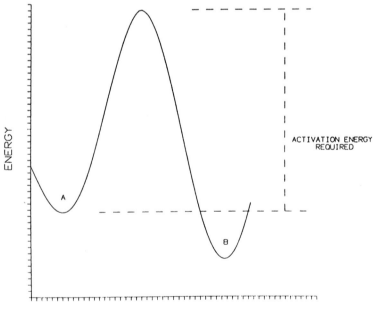

(b)   ATOMIC OR MOLECULAR POSITION

the curves for the other temperatures are shifted horizontally until one smooth curve is obtained.

This formulation implies that the viscoelastic properties of the plastic are essentially controlled by the fluidlike elements, since the strain rate alone determines the required stress for deformation (flow). A more complete specification would include the recognition that the amount of prior deformation also affects the yield stress. This formulation was first given by Inouye (1954) and expanded by Tate et al. (1988). The Arrhenius formula for the effect of temperature was included to yield

$$\sigma_y = K e^n + \hat{e}^m \exp(-CT)$$

The unknown constants K, n, m, C must be obtained by multilinear regression of experimental data.

The Eyring Approach to Yielding

One approach to evaluating the dependence of the yield strength on temperature is modeled by using the Eyring concept that an energy barrier prevents easy motion of an atom or molecule from one equilibrium position to another. This concept is demonstrated in Fig. 6.9a. This diagram shows the energy of an atom, molecule, or molecular kink vs. position. In its initial equilibrium position, the atom, molecule, or kink has energy E, and small movements tend to increase its energy because its chemical bonds with nearest neighbors become stretched or weakened. Another relatively stable position B exists when the particle has moved to become bonded to other neigboring molecules. As a result of random thermal vibrations, some atoms will gain sufficient kinetic energy to overcome the barrier, break the chemical bonds to its neighbors, and move from A to B. The additional energy required to overcome the barrier is called the activation energy E^*. Similarly, an equal number statistically will move from B to A, so there is no net diffusion or change in shape of the sample.

Boltzmann statistics provide the energy distribution of atoms. The probability of an atom having sufficiently high energy E^* to overcome the barrier and move to a new location is

$$P = A \exp\left(-\frac{E^*}{kT}\right)$$

Figure 6.9. Eyring theory of molecular motion producing strain. (a) Energy of an atom as a function of position in the absence of stress. (b) Energy of an atom as a function of position when stress is applied.

where P = probability of an atom having energy E^*
        T = absolute temperature
        k = Boltzmann constant
        A = constant dependent on the material

If an applied stress σ exists, the energy E of state A is raised by

E = force·distance = stress·cross section of atom·distance moved

These last two terms have units of volume and are lumped together and called the activation volume v^*. However, no acceptable correlation has been found for the value of v^* and the dimensions of atoms or moving segments of molecules, and it therefore remains an adjustable constant. The effective barrier for forward motion in the direction of the applied stress is therefore reduced to

$E^* - \sigma v^*$

and the barrier for reverse motion is raised to

$E^* + \sigma v^*$

The probability of a forward transition is therefore

$$P = A \exp - \left(\frac{E^* - v^*\sigma}{kT} \right)$$

Conversely, the probability of backward transition is

$$P = A \exp - \left(\frac{E^* + v^*\sigma}{kT} \right)$$

The net forward rate of reaction is proportional to the difference between these two terms:

$$P = A \exp - \left(\frac{E^* - v^*\sigma}{kT} \right) - A \exp - \left(\frac{E^* + v^*\sigma}{kT} \right) \qquad (6.24)$$

or

$$P = 2 \exp - \left(\frac{E^*}{kT} \right) \sinh \left(\frac{\sigma^* v^*}{kT} \right)$$

The shift of energy of the two possible positions for the molecule is shown in Fig. 6.9b. This general approach has been applied to various types of deformation processes, including creep of different classes of materials, as well as to yielding of plastics. Since different jump heights and distances between atomic positions exist between materials, this equation can give only the form of the expected variation of creep strain with temperature. Although any excellent fit to the temperature dependence of some creep data has been found with this equation, it has not received universal recognition as a good formula for accurate fitting of experimental creep data.

For yielding, it can be considered that the applied stress is sufficiently large that the probability of atoms moving in the reverse direction is negligible, and only the forward direction need be considered. The strain rate can therefore be taken as proportional to the probability of atoms jumping in the direction of the applied stress. For yielding the stress is the yield stress of the sample, and the probability of a transition is replaced by the strain rate. Solving the equation for the yield stress:

$$\sigma_{yield} = \frac{E^*}{v^*} + \left(\frac{kT}{v^*}\right)\ln\left(\frac{\varepsilon_r}{A}\right)$$

where ε_r is the strain rate applied during the test.

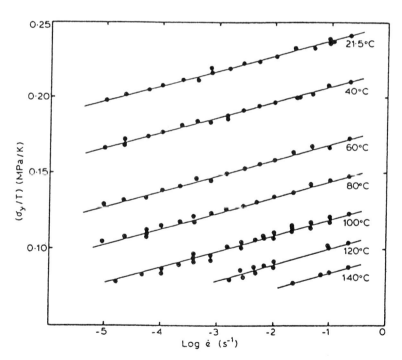

Figure 6.10. Ratio of yield stress to temperature for various strain rates for PC. [From C. Bauwens-Crowet, J. C. Bauwens, and G. Homes, *J. Polym. Sci. A2* (7), 735 (1969). Reprinted by permission of John Wiley and Sons, Inc.]

Based on this equation, a plot of yield σ/T vs. log ε_r should give a series of parallel lines for different temperatures. The data of Bauwens-Crowet et al. (1969) shown in Fig. 6.10 demonstrate this behavior remarkably well. However, these data are for a limited temperature range, and for extended temperatures the agreement is not as precise. To maintain the same terminology and model, various authors have included more than one activation energy, indicating that several different types of flow or jump processes are occurring. The activation energy changes with temperature and may be correlated with secondary transitions. Such models require experimental data to permit determination of the activation energies and do not actively assist in predicting the behavior of the materials.

Effect of Hydrostatic Pressure

The effect of hydrostatic pressure on the yield point of a plastic can be rationalized by using these concepts. Sasabe and Saito (1968) found that molecular motion in plastics is affected exponentially by hydrostatic pressure. This pressure dependence can be included by recognizing that the rate constant A in Eq. 6.24 depends on the hydrostatic pressure:

$$A = A' \exp(-xP)$$

where A', x are material constants to be experimentally determined, where P is the applied stress. Bowden and Jukes (1972) suggested substituting this equation into Eq. 6.24 (neglecting the reverse reaction of molecules moving backward). This yields for the rate or molecular motion v

$$v = A' \exp(-xP) \exp\left[\frac{-(E^* - v^*\sigma)}{kT}\right]$$

At a constant strain rate at yield, the probability of molecular motion is independent of material, and v is a constant.

Solving for the yield stress σ gives an equation of the form

$$\sigma = A + BP \tag{6.25}$$

indicating a linear relation between the yield stress and the applied hydrostatic pressure, in agreement with the conclusions reached in the section on the maximum distortion theory of yielding. Reasonable agreement can be found between the values of A and B in Eq. 6.25 and the values of the corresponding $\tau_{crit\ 0}$ and μ in Table 6.1.

6.5. NECKING (DRAWING)

During mechanical deformation, the molecules are brought closer to each other, and the intermolecular forces may increase to further induce molecular alignment. If the polymer is partly crystalline, it may further crystallize during deformation by alignment of the molecules as a result of the deformation and flow of the material. If the polymer is amorphous but capable of crystallization, the deformation may initiate crystallization. Such self-reinforcing molecular alignment may continue until a neck forms in the specimen. This ordered material in the neck is significantly stronger than the undeformed parts of the test specimen. The remaining material then deforms preferentially, inducing a spread of the necked region along the specimen's length until the whole specimen is reduced in cross section. The entire specimen may then thin further before failure.

The appearance of the engineering stress-strain diagram of such crystalline polymers is deceptively similar to that of low-carbon steels which show a yield point phenomenon. However, in metals, the neck begins well after yielding has occurred in a region of weakness, and the neck continues to narrow and failure occurs at that point (Fig. 6.11A). It is noteworthy that the stress appears to decrease when necking begins only because it is calculated based on the original cross-sectional area. The material in the neck area is actually increasing in real strength considerably if the strength is calculated based on the true cross-sectional area of the neck at any time (true stress).

Drawing is frequently observed in easily crystallizable plastics; polyethylene is a prime example. The behavior of a plastic sample during drawing during a tensile test as shown in Fig. 6.11B can be summarized as follows:

1. The specimen becomes smaller in one location, usually near the shoulder of the sample.
2. Rather than the sample continuing to neck at that point, the deformation at that point increases the crystallinity and aligns the chains to a greater extent. This region therefore becomes stronger.
3. The thin region extends down the sample, becoming longer.
4. After the entire sample has decreased in diameter, the sample continues to shrink in diameter.
5. The sample eventually fails, showing long fibers of drawn molecules in the failure zone.

It should be noted that this extensive necking provides a mechanism for exceptional energy absorption during deformation of plastics, as the entire sample is deformed, unlike the localized region in metals. In applications where such large deformations may occur, as in parts where puncturing may occur, this mechanism provides extreme resistance.

PET (polyethylene terephthalate) is another easily crystallizable polymer, which may be obtained in the amorphous form readily in copolymers. The behavior of a copolymer with respect to drawing depends on the temperature of deformation relative to T_g. If the drawing is above T_g, the drawing

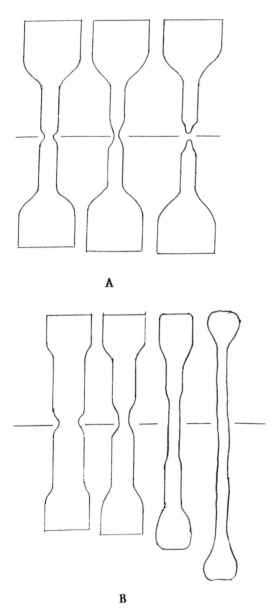

A

B

Figure 6.11. Behavior of a metal during necking compared to a plastic during drawing. (A) Metal necking. (B) Plastic drawing.

and decrease in diameter of the sample are uniform, generating a steadily thinning sample, similar to the behavior described above for PE. If the drawing is below T_g, necking occurs and the deformation does not produce crystallization, since large molecular motions are not possible. Rather, the sample remains amorphous. As with many amorphous polymers, despite the fact that the load is applied below T_g, large extensions can be obtained and a strain of 200% has been observed (Benavente and Perena, 1987).

Strain Hardening

Strain hardening is the increase in load required to continue deformation past the yield strength of the material. Although some metals strain harden significantly, virtually all polymers exhibit almost no strain hardening. Once the maximum load is reached, the material will elongate (usually by the neck traversing the sample) with essentially no increase in load. Since the cross section is decreasing, the load required to continue the deformation may actually decrease somewhat with continued deformation. The true stress tends to remain constant until extensive deformation has occurred and then begins to increase.

Triaxial Stress in Necking

Shear stresses are responsible for yielding and extensive deformation in plastics, as in metals. In a simple tensile test, shear stresses are developed on planes not perpendicular to the tensile axis. The generation and magnitude of the shear stresses produced by the axial stresses can readily be visualized by the Mohr circle representations in Fig. 6.6.

As the three tensile stresses approach each other in magnitude, the shear stress becomes continually smaller, and in the limiting case of equal tensile stresses applied along the three axis, the shear stress becomes zero and no yielding is possible.

Although a tensile test might be considered to produce only normal stresses in the direction of the applied load, necking or inherent defects and notches within the sample generate multiaxial stresses near the change in section. When necking occurs, the material within the neck has reduced in cross section to such an extent that it can no longer withstand the load, and all further permanent extension occurs exclusively within the neck. The material within the neck therefore tends to elongate and, as a result, must also decrease in cross-sectional area. However, the material just outside the necked region experiences a much lower stress and therefore tends to resist the further contraction of the necked material. This resistance generates tensile stresses in the radial and circumferential directions. The result is the generation of a triaxial stress field within the necked area, as opposed to the simple tension previously existing. As discussed previously, plastic deformation is produced only by shear stresses, and the triaxial tension reduces the shear stress within the necked region. This results in a greater applied load being required to continue the deformation. The greater the curvature of the necked region, the greater the plastic constraint generating triaxial stresses and the greater the additional stress required. The equation relating the applied stress and the true stress within the necked region was calculated by P. W. Bridgman (1954) as

$$\frac{\sigma_{neck}}{\sigma_{applied}} = \frac{1}{(1 + 2R/a)\ln(1 + a/2R)}$$

where R is the radius of curvature of the necked region and a is the radius of the necked region. For most plastics, necking occurs over a large fraction of the sample, generating a region with a large radius of curvature. Since $\ln(1 + x) = x$ for small x values, for plastics where a/R is small, this difference is usually insignificant. For intentionally introduced notches, the term can become important.

6.6. TYPICAL STRESS-STRAIN DIAGRAMS

Glassy Polymers

Brittle failure is usually associated with glassy polymers below their T_g, but notable exceptions occur, with many glassy materials showing good ductility (polycarbonate (PC) is a prime example). The stress-strain diagram of PC at room temperature, well below the glass transition temperature of about 140°C, is shown in Fig. 6.12.

Effect of Temperature

The effect of increasing the temperature on the mechanical properties of a polymer depends on the extent of crystallinity of the material. For a partly crystalline sample, the principal effects occur within the amorphous regions. These regions become increasingly rubbery with increasing temperature. The crystalline regions are affected to a much smaller degree. The softening of the amorphous regions permits independent motion of the lamellae, such as sliding and rotation, to occur more easily. Thus the deformation process of crystalline polymers in which the individual lamellae become drawn into microfibrils can occur more readily. As a result, the amorphous regions tend to control the temperature dependence. As the temperature approaches the melting point, the ready movement of the crystallites within the amorphous matrix becomes so great that the yield strength falls toward zero. The yield

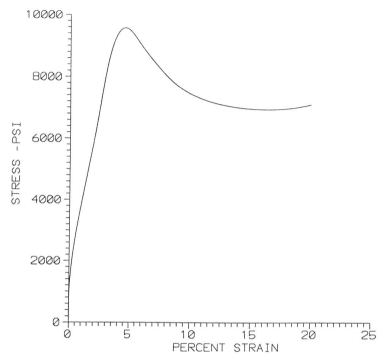

Figure 6.12. Stress-strain behavior of PC.

stress of glassy polymers decreases with increasing temperature much more rapidly than for other structures. At the glass transition the material becomes rubbery, and the yield stress approaches zero. The relationship between the yield stress and the temperature can be given by Brown (1988)

$$\tau_{\text{yield}} = 0.08 \cdot G \cdot \left(1 - \frac{T}{T_{\text{g}}}\right)$$

where
τ_{yield} = shear yield stress
G = shear modulus
T_{g} = glass transition temperature
T = temperature of the test

Crystalline Polymers

Crystalline polymers frequently fail with extended ductility, as a result of molecular reorientation during deformation. The degree and morphology of the crystallinity within the plastic depend on the prior history of the sample.

However, the crystalline structure tends to become disordered to some extent as a result of the deformation. This breakdown of the structure weakens the intermolecular bonding, and the sample may exhibit a stress-strain diagram that demonstrates a region in which the material elongates under reduced applied load. Continued deformation produces a more strongly ordered molecular arrangement, and work hardening or strengthening of the sample begins, producing a stress-strain diagram that demonstrates an increasing load for continued deformation. The color of transparent crystalline plastics shows definite changes during these processes. The sample becomes milky during the initial deformation due to the breakdown of the internal structure into smaller regions which scatter light, and then becomes clearer again as the new structure forms.

Data for PE are shown in Fig. 6.8 at various strain rates (Popelar et al., 1993). The typical increase in yield strength and decrease in ductility with increasing strain rate can be seen. Note that the unloading curve does not reproduce the loading curve. As discussed previously, this does not necessarily mean that this loading results in permanent deformation, as much of this strain is due to kinks in the molecules being straightened and will be recovered on standing for extended periods.

Rubbery Polymers

A typical stress-strain diagram for a rubber polymer is shown in Fig. 6.13. Polymers in the rubbery region exhibit stress-strain diagrams with very low initial stresses generating some deformation. If the load on the sample is released prior to failure, the sample should return to its original dimensions, since the entire deformation is due to kink straightening and elastic deformation of the C—C bonds. Although the diagram looks deceptively similar to that of crystalline and amorphous plastics, the extension, which is substantial, is all recoverable, no permanent deformation or yielding occurring. The initial rapid rise in stress for relatively small elongations is considered to be due to stretching of the C—C bonds within the chains and the large viscous forces required to begin straightening the long chain molecules. As the stress is increased, the kinks within the molecular chains begin to straighten, tending to bring the molecules within close proximity. This produces stronger intermolecular attractive forces, which assist in further straightening the molecules and lengthening the sample. The applied external force required to produce further extension therefore decreases, and the marked dip in the stress-strain diagram results. Eventually all the kinks become straightened, and the rapid increase in stress required for further

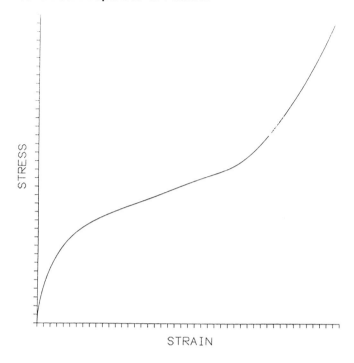

Figure 6.13. Typical stress-strain diagram for a rubbery plastic.

deformation is again a result of the C—C bonds being stretched. Obviously, not all the molecules are perfectly aligned, as cross-links or intermolecular entanglements prevent complete movement of the entire molecule. The strength of the intermolecular attractive forces affects the behavior of the rubber. If these forces are sufficiently large (due to polar groups or simple molecular configurations), then the attractive force may be sufficient to develop some crystallization during this stretching.

Hysteresis

Rubbery materials all exhibit significant hysteresis in their stress-strain diagrams on application and release of the load (see Fig. 6.14) as a result of their molecular configurations. The amount of hysteresis is determined from the energy dissipated in one loading cycle. Since the area under the stress-strain diagram is the energy absorbed, the area within the hysteresis

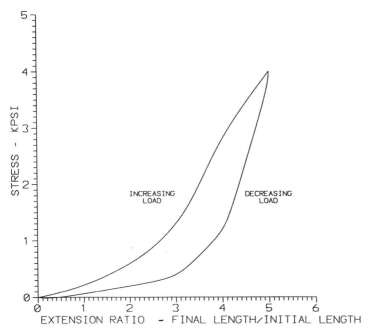

Figure 6.14. Typical hysteresis curves for carbon-filled rubber.

curve is the energy dissipated during a cycle. The hysteresis ratio h is

$$h = \frac{\text{energy absorbed}}{\text{energy during initial loading}}$$

$$= \frac{\text{area within hysteresis curve}}{\text{area under initial stress-strain diagram}}$$

The amount of hysteresis will depend on the strain rate and temperature. If the molecular motion could occur rapidly and in an unimpeded
fashion, no hysteresis should occur. However, the sliding of portions of
the chain molecules past each other during straightening requires time
and requires breaking the attractive forces between the molecules, producing
internal friction. Hysteresis is a result of the molecules being incapable
of responding instantaneously to the applied load. A reduction in temperature will decrease chain mobility and increase hysteresis. Similarly,
an increase in strain rate will make it more difficult for the molecules
to follow the applied load and should also increase the amount of
hysteresis.

As with mechanical properties, increasing strain rate results in changes similar to those resulting from lowering the temperature. If the temperature is lowered to near T_g, however, the chains become frozen, kinking stops, and the extent of hysteresis decreases. Similarly, at large strain rates the molecules do not have time to begin to move and behave like glassy plastics with reduced hysteresis. The effect of temperature on the flow viscosity can be determined by the Williams–Ferry–Landau (WFL) equation to be discussed later, and the same equation should therefore relate the effect of temperature on internal friction. Although this should theoretically apply only to uncross-linked rubbers, it should work reasonably well for all rubbery materials. Many rubbery products are filled with some strengthening additive, carbon black being the most common. The strengthening by carbon black is due to adsorption of part of the molecular chain of different molecules on the surface, resulting in effective cross-linking or entanglement sites, which would increase the internal friction and hysteresis. Partial crystallization during stretching due to large intermolecular forces also increases the amount of hysteresis, since decomposition of the crystalline regions occurs at a lower rate than their formation. Natural rubbers tend to crystallize to a greater extent than artificial ones and exhibit significantly higher levels of hysteresis.

The Mullins Effect

If a stress is applied to a rubbery sample and then released, on reapplication of the load the stress-strain diagram falls below the original one until the maximum stress originally applied is reached. The continuation of the test then follows smoothly the curve for the original test or the curve for a different sample tested to failure without load release. If the sample is further loaded, released, and then retested, the retest again falls below the previous loading curve. However, retesting repeatedly to the same stress value will within a few repetitions produce a reproducible hysteresis curve. For carbon-filled rubbers, this is considered to be due to the adsorbed chains being dislocated from the carbon particles during the loading and then reattaching at some other point which would permit easier kink reduction on reloading. In unfilled rubbers, it may be due to the breaking of highly stressed cross-links, producing a more compliant sample.

6.7. FLEXURAL TESTING

Flexural Modulus

Flexural testing by three-point bending of rectangular specimens is a simple test frequently performed and specified in ASTM D790. The specimen does

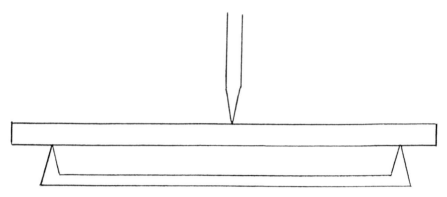

Figure 6.15. Flexural testing.

not have to be machined to a complicated form, and gripping problems and breakage of the specimen in the grips are eliminated. In the flexural test, as shown in Fig. 6.15, the specimen is supported at two locations 4 inches apart. The load is applied at the center of the specimen. As in any test of a plastic, the mechanical properties are a strong function of the loading rate. The loading rate is therefore specified in the standard. The deflections measured are much larger than the tensile elongations observed in a typical stress-strain test and therefore require less expensive test equipment. For three-point bending, the modulus is related to the deflection of the mid-point by

$$E = \frac{PL^3}{4BD^3 q}$$

where E is the modulus, P the load, B the width, D the thickness, L the length, and q the deflection. In bending some deflection is a result of shear stresses generated. For long beams, this contribution is small, and tests with $L/D > 15$ reduce this error to a negligible value.

The modulus is found by drawing the load vs. deflection curve, and, as for the tensile test, either a secant or a tangent modulus can be computed. The compressive modulus is greater than the tensile modulus. Since essentially half the sample is in compression (due to the difference in tensile and compressive moduli, the neutral axis is not exactly at the center), the flexural modulus tends to be slightly greater than the tensile modulus. Although many plastics fail in this test, giving the flexural strength directly, rubbery materials do not and their flexural stress is defined as the stress at 5% strain.

6.8. RECOVERY AFTER LOADING

The behavior of plastic samples is not truly elastic at any stress level or time independent. On unloading within the region both above and below what is considered the yield strength of the material, the sample closely returns to its original dimensions as long as the applied strain is not excessive. The limiting strain varies with the polymer, but 2% strain is usually considered within the range in which complete recovery can be expected. The rate of recovery depends on the time under the applied load, the time after the load has been reduced, and the applied strain. Recovery data are not available for all polymers. The most informative method of presenting such data is to plot fractional recovery vs. reduced time suggested by Turner (1966).

$$\text{Fractional recovery} = \frac{\text{strain recovered}}{\text{total initial strain}}$$

$$\text{Reduced time} = \frac{\text{recovery time}}{\text{total time under load}}$$

A typical curve is shown in Fig. 6.16. Note that recovery is slower with increasing time of application of the load and with increasing resultant strain.

6.9. STRESS RELAXATION

Because molecular kinks tend to straighten under applied load, samples which are loaded under conditions of fixed strain slowly relax, reducing the internal stress. Such conditions are of major significance in designing plastic bolts. However, stress relaxation and creep testing are both time consuming, and therefore expensive, and both sets of data are not available for most polymers.

In a stress relaxation test, the initial strain is instantaneously applied and then fixed. Experimentally, providing a step function of strain is quite impossible and difficult to approximate in practice. Furthermore, continuous monitoring of the stress within the sample requires a load cell to be utilized continuously for each sample, and an extensometer is required to measure the initial deflection. Thus this test involves twice the instrumentation and cost of the components of a creep test. The general behavior shows an exponential decay of the stress with time, but the rate of decay tends to decrease less rapidly with increasing time. The general variation of stress with time is shown in Fig. 6.17.

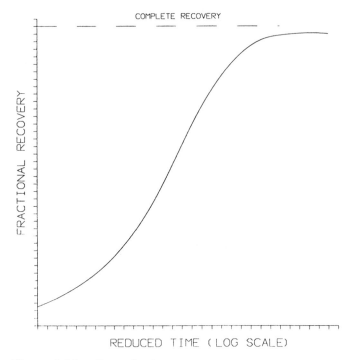

Figure 6.16.   Generalized recovery curve.

6.10.   IMPACT TESTING

Impact, although not precisely defined, clearly refers to rapid loading. Plastics are more sensitive to the rate of loading than other types of materials. A plastic sample may be very ductile when loaded slowly and fail in a brittle manner with negligible ductility when the loading rate is increased. The standardized test methods to identify the resistance of a material to impact loading are mainly qualitatively informative and do not give information which can be directly applied to the design of a part that will be subjected to such loading.

Toughness

The area under the tensile diagram corresponds to the energy per unit volume required to break the sample:

$$\text{Energy} = \text{force} \cdot \text{distance} = \int \sigma \, d\varepsilon$$

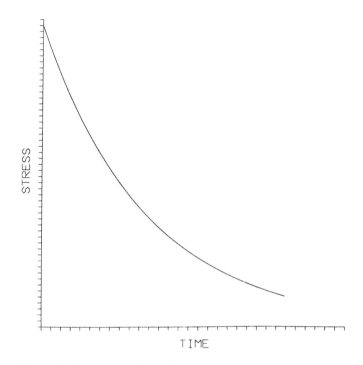

Figure 6.17. Stress relaxation data.

where σ = applied stress
ε = resultant strain
ε_f = strain at failure

Since the tensile diagram depends on the rate of loading, the speed of testing is critical. Evans (1960) found that the area at a strain rate of about 0.3 in/sec gave good agreement with other impact tests, but the data obtained at the usual rates of testing did not.

Izod Impact Testing

Izod impact testing of plastics is performed in the same type of equipment as utilized for metals. The thickness of the sample can vary from 1/8 to 1/2 inch. A value of 1/8 inch is most commonly used, as this corresponds to the typical wall thickness of a molded part. Samples should be conditioned as in all mechanical testing. The energy to break is calculated as ft-lb per inch of notch (same as inch of width).

In plastic samples, the corrections are much more important than for metals. During breaking the broken part may be thrown a significant distance, and the energy associated with this process should be subtracted from the energy absorbed to yield the true absorbed energy to cause failure. This energy can be determined by replacing the broken pieces in the vise and dropping the hammer again. The energy absorbed to strike and throw the part can be subtracted from the energy obtained in the impact to failure test. However, the piece spins on being broken, since a torque is generated during the bending and fracture of the sample. During the retesting, the thrown part does not spin, and the absorbed energy is not exactly that absorbed during the test. The clamping pressure has some effect on the impact strength, as the samples are soft and can be deformed by the gripping mechanism.

As with metals, the Izod test is a useful test to indicate qualitatively the toughness and susceptibility to notch effects. It is very useful for quality control, but the test does not correspond to service conditions and the values of toughness cannot be utilized in any calculation. Fracture mechanics testing is of much greater value for these reasons but requires much more expensive machinery and analysis of data.

6.11. EFFECT OF ANNEALING

As described in Chapter 2, annealing a glassy polymer below T_g permits some molecular motion and reduction in the free volume. This annealing may also partially relieve the residual stresses generated by rapid cooling or by deformation. Broutman and Krishnakumar (1976) discussed the variation of Izod impact strength of PC with sheet thickness. A ductile-brittle transition occurs in this material at a thickness of the order of 0.25 inch, with thicker materials being ductile. Figure 6.18 shows this variation. They showed that quenched samples of the thin material were ductile, but samples annealed at 140–150°C and then slowly cooled became brittle. The ductility and good impact resistance of the quenched samples were attributed to the residual stresses generated during quenching. The stress distribution was found to be of the order of 3000 psi compression at the surface, with corresponding tensile stresses in the interior. These surface compressive stresses reduced the triaxial tensile stresses at the base of the notch in the Izod specimens and prevented craze growth and brittle fracture. All these samples were notched prior to heat treatment. If the samples were notched after quenching, low impact strengths were observed, since the material with residual compressive stresses was removed by the machining operation.

Because molecular motion increases with increasing temperature, to be effective, annealing must be performed close to T_g. Thakkar and Broutman

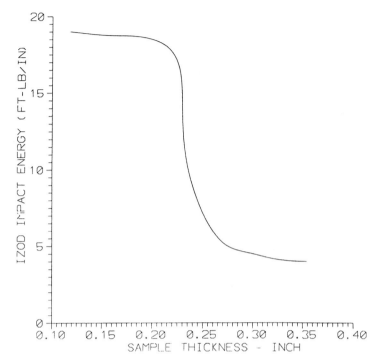

Figure 6.18.   The ductile-brittle transition of PC at various temperatures.

(1981) demonstrated that annealing PC ($T_g = 150°C$) at temperatures between 125 and 85°C returned the PC to the brittle condition as a result of the reduction in residual stresses and molecular orientation. Annealing close to T_g is also required to eliminate the molecular orientation. Since molecular motion increases exponentially with temperature, the rate of reduction in orientation and residual stress is given by an Arrhenius-type equation:

$$r = Ae^{-Q/RT}$$

where r = rate of disappearance of molecular orientation
　　　Q = activation energy for molecular motion and is a function of the applied stress level which generated the orientation
　　　R = gas constant
　　　T = absolute temperature

For 2,6 dimethyl polyphenylene oxide (2,6 MPPO), the data shown in Fig. 6.19 indicate that the as-cast material has some ductility (17%), probably due to the presence of some free volume in the glassy material, which is fairly

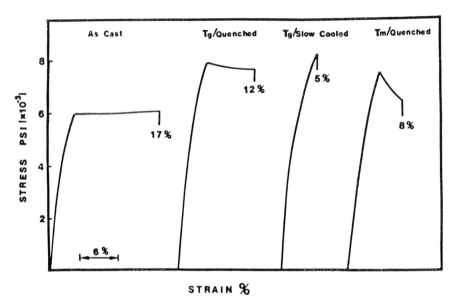

Figure 6.19. Stress-strain behavior for MPPO. (1) As cast; (2) quenched from 200°C; (3) slow cooled from T_g; (4) quenched from T_m. [From V. F. Dalal and A. Moet, *PES 28*, 544 (1988). Reprinted by permission of the publishers.]

rapidly cooled during normal casting. It may be that the free volume at the casting temperature was frozen in on rapid cooling from the casting temperature. When the sample is cooled slowly to T_g and then quenched, the free volume is reduced to the equilibrium value at T_g, resulting in less molecular mobility. If the sample is slowly cooled below T_g, the free volume may approach the smaller equilibrium value, again lowering the ductility. Why quenching from above the melting point produces less ductility is unclear.

The mechanical properties of a semicrystalline polymer depend on the percentages of crystalline and amorphous material. Annealing a partly amorphous polymer increases the crystallinity, which should increase the magnitude of the elastic modulus and the yield and tensile strength. Figure 6.20 (from Arzak et al., 1992) shows the effect of annealing on poly(ether ether ketone) (PEEK). The modulus and strength increase with increasing annealing temperature and annealing time. The T_g of PEEK is 140°C, and while extended annealing below this temperature should decrease the free volume toward the equilibrium value at that temperature, molecular mobility is so small that no additional crystallization will occur. This is demonstrated

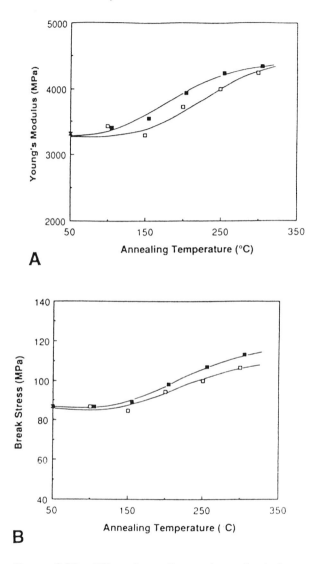

Figure 6.20. Effect of annealing on the mechanical properties of quenched PEEK. (A) Effect on elastic modulus. (B) Effect on breaking stress. Open squares represent 1 hr anneal, solid squares 24 hr anneal. The increase in mechanical properties can be attributed to an increase in crystallinity and in crystal perfection with increased annealing. [From A. Arzak, J. I. Eguiazabel, and J. Nazabal, *PES 31*, 586 (1992). Reprinted by permission of the publishers.]

by the very small change in properties on annealing below T_g. Increasing crystallization decreases ductility and toughness in this polymer at room temperature. There are apparently more mechanisms for deformation in the glassy than in the crystalline material.

6.12. TENSILE IMPACT

In Izod testing, the specimen is notched and bent during the test prior to failure, so the stress distribution is uncertain. Tensile impact is designed to apply a uniform tensile load rapidly to the specimen, avoiding uncertainty in stress distribution and also permitting testing of unnotched samples. In the test the sample in the form of a tensile specimen is held between a fixed grip and a grip which is struck by a rapidly moving pendulum. By adjusting the height of the pendulum, failure at different strain rates can be effected. This test has not received wide usage but is very useful in determining the effect of loading rate at high rates, above those attainable on a standard tensile machine. A technique for applying tensile impact is shown schematically shown in Fig. 6.21.

The energy absorbed to cause failure is a very strong function of the applied strain rate. At low rates of applied loading, the energy is relatively mildly affected, as the molecules have sufficient time to adjust to the applied load. Kink opening resulting in straightening of the chains can occur, and

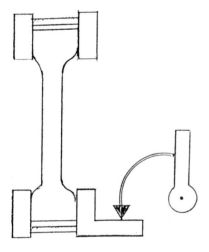

Figure 6.21.   Tensile impact testing.

the specimen absorbs a large amount of energy before failure. At a critical strain rate, the molecules can no longer respond to the rapidly applied load, and the material fails with little kink motion, with little deformation and energy absorption. In materials with secondary transitions, the side chains can still move rapidly at high deformation rates to provide some mechanism for energy absorption. At sufficiently high loading rates, these small segments of the molecule cannot follow the deformation either, and a second drop in energy absorption occurs.

Falling Weight Methods

The Izod test determines the energy absorbed during failure from an object containing excess energy (the pendulum). In the falling weight test, increasingly heavy weights are dropped from a fixed height until a crack appears in the sample. The striking surface is a 1/2-inch-diameter hemispherical nose. Since the drop height is fixed, the strain rate is also fixed. Because some damage may occur without evidence of failure, a new sample is required for each drop. A minimum of 20 pieces is required, and multiple samples are required at each drop height. A plot of the fraction of samples broken vs. energy is obtained, and the energy at which 50% of the samples fail is defined as the impact strength of the material. Although this test may correspond to failure modes during usage, it is a very time consuming and elaborate procedure and therefore rarely used.

6.13. HARDNESS

The Rockwell test for hardness is so well recognized and standardized that it is frequently used for plastics. In the Rockwell test, the indenter (which is always a steel ball for plastics) is impressed into the surface under a minor load, and the further depression into the surface under an additional load is measured. Due to the time dependence of the mechanical deformation of a plastic, the load application time must be carefully specified for a plastic (ASTM D785).

For softer plastics, the durometer, which has a needlelike indenter extending below a flat plate, is commonly used. The plate is lowered onto the surface of the specimen, and the extent to which the indenter extends into the sample is read on a dial indicator. The time of application of the load also affects the reading of this instrument (ASTM D2240).

6.14. HEAT DEFLECTION TEMPERATURE

Theoretically, crystalline plastics should not distort significantly below their melting points and glassy polymers below their glass transition temperature; in practice, the temperature at which a polymer begins to distort is usually significantly different from these two values. The heat distortion temperature is determined by placing a sample on two supports 4 inches apart and applying a load which corresponds to 66 or 264 psi at the center. The assembly is placed in a furnace and the temperature raised 2°C per minute. The temperature at which the center of the bar deflects 0.01 inch is reported as the heat deflection temperature. The stress should also be reported. As with many such practical tests, the value reported cannot be used in calculations to find the actual deformation at different temperatures at various loading times. The heat deflection temperature is primarily a guide for quality control and rough design material choice. It also cannot be used in determining the effectiveness of a material held for extended periods at elevated temperatures. More complete data on the degradation over time and the creep data at the use temperature are required for any meaningful evaluation.

6.15. CREEP

It is helpful to emphasize that stress-strain diagrams have little usefulness in design for plastics. All plastics have properties which are very time dependent, and therefore the basic data for design of plastic components are creep data. Creep data are experimentally determined by observing changes in length with time under constant load. Creep in metals is associated with permanent internal changes in structure, such as dislocation movement or grain boundary sliding. In plastics, the predominant mechanism for creep is the opening of kinks existing in the amorphous regions of the sample. These kink motions occur slowly under the applied stress, resulting in general chain straightening. As the chains become elongated, the rate of creep tends to diminish. Chain sliding or permanent changes in length due to changes in molecular neighbor positions also occur, but in many plastics this is negligible compared with the kinking. If the load is removed, the kinks slowly begin to return to their original configuration, and the sample slowly returns to its original length. Thus creep in plastics (for small strains) may be fully recoverable when the load is released. The strain level at which the strain is not recoverable varies with loading rate, loading time, and total strain; for most polymers strains below 2% may be recovered fully.

6.16. THE CREEP TEST

According to the definition of the creep experiment, the load should be applied as a step function of time: the load is initially zero, then increases instantaneously to the desired value. This implies impact loading. If such loading were applied, a shock wave would be transmitted to the sample, and ringing or oscillation of the stress as a function of time would occur. To avoid such oscillations of stress, the load must be applied smoothly over a period of time. Too slow a loading leads to uncertainty about the exact time of the start of the testing; too rapid a loading approximates impact loading, and the stress within the sample oscillates. Figure 6.22 shows the two extremes. Careful experimental design is required to obtain the optimum loading rate. In either case, the exact definition of the time at which loading is applied is somewhat uncertain, and the initial few seconds of the curve are inaccurate or the data are unavailable. Although this is often unimportant for long-term loading applications, as we shall see, it complicates the analysis of situations in which loads are varied over small time intervals.

Tensile and Compressive Stresses

Tensile creep is almost universally investigated. Creep strains in compression are somewhat lower than in tension. The tensile data therefore give conservative values for design. The differences depend on the material and the strain, becoming more severe at higher strains, usually well above those of interest in design.

Changes in Morphology

During an extended creep test, aging phenomena occur and can be accelerated by the applied stress. Semicrystalline polymers tend to continue crystallization and have enhanced postmolding crystallization. Amorphous polymers tend to approach their equilibrium free volume at a more rapid rate. These changes are usually small, however, and if the strains are not in excess of 2%, morphological changes are usually small enough that on release of the load, the specimens may eventually recover the applied strain completely. When morphological changes occur, however, complete recovery is not possible.

Creep Data

Creep data are usually presented as a plot of the resultant strain as a function of time for different applied stresses. If the strain at any time is a linear function of stress, the material is said to be linear viscoelastic. Some polymers

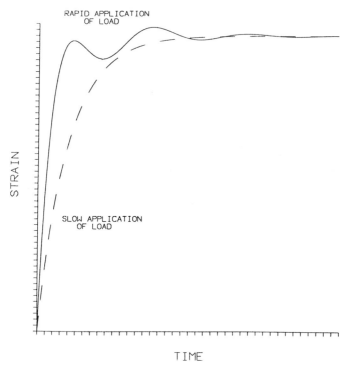

Figure 6.22.   Effect of loading rate on behavior during creep testing.

exhibit this property at low strains ($<2\%$), but deviate at higher stress values. Other polymers show deviations from linear behavior at all times and strains. However, the simplification of the presentation of data and the calculations possible with linear materials often lead us to neglect small deviations.

A schematic illustration of typical creep data is shown in Fig. 6.23. Three regions can be identified. Initially, the material deforms very rapidly. Data in this region often are not shown, since they may be inaccurate due to the difficulties in starting the step loading correctly. After the first rapid increase in strain, the strain continues to increase at a steadily decreasing rate for extended time periods. Plots of strain vs. log(time) tend to obscure this decrease in creep rate. At large strains the cross section of the material may be so reduced and the true stress in the reduced area so large that the rate of extension increases rapidly. The material is well beyond the proper design region when this occurs.

The creep compliance is a function of time and is defined as the strain

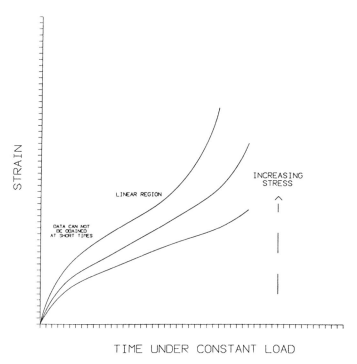

Figure 6.23.   Typical creep data.

at a given time per unit stress, where the stress is considered constant over the life of the test:

$$J(t) = \frac{\varepsilon(t)}{\upsilon}$$

If the material is linear viscoelastic, a plot of $J(t)$ vs. time will exhibit superposition of all the creep curves obtained at different stresses on one curve. The extent of deviation from linearity of behavior is evident in the spread of the $J(t)$ curves with stress (Fig. 6.24).

Some polymers may exhibit normal creep extension, followed by the development of a crack and brittle rupture in a relatively short time after the observation of the crack. There is no correlation between the appearance of the creep data and the rupture data. This exemplifies the danger in extrapolating creep data. Examination of the tensile data for plastics might suggest that there is an elastic range in which creep can be neglected. This is not true, and any stress level will be sufficient to produce creep strain.

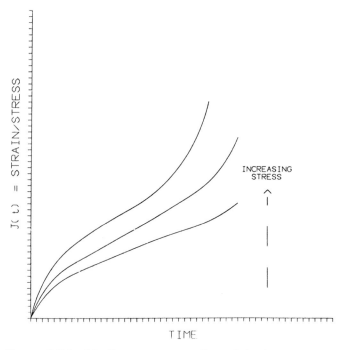

Figure 6.24. Nonlinear creep compliance behavior.

Often the weight of the article alone may be sufficient to cause significant creep deformation.

6.17. EQUATIONS FOR CREEP CURVES

Although creep data are most accurately presented as the plot of strain vs. time for various stresses and temperatures, many theoretical and empirical relations have been suggested for the dependence of creep strain on stress and temperature for plastics. Most of these are identical to the equations found to apply to metals. These general expressions do not take into account any changes in morphology which may occur during the aging of the polymer or the occurrence of crazing or stress rupture, and therefore they should not be used for extrapolation beyond the region of experimental data. If the polymer is linear viscoelastic, then the creep strain should be directly proportional to the stress. Some equations assume linearity, but most include the dependence of stress. Among the more commonly used expressions are those shown in Table 6.2.

Table 6.2. Common Expressions for Creep Strain as a Function of Time[a]

$$\varepsilon = a + bt^c$$
$$\varepsilon = a + bt^{1/3} + ct$$
$$\varepsilon = at/(1 + bt)$$
$$\varepsilon = 1/(1 + at^b)$$
$$\varepsilon = a + \ln(b + t)$$
$$\varepsilon = a(b + \ln t)/(c + d \ln t)$$
$$\varepsilon = a + b(1 - e^{-ct})$$
$$\varepsilon = \sinh(at)$$
$$\varepsilon = a + b[\sinh(ct)]$$
$$\varepsilon = a + b[\sinh(ct^{1/3})]$$
$$\varepsilon = a(1 + bt^{1/3})e^{-ct}$$

[a] a, b, c, d are determined from experimental data.
From T. Sterrett and E. Miller, *J. Elastomers Plastics* 20, 346 (1988).

The hyperbolic sine equation was derived previously from the Eyring theory of activated states and is favored for theoretical studies. For calculations the most convenient expression is the first one, and it has been found to best fit the data for many plastics. The value of c for most polymers lies in the range of 0.5 to 5. If no value of c is available and the stress level is low, linear viscoelasticity might be assumed with $c = 1$.

The exponent c has been considered to be independent of stress, but this is not precisely true, as it increases slowly with increasing stress level. If c is truly a constant, a log-log plot of strain vs. time should be a straight line. This gives a simple means of extrapolating data to longer times. In such extrapolation the danger always exists that the other forms of failure mentioned above may suddenly and catastrophically occur. To help decide whether such extrapolations are valid, examination of creep data at higher temperatures and stresses is helpful. In general, strain extrapolations greater than 2% should never be made.

6.18. CROSS-PLOTTING OF CREEP DATA

Isometric Stress Versus Time

Creep data can be cross-plotted to present the data in a variety of useful forms. By taking data at numerous constant strain values, plots of stress vs.

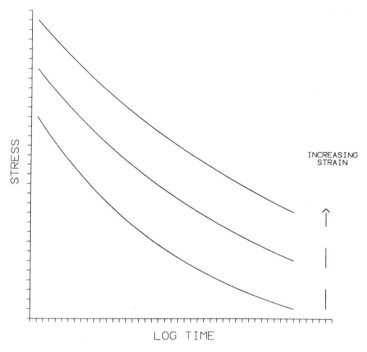

Figure 6.25. Cross-plotting creep data to obtain isometric stress vs. time plots.

time at constant strain are obtained (Fig. 6.25). These have the appearance of stress relaxation curves and are sometimes used for estimating such relaxation, but they are obtained from creep curves (constant stress) and do not correspond accurately to constant strain data. These curves are valuable primarily for quick visual identification of the time that a stress can be applied if the maximum strain is the material specification.

Creep Modulus Versus Time

The creep modulus obtained from a creep curve is defined as

$$E(t) = \frac{\sigma}{\varepsilon(t)}$$

These values computed from the creep data are shown schematically in Fig. 6.26. If the material is elastic, the strain should be independent of time and the modulus should be independent of time. If the material is linear viscoelastic, $E(t)$ should be independent of the stress level, and curves

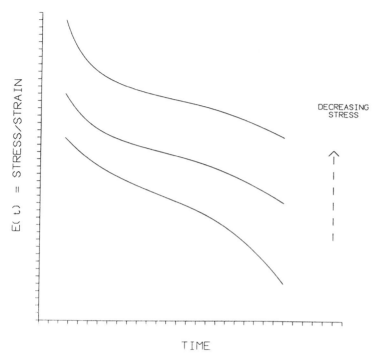

Figure 6.26.   Nonlinear creep modulus behavior.

obtained at different stress levels should all superimpose on one curve. For all polymers, the creep modulus decreases with increasing stress. At a given time of loading, the stiffness of the material will be smaller the greater the applied load. The creep modulus is very useful in design calculations. As with other mechanical strength properties, the creep modulus decreases markedly with temperature. The dependence on temperature is a result of the changes in molecular motion discussed previously:

1. In the glassy region, the molecular configurations are frozen, and the response to an applied stress is mainly due to changes in the C—C bond distance.
2. In the rubbery region, the kinks can change shape, with straightening of the molecules occurring readily. At elevated temperatures close to the melting point, sliding of one molecular chain past another occurs readily and permanent deformation occurs in the material.
3. Since the molecules are essentially frozen in the glassy region, the molecular length has only a small effect on the modulus. At elevated

temperatures at which flow molecular sliding occurs, the higher the molecular weight, the higher the modulus. As expected, increasing crystallinity increases the modulus.

Isochronous Stress Versus Strain

By cross-plotting the creep data at specific time intervals, stress-strain diagrams can be obtained. These diagrams appear similar to the tensile test data. However, in a tensile test the time varies throughout the test, and the two sets of data do not coincide. This plot is useful for quickly identifying the permissible stress to be applied for a known lifetime and specified maximum strain. If the material is linear viscoelastic, the stress varies linearly with strain independently of time, and this diagram degenerates into a series of straight lines passing through the origin. Isochronous stress-strain diagrams eliminate the effect of pulling rate on the properties of the material and are frequently drawn at times of 1 second to permit comparison of various materials. Figure 6.27 shows these plots for various polymers. The larger

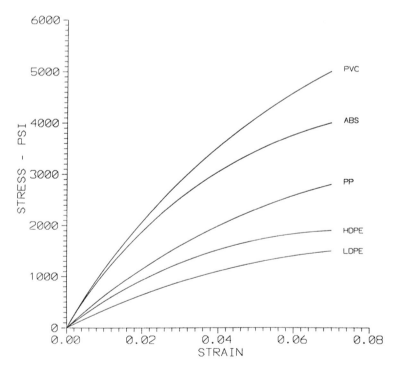

Figure 6.27. Cross-plot of creep data to obtain isochronous stress-strain data.

deviation from linearity for acrylonitrile-butadiene-polystyrene (ABS) as compared to PE is apparent.

6.19. CREEP RUPTURE

The mode of failure expected as a result of extended creep is a reduction in cross-sectional area to such an extent that the resultant material is incapable of supporting the applied stress, and the material necks down and fails in a ductile manner. However, some polymers begin to develop cracks after extended creep. These cracks propagate rapidly and cause failure after a relatively small additional time. The type of failure is a function of temperature. Figure 6.28 (Gedde, Ifwarson (1990)) shows the sudden change from ductile failure to brittle failure at extended times for PE. Whether failure occurs by general yielding or by crack propagation is a function of the applied stress and service temperature. Lower stresses tend to produce a crack propagation type of failure. The data in Fig. 6.29 indicate that the temperature dependence for crack propagation in PE film is much less than for ductile failure, and a transition therefore occurs, failure by crack propagation controlling at low temperatures and failure by shear deformation at higher temperatures. Failure occurs at longer times at lower temperatures for both branches, since crack propagation rates increase with temperature and the yield strength decreases with increasing temperature, shifting both branches of the failure curve to shorter times at higher temperatures.

6.20. MOLECULAR ORIENTATION AND RESIDUAL STRESSES

The mechanical properties of many amorphous and crystalline polymers are improved by mechanical deformation procedures. These improvements are a result of the molecular orientation produced, as well as the resultant residual stresses introduced into the product. The particular orientation and residual stress pattern depend on the specifics of the deformation, and different properties will be observed in finished products manufactured by rolling, extrusion, drawing, etc. In these deformation processes, most of the orientation effects occur in the surface layers, where the deformation is concentrated. The direction of the molecular orientation or residual stresses relative to the applied stress direction is of particular importance. Unoriented polystyrene is a typical brittle glassy polymer. However, uniaxially oriented

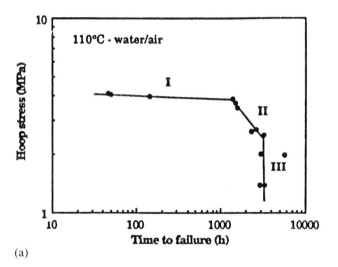

(a)

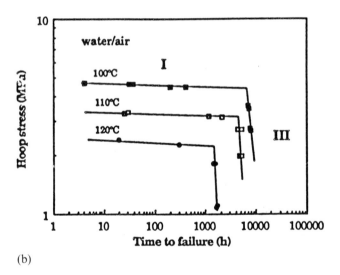

(b)

Figure 6.28. Transition from ductile to brittle failure in cross-linked PE during creep testing. (a) Low cross-link density; (b) normal cross-link density. Three stages of fracture can be observed. Stage I is ductile failure. Stage II is the onset of brittle fracture at longer times from creep rupture. Stage III is almost stress independent (note the almost vertical line for this stage) and occurs due to chemical defects and degradation of the PE. [From U. W. Gedde and M. Ifwarson, *PES 30*, 202 (1990). Reprinted by permission of the publishers.]

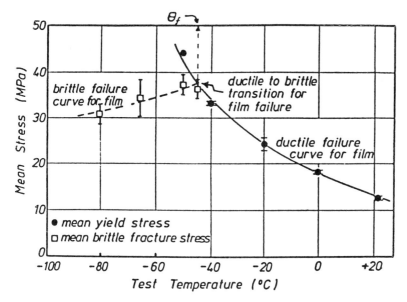

Figure 6.29. Transition from ductile to brittle failure in PE film as a function of test temperature. At temperatures above $-40°C$, the yield and breaking strength increase with decreasing temperature, since the behavior is ductile. Below this transition temperature, brittle behavior is observed, with the brittle failure stress being well below the extrapolated ductile failure stress. [From M. Simpson and J. Bowman, *PES 31*, 487 (1991). Reprinted by permission of the publishers.]

PS has a much higher tensile strength, its elongation is good, and its resistance to environmental stress cracking is greatly improved in the direction of the orientation (Alfrey, 1974). However, in the direction transverse to the orientation, the properties are degraded.

Cold rolling of 6.4 mm (0.25 inch) thick PC samples to 2% reduction in thickness increases the impact strength by a factor of 10, from 100 J/m (2 ft-lb/in) to 1000 J/m (20 ft-lb/in) (Broutman and Krishnakumar 1976). Thicker samples exhibited smaller increases in impact strength, probably due to the inability of the deformation to extend throughout the sample. This increase is due to the generation of a surface and interior residual stresses. At larger reductions, molecular orientation further improves the toughness. Their experiments indicated that at high strains, the molecular orientation produced is much more significant than the residual stresses generated. Rapid cooling of thin samples also improves impact resistance due to the generation of residual stresses (Mills, 1976).

6.21. STRESS-STRAIN HYSTERESIS LOOPS

Stress-strain diagrams of samples obtained continuously during fatigue loading provide information regarding the changes in mechanical properties resulting from the deformation. Since the stress and strain are not in phase during the deformation, the resultant curves are called hysteresis loops. Such curves have been studied extensively for metals, most commonly obtained under constant stress amplitude, since the strains generated in metallic samples are so small that strain control is difficult to achieve accurately. Annealed metallic samples tend to work harden during cyclic deformation, and the stress required to produce the predefined strain increases slowly as cycling continues. Metallic samples which are work hardened prior to cyclic deformation tend to soften as cycling progresses; the hysteresis curve showing a continually lower required stress to reach the defined strain. A stabilized condition tends to be approached for both annealed and work hardened metals.

If the metallic samples were perfectly elastic, the stress-strain curve generated on increasing stress should be exactly matched on decreasing the stress, and the hysteresis loop should be a single straight line passing through the origin, with slope corresponding to the elastic modulus of the material. For real metallic samples, the generation of a hysteresis loop is a result of a small component of permanent strain produced during each cycle. Elastic strain is also generated but is fully recovered when the stress returns to zero. The width of the hysteresis loop along the strain axis at zero stress is therefore a measure of the permanent strain generated in the sample during the cycle. The accumulation of this plastic strain eventually results in crack generation and fatigue failure. The cyclic life has been related to the amount of plastic deformation per cycle by

$$e = MN^a$$

where e is the plastic strain per cycle, N is the number of cycles to failure, and M and a are material constants (Manson, 1954).

Few data are available regarding these hysteresis loops for plastics. The most comprehensive studies of the cyclic properties of polymers are those of Beardmore and Rabinowitz (Beardmore and Rabinowitz, 1974). They performed experiments in fully reversed tension-compression under strain control. They observed only cyclic stress softening in all the polymers they investigated. Their results for ductile amorphous polycarbonate were particularly interesting. Initially, the stress-strain curve had a distinctly propeller-shaped hysteresis curve, which rapidly widened during cycling (Fig. 6.30). They observed four different regions in testing at constant strain amplitude. The first or incubation stage was one in which relatively small

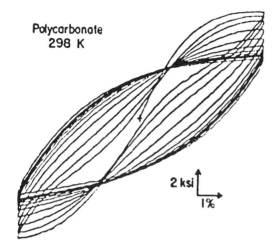

Figure 6.30.   Hysteresis curve for PC. [From S. Rabinowitz and P. Beardmore, *J. Mater. Sci.*, *9*, 81 (1974). Reprinted by Permission of Chapman & Hall.]

amounts of stress softening were observed. As the strain amplitude increased, the number of cycles associated with the incubation stage decreased. The second (or transition) stage, in which the material significantly stress softened, was of relatively short duration. The third stage of cyclic stabilization occupied the major portion of the lifetime, as in metals. Crack initiation followed this stage.

For brittle plastics (epoxy, PMMA) small amounts of continuous softening occurred throughout the fatigue life. This softening is small compared to that observed in the ductile materials. Semicrystalline polymers (nylon, polyoxymethylene) did not exhibit an incubation period. ABS exhibited cyclic softening throughout the life of the specimen. Surprisingly, the softening of the specimens was associated with an increase in density of approximately 1%. Since such an increase in density would imply greater packing and more internal order, the softening reported must be due to the removal of molecular entanglements which strengthen the polymer. Beardmore and Rabinowitz suggested that for plastics, as for metals, the width of the loop at zero stress is the plastic strain in the cycle. The viscoelastic response of the sample requires stress-strain diagrams to be loops, since the stress and strain are not in phase. In Chapter 9 the relation between strain and stress during cyclic loading of a four-parameter model is given, and the concepts of modeling the times for molecular relaxations are discussed in detail. At this time it is convenient to recognize that the

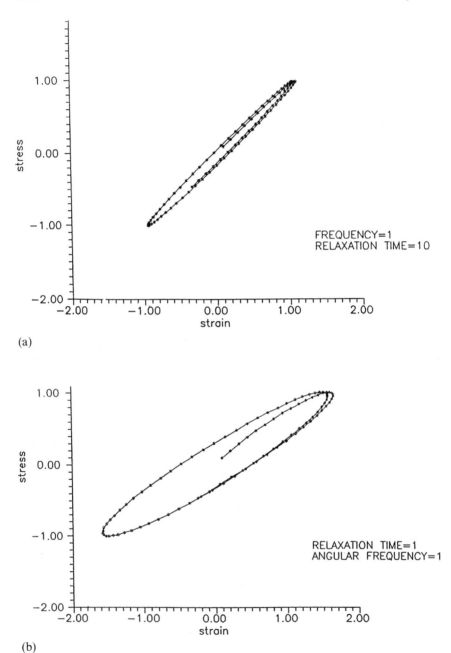

(a)

(b)

Figure 6.31. Hysteresis curves for different relaxation times.

relaxation time of a polymer represents a measure of the time for molecules to return to their original configuration after being disturbed by the application of a stress. Figure 6.31 shows hysteresis loops for two different relaxation times. When the relaxation time is orders of magnitude larger than the period of the applied stress, the polymeric long-chain molecules have insufficient time to change their kink configurations, and the material behaves elastically, the strain being virtually linear with applied stress. As the relaxation time approaches the time of the load cycling, some lagging of strain behind the stress occurs, and the hysteresis loop begins to open. When the relaxation time becomes much shorter than the cycling period the molecules can follow the applied stress, and the width of the hysteresis loop begins to close again. The shape or openness of the hysteresis loop in plastic can therefore not be taken as a measure of the amount of permanent strain for polymers, as it can for metals. Rather, it is likely a measure of the lag angle between the strain and stress and a strong function of the cycling rate and the relaxation time of the material. It is therefore unlikely that predictions of fatigue lifetime can be made from these data for polymers, as they can for metals.

7
Fracture Properties

7.1. CRAZING

In ceramics, a fine network of cracks on the surface is referred to as crazing. A craze appears in a transparent amorphous polymer as a thin white line, similar to a crack. Not surprisingly, in polymers crazes were confused with cracks for many years, until it was determined that the craze was a region of material with lower density, occupied by deformed, extended polymer chains. Crazing is one of the important deformation mechanisms observed in glassy polymers. Crazing is an important energy-absorbing mechanism, and the fracture energy of a plastic is increased by the generation of crazes. In particular, if multiple crazes occur throughout the material, its toughness can be increased considerably over that of a material which exhibits few crazes. Crazes occur when a dilatational stress which tends to increase the volume of the sample exists. Under the dilatational stress, microvoids are produced, usually nucleating at a defect which generates a local stress concentration. Such defects can be surface scratches produced during molding, machining, or handling; small air bubbles or flaws introduced during molding; or internal structural defects such as a region where molecular entanglements, unreacted catalyst, or excessive additives exist. The material between the microvoids plasticly deforms under tensile stresses, and the microvoids are separated by material with large elongations and resultant molecular alignment, similar to the material in a necked tensile sample.

In window materials or light fixtures, crazing produces light scattering and unsightly appearance and can be considered as a failure of the material for those applications. In structures, the generation of a craze may not produce instantaneous failure, but the crazes are defects from which eventual failure will occur, by a true crack propagating through the crazed material. In general, the formation of crazes within a material should be considered unacceptable, and the material should be chosen so that no crazing occurs for the loads and environments encountered.

❡ The tendency of a glassy plastic to craze is strongly affected by liquid environments and to some extent by gaseous ones. Some liquids promote the formation of crazes under very low external loads, although internal residual stresses contribute strongly to the crazing of many products. Liquids that promote crazing plasticize the surface of the polymer, increasing the ease of deformation. The liquid then flows into the craze through the voids once initiated, and the process continues. For this to be so, the solubility parameter of the liquid should control the ease of crazing. Figure 7.1 shows the time for crazing of polymethyl methacrylate (PMMA) at different stress levels. The higher the stress, the faster crazing is initiated. Figure 7.2 shows the minimum specimen strain (the critical strain) at which crazes ❡

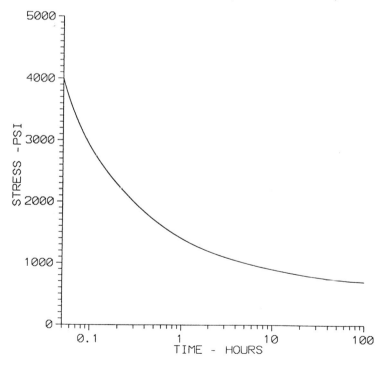

Figure 7.1.   Time for craze initiation in PMMA in air.

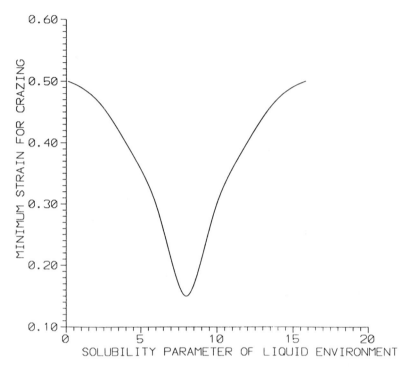

Figure 7.2.   Critical strain for crazing in various solvents.

⟍ form in polyphenylene oxide (PPO) in various liquids. This polymer has a solubility parameter of 8. It can be seen that the ease of crazing is greatest for liquids of identical solubility parameter, and the strain required to initiate a craze increases very rapidly as the solubility parameter difference between the polymer and the environmental liquid increases. ⟍

Density of Crazed Material

⟍ Crazes are regions of lower density within the polymer. The craze consists of a long thin region which contains small voids and interpenetrating fibrils of elongated chains. Due to the lower density of material within the craze, the refractive index is much lower than that of the bulk material. This change in refractive index gives the craze its characteristic milky appearance. Since the refractive index is a function of the density, the determination of the refractive index permits an evaluation of the concentration of microvoids within the craze. The density of material within the craze decreases with increasing strain as a result of increasing void formation within the craze.

The volume fraction of material within the craze has been found to be related \ to the strain e by

$$V_f = \frac{1}{1 + e} \tag{7.1}$$

The strain within the craze is of the order of 50%, resulting in the molecular fibrils connecting the two halves of the craze being quite elongated and ordered. This strain leads to a volume fraction or density within the craze of approximately 40–60% of the bulk material. Similarly the elastic modulus of the material within the craze is approximately one half that of the bulk material. Edge grooves are observed where the craze intersects the surface. Since extensive strain has occurred within the craze, lateral contraction would be expected and should be computable by Poisson's ratio. However, since the density of the material is so low compared to the normal material, the lateral contraction is actually negligible compared to that which would occur if the material maintained its true density.

Lauterwasser and Kramer (1979) studied the true stress-strain curve for the material within a craze in PMMA. They determined that the strain in the fibrils near the craze tip is far higher than that at the craze base, indicating that as the craze forms the fibrils are generated by marked drawing and elongation. In fact, the elongation measured at the craze tip correspond to about 500%, a value corresponding to fully elongated molecules. The material farther from the craze tip, where thickening of the craze width has occurred, must have been somewhat reduced in overall strain by the pulling in of additional material from the bulk material-craze interface, resulting in less overall extension of the molecules. Since the molecules at the crack tip are fully extended, any additional strain can result only from separation of molecules by breaking the intermolecular van der Waals forces or by chain scission. Either of these mechanisms would result in the generation of a crack within the craze.

Crack Propagation Along Crazes

As the stress increases, the fibrils become more oriented and the size of the voids increases. As deformation proceeds, a crack develops down the center of the craze, and the craze propagates through the sample, with the crack progressing behind the craze front until failure occurs. At low crack velocities, the crack penetrates through the center of the crazed material. As the velocity increases, the crack jumps from one craze–bulk material interface to the other. At very high velocities, the kinetic energy of the crack (actually of the material on either side of the crack) can assist in the crack extension, resulting in an acceleration of the crack. The crack therefore overruns the craze production and runs to the end of the previously formed craze, and new crazed regions develop in bursts, through which the crack then follows. This leads to interesting variations in the appearance of the fracture surface. In the initial stages of crack propagation, where the crack velocity is low, both fracture surfaces are coated with a thin hazy material, the remains of the crazed material which is partly attached to each fracture surface. Farther on in the direction of crack propagation, islands of crazed material appear, since the entire craze remains attached to one surface in localized regions. Toward the final stages of crack propagation, the crazes are bundle shaped (since they formed during the crack movement), and the resultant surface takes on the appearance described as *hackle bands*, in which clumps of hazy, crazed material appear, pointing in the general direction of crack propagation (although sometimes much imagination is required to identify the direction).

Crazes do not form as isolated flow markings. Near the tip of a craze the stress concentration is increased, much as at the tip of a crack, and additional crazes are likely to be initiated in the surrounding material. Surprisingly, high molecular weight material appears to be more susceptible to such secondary crazing than low molecular weight material. This effect is desirable, however, since the greater the number of crazes formed, the greater the energy absorption and the greater the fracture toughness. In addition, the strength of the crazed material tends to increase with increasing molecular weight.

Crazes also form during fatigue loading. The fatigue crack propagates by advancing through the crazed material, and the crazed material at the crack front also propagates through the sample. The advancing craze can therefore be considered to be a front weakening the material, thus permitting the crack to grow. However, it should be recognized that the craze involves deformation, and therefore the energy required to cause failure is increased by this mechanism. Huang and Brown (1990) have shown that increasing the density of molecular branches improves the resistance to crack propagation if the branches are on the polymer main chains that are responsible for tying the various crystallites together. A relatively uniform spacing of the branches on the tie molecules is more effective than a cluster of such branches.

Craze Initiation

As stated a craze is a flat thin region of the material which has developed a commingled area of voids and elongated packets of molecules. The fibrils of elongated molecules within the craze are of the order of 400 angstroms in diameter. The molecules within the bulk of the sample are disordered, but within the craze the fibrils have bceome ordered as a result of (or producing) the extensive strain within the craze.

The conditions required for craze initiation have been discussed extensively in the literature. Crazes tend to be generated at discontinuities such as surface notches, machining nicks, or dirt particles. However, the cleanest specimens still exhibit crazing under the proper conditions. Since the craze has a lower density than the bulk material, a dilational stress field must exist for its formation. Most commonly the stress is uniaxial tension, and in this case the crazes usually develop perpendicular to the tensile stress direction. Various criteria have been proposed for the initiation of crazing. These have been presented in terms of a critical minimum stress, minimum strain, or minimum stress intensity factor. Since voids are formed within the craze, a dilatational stress must exist, and a logical criteria for craze formation should therefore include the dilatational stress as defined by I:

$$I = \sigma_1 + \sigma_2 + \sigma_3 \tag{7.2}$$

Sternstein and Ongchin (1969) applied biaxial tension to specimens of PMMA and proposed the criterion for craze initiation of

$$\sigma_1 - \sigma_2 = A(T) + \frac{B(T)}{I} \tag{7.3}$$

where $A(T)$ and $B(T)$ are functions of temperature.

Another fruitful concept is that of the requirement for a critical strain (dependent on environment) at which crazing initiates. Wang et al. (1971) and Mitz et al. (1978) embedded balls of different elastic moduli in polymeric samples to generate principal stresses and strains along different directions. Their results indicated that the magnitude of the principal strain determined the initiation of crazing. Although the critical stress depends on the material and the environment, they found that typically, the uniform strain had to reach a value of about 0.75% for crazing to be initiated.

Bowden and Oxborough (1973) converted the concept of Sternstein and Ongchin to that of a critical strain by restating Eq. 7.3 in the form

$$e_c = A(T) + \frac{B(T)}{I} \tag{7.4}$$

Since

$$e_1 = \frac{\sigma_1 - \mu\sigma_2 - \mu\sigma_3}{E} \tag{7.5}$$

then

$$\sigma_1 - \mu\sigma_2 - \mu\sigma_3 = A(T) + \frac{B(T)}{I} \tag{7.6}$$

where $A(T)$, $B(T)$ are material constants which must be determined experimentally. Excellent agreement was observed between the stress at which crazing begins and that computed from this equation.

Effect of Temperature

The equation proposed by Gent (1970) is

$$\sigma_{\text{critical}} = \frac{b(T_g - T) - P}{k} \tag{7.7}$$

where k = stress concentration factor
         b = a material constant
         P = the hydrostatic pressure = $\dfrac{\sigma_1 + \sigma_2 + \sigma_3}{3}$

Under a normal tensile load, the stress to produce crazing should decrease with increasing temperature until it approaches zero at the glass transition temperature. This equation cannot be utilized close to the glass transition temperature, as the crazing stress does not drop to zero at the glass transition temperature. Furthermore, most glassy polymers exhibit a relation between temperature and σ_{critical} which approaches a hyperbolic one more closely than the proposed linear one. The increase in critical stress produced by a hydrostatic pressure tending to close the crazes and cracks present is noteworthy. Above T_g, the material is rubbery, and crazing does not occur.

Craze Growth

Although a hydrostatic dilatational stresss is generally required to generate microvoids and a craze, Argon and Salama (1977) showed that craze growth could occur when only a pure shear stress existed. Once crazes form, they tend to increase in size at constant stress or strain with time. Crazes grow both in thickness and in length, although the increase in thickness is rather small. Most studies have examined only the variation of craze length with time. Since most viscoelastic effects depend on time according to a power law, the time dependence of the craze initiation stress can be expressed as

$$\sigma = At^{-n} \tag{7.8}$$

where the exponent n is usually less than 1. Similarly, the length of a craze might be expected to obey a similar time dependence.

$$l = At^n \tag{7.9}$$

Taking logarithms of this equation yields

$$\log l = \log A + n \log t \tag{7.10}$$

where l is the length of the craze. This can be rewritten in the form

$$\log l = n \log\left(\frac{t}{A_0}\right) \tag{7.11}$$

Regel (1956–7) reported the increase in craze length l with time t at constant stress to be of a somewhat similar form:

$$l = k \log\left(\frac{t}{t_0}\right) \tag{7.12}$$

He associated t_0 with the initiation time for the craze to start. l

For polyvinyl chloride (PVC), at stresses below 16 MPa crazes do not form. The time for crazing to occur decreases with increasing stress above this level. These stresses are all well below the yield stress of the material, and the instantaneous strain can be computed directly from the modulus. If critical minumum strain criteria were used instead of the applied stress, then the critical strain of 0.35% would be reported. As the stress or the holding time at constant stress increases, the strain and the void fraction of the craze increase, decreasing the density further.

Jasver and Hsaio (1953) reported that the craze growth rate increases linearly with increasing tensile stress above the critical minimum stress to produce crazing σ_0:

$$\frac{dl}{dt} = k(\sigma - \sigma_0) \tag{7.13}$$

where k is a material constant and σ_0 is the minimum stress at which crazing starts. Note that this equation implies a critical stress rather than a critical strain for craze initiation. For uniaxial tension, these criteria are obviously related.

Williams and Marshall (1975) applied the concepts of fracture mechanics to craze growth. This is discussed in Section 7.11.

7.2. FRACTURE MECHANICS

Fracture

Fracture is the separation of a part into two or more pieces. Although failure may take place as a result of improper design so that the ultimate strength is less than the actual load on the sample, these type of failures are uncommon with our improved understanding of stress analysis and material properties. More commonly, such failures are initiated at preexisting flaws or cracks in the material; then, as a result of environmental factors or fatigue loading, the crack extends until a critical size is reached, at which time the crack propagates rapidly to cause failure. Due to the stress concentration at the crack tip, this failure is likely to be at a load well below the nominal failure load of an ideal uncracked sample. It is an axiom of fracture mechanics that all materials contain defects, generated during molding, machining, or handling, and that the ease of propagation of these cracks determines the resistance of the material to fracture. The significance of a crack in controlling the properties of a sample depends on the ability of the material at the end of the crack to deform plastically to blunt the crack tip, thereby lowering the stress concentration.

This approach indicates that the loading-bearing ability of a member which contains a crack depends on:

1. The material's ability to blunt a crack or withstand its propagation (ability to deform in the localized region of the crack tip).
2. The size of the preexisting crack. The greater the crack size, the less the critical stress to cause failure. Observation of plastics have indicated that various sizes of cracks are present, and Benham (1973) proposed that polystryene (PS) contains an inherent crack size due to its morphology, and the material cannot be prepared without a crack size lower than 5 µm.

Fracture usually results in its separation into two parts, although in bending, brittle materials may shatter, breaking into three or more pieces. The common terminology is that a *brittle fracture* is one in which the material behaves essentially elastically up to the point of failure. The area under the stress-strain diagram is the energy required to break the specimen. In brittle materials the strain is primarily elastic. Because elastic strains are usually quite small, the energy to cause such failures is small and is stored in the material prior to failure as elastic strain energy. Since no permanent deformation has occurred, the fracture surfaces are relatively smooth and largely perpendicular to the direction of the applied stress. The two fracture surfaces may be fit together quite accurately. Fitting of mating parts, such as a connecting rod cap and the connecting rod, is performed by forging the two as one unit and then separating them by fracture at an induced notch, resuting in excellent reproducible contact when fitted over the bearing. Data show a decrease in brittle fracture stress with decreasing temperature.

Ductile fracture implies that large permanent deformation has occurred before failure, requiring a significantly larger amount of energy absorption by the part before failure. As a result of the permanent deformation, the two fracture surfaces do not match, and the cross-sectional area at the loaction of the fracture is reduced from the original value.

The separation of materials into ductile and brittle ones is not necessarily obvious. We expect a change from ductile behavior above the glass transition to brittle behavior below the glass transition for amorphous materials. However, materials which exhibit reasonable ductility during a tensile test may fail in a brittle fashion during service. The theoretical tensile strength of an ideal elastic material is of the order of one-tenth the elastic modulus, orders of magnitude higher than that experimentally observed for most materials.

The yield strength of crystalline polymers increases with decreasing temperature. Crystalline polymers in general exhibit some ductility. At sufficiently low temperatures in semicrystalline polymers containing a large percentage of amorphous material, a transition may occur to brittle behavior as the glass transition of the amorphous region is approached. Brittle failure occurs at stresses below that at which yielding will occur and is dependent on surface smoothness and small nicks and scratches on the surface. As a result, the strength of the sample decreases once brittle behavior is encountered.

Stress Concentrations at Cracks

If a sample is ostensibly uniformly loaded, the actual stress within the material may vary due to the presence of imperfections and cracks in the material. If the stress field is illustrated by lines running from one end of the sample to another, then the stress lines must be concentrated around the edges of any notch or internal discontinuity to transfer the stress across the sample. The magnitude of the stress concentration depends on the location and the sharpness of the crack. The stress concentration can be computed for the condition of a perfectly elastic material. For an elliptical crack in the center of a plate with the crack running perpendicular to the applied stress, the stress concentration was computed by Inglis (1913):

$$k_t = \frac{\sigma_{max}}{\sigma_{nom}} = 1 + 2\frac{a}{b} \tag{7.14}$$

where a = half major axis of the elliptical crack
   b = half minor axis of the elliptical crack
   σ_{max} = maximum stress
   σ_{nom} = applied stress considered to be uniform
   k_t = stress concentration due to the crack

For a material considered to contain cracks, it is common to assume that the crack is elliptically shaped, with the radius of curvature ρ at the crack tip being very small. Equation 7.14 is normally expressed for such cracks in terms of the radius of curvature:

$$\sigma_{max} = 2\sigma_{nom} \left(\frac{c}{\rho}\right)^{0.5} \tag{7.15}$$

where ρ = radius of curvature of the elliptical crack
         c = crack length

The normal convention for defining the length of a crack in fracture mechanics is that the crack length c (the standard symbol for crack length) is one half the crack length if the crack is internal or the total crack length if the crack runs through to the surface. Note that as the crack becomes narrower and sharper (as a/b increases and ρ decreases) the stress at the tip of the crack rapidly increases.

Tables of stress concentration values k_t are available (Peterson, 1974) for various crack configurations. Regardless of the position of the crack and its general appearance, the concentration increases with increasing crack length and increasing sharpness (or decreasing radius of curvature at the crack tip). The sharpest crack can be considered to be of atomic thickness, resulting in very large stress concentrations. The effect of cracks and discontinuities is much more significant in materials with no ductility. A ductile material will plastically or viscoelastically deform in the region of high stress concentration, increasing ρ and decreasing the stress levels until yielding or deformation no longer occurs.

Crack Initiation in Rubbery Polymers

Filler particles such as carbon in rubbery polymers are likely to be the regions in which crack initiation and crack growth are most likely to develop. As a result of the particles' presence, triaxial stresses are generated directly above and below the filler particles, restricting shear distortion and encouraging crack initiation. The equation developed for the stress to initiate these cracks (Oberth and Brenner, 1965) is

$$\sigma = \frac{E}{\eta} - P \tag{7.16}$$

where η is the stress concentration factor for the inclusion, and P is the external hydrostatic pressure. For a spherical particle η is close to 2, and the external pressure can be neglected, yielding

$$\sigma = \frac{E}{2} \tag{7.17}$$

Good agreement with this equation was obtained for the stress to initiate cracking in polyurethane rubbers.

Besides introducing triaxial stresses, the filler particles usually have low surface adhesion to the matrix, and separation from the matrix will generate surface cracks around the particles. Gent (1980) considered the energy required to create the new surface at the filler-matrix interface and developed the relation for the energy required to separate the filler from the matrix:

$$\sigma = 2\left(\frac{\pi G_i E}{3r_p}\right)^{0.5} \tag{7.18}$$

where G_i is the interfacial fracture energy, E the modulus of matrix material, and r_p the radius of the filler particle. The larger particles are therefore more likely to develop separation from the matrix.

7.3. CRACK PROPAGATION

Cracks in linear elastic materials were the original defects studied. These studies developed into the field of linear elastic fracture mechanics (LEFM). The original concepts were applied to cracks in completely brittle materials by Griffith (1921). These concepts were then expanded to materials which exhibit plastic deformation at the crack tip and then to materials which exhibit extensive plastic deformation throughout the sample as well. The application to such materials is currently being actively pursued. Application to viscoelastic materials is the most difficult and at this point most inconclusive, with few useful engineering approaches yet developed. Most commonly, the viscoelastic behavior is neglected in applying fracture mechanics to polymers, and most experimental data can be satisfactorily analyzed by techniques which do not include the viscoelastic effects.

Whether preexisting cracks within a material will extend under an applied load depends on a variety of controlling factors. These will be discussed in more detail in later sections. It should be helpful at this stage to summarize the general behavior of cracks in any material. The configuration most commonly considered is a wide plate with a central crack.

1. For materials primarily elastic-plastic in behavior, if the sample is under plane strain, the cracked material will be stable; the crack will not increase significantly in length until a critical stress is reached. Some stable crack growth can be observed under these conditions, but it is usually not sufficiently large to warrant inclusion in the

analysis. At the critical stress the crack will propagate rapidly, increasing in length. Since the stress concentration at the base of the crack increases with increasing crack length, in most cases the crack will continue to extend until failure occurs. This extension will usually be very rapid under constant load. However, some loading and sample configurations exist which result in a decrease in stress intensity at the crack tip with increasing crack length, and in these cases the crack will extend slightly and then become stable again, requiring an increase in applied stress to further extend the crack. The most common example is in testing, in which the sample's length is fixed by maintaining the grip position fixed and/or using a tapered sample with increasing width in the direction of crack growth to increase the sample's resistance to crack growth as the crack progresses. Under those conditions the crack will begin to grow at a certain applied stress but will not continue to extend at that fixed load. For viscoelastic samples, the behavior under plane strain conditions suggests that at a certain minimum stress the crack begins to increase in length, but rapid propagation does not occur. Rather, slow growth occurs until failure at higher loads or later times.

2. If the sample is under plane stress, the crack will not extend until a critical stress is applied. Unlike the case of plane strain, rapid crack propagation will not occur. Rather, the crack will stop growing if the load is kept constant. For continued extension, the load must be increased. When the applied stress reaches a second high critical value, the crack will suddenly and catastrophically propagate. The longer the initial crack, the lower the stress at which crack growth starts (as expected for the higher stress concentration at the base of a longer crack), but the amount of crack growth before failure is greater than for shorter initial cracks, since the longer crack will fail at a lower stress.

3. The behavior cited above depends on the ambient conditions and generally is correct for dry air. However, many environments cause environmentally assisted cracking, and slow crack growth at stresses below the critical one for rapid crack extension will occur at constant stress due to the interaction of the environmental fluid and the molecules at the crack tip. This will cause slow crack growth until the critical crack size for rapid extension is reached.

4. The stability of a crack is also affected by varying loads. Under fatigue loading at sufficiently high stresses, crack extension occurs during the fatigue loading until the critical crack size for rapid extension is reached.

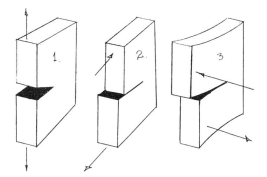

Figure 7.3. Modes of crack propagation.

Modes of Crack Extension

A crack in a sample as shown in Fig. 7.3 can extend in several modes, depending on the stress state existing in the material. Mode I results from normal stresses, mode II from in-plane shear, and mode III from out-of-plane shear, commonly called tearing. The most common form of crack extension in other than very thin sheets is mode I, and this will be considered in the following discussion.

Two general approaches have been taken to evaluate the effect of cracks on ultimate strength. The *energy balance approach* considers the changes in energy that occur within the sample as the crack increases in length. The *stress intensity approach* examines the details of the stress field around the crack and predicts crack extension based on a critical stress field intensity.

7.4. ENERGY BALANCE OF A CRACKED SAMPLE

The evaluation of energy and stress fields by the techniques of linear elastic fracture mechanics in samples containing cracks involves the implicit assumption that the response of the sample is controlled by the behavior at the crack tip. Plastic deformation processes are considered to occur only at the crack tip, and the size of the plastic zone at the crack tip is small compared to the crack size. As a result, the energy stored within the sample can be computed by considering that it is purely elastic in behavior, and growth of the crack will change the amount of elastic energy stored in the sample. In cases in which energy dissipation and plastic deformation occur through regions large compared to the region at the crack tip, other forms of analysis are required, leading to the *J*-integral discussed in later sections.

The fundamental approach is to calculate the energy input to the system (the material of the specimen) when an applied loading system increases a preexisting crack of length c by an infinitesimal amount dc. This energy can be compared with the energy absorption mechanisms within the sample which are activated by the crack growth. If the total energy input to the crack as it increases in length is greater than the energy absorption mechanisms within the sample, crack growth will be favored and failure will occur.

In its general form, if we consider a specimen containing a crack of length c, an extension of the crack will result in:

1. Work being done by the externally applied load if the sample elongates during crack propagation. Specimen elongation will occur under conditions in which the load is maintained constant during crack growth. Many configurations exist in which the ends of the same are fixed, resulting in no net work being done. The work per unit volume is given the symbol W.
2. A change in the elastic strain energy stored within the material per unit volume (S). In general, under fixed grip conditions, crack extension will relax the material around the crack, lowering the stored strain energy.
3. An increase in the kinetic energy (KE) of the parts of the sample surrounding the crack if the crack moves rapidly. This is frequently referred to as the crack kinetic energy. For slow-moving cracks, this term can be neglected.

These energy input terms are balanced by the resistance of the material to crack growth. This resistance is a result of the energy absorption mechanisms within the sample. This energy is the sum of the increase in surface energy of the sample and the energy irreversiby dissipated by viscoelastic or plastic deformation within the sample. This energy absorption per unit crack area increase is commonly called the fracture toughness or the fracture resistance of the sample (R). The second term will be used in this text.

The total energy change dU required for an infinitesimal crack extension is therefore the sum of three terms:

$$dU = dW - dS - dKE \tag{7.19}$$

The kinetic energy associated with the motion of the material around the crack if the crack velocity is high may be significant. It affects the ability of a moving crack to drive through obstructions such as the passage from one spherulite to another, or its may be converted into the formation of additional cracks, generating crack branching. However, the analysis is usually concerned with the behavior of the part while the crack is still stable and does not extend catastrophically. Therefore, in this case,

$$dU = dW - dS \tag{7.20}$$

When a crack propagates a change in energy results from the load moving and the change in stored internal energy. This energy release per unit of crack area increment, is defined as G, called the *energy release rate*. The rate is with respect to crack length, not time. Calculations are usually made based on a unit thickness of sample. Then G can be formally defined as

$$G = \frac{dU}{dc} = \frac{d(W - S)}{dc} \tag{7.21}$$

G is related to the energy available to cause the crack to propagate.

The fracture resistance R can be defined as the energy required to form an additional increment of crack:

$$R = \frac{dU}{dA} = \frac{d(U_{\text{surface}} + U_{\text{volume}})}{dc} \tag{7.22}$$

where A is the area of the crack.

The concepts of G and R are closely related, and the difference should be clearly understood. Physically, R refers to the resistance of the material to crack propagation, whereas G refers to the concept that in order for the crack to propagate, the energy of propagation must come from sources outside the crack. The usefulness of concentrating our attention on G rather than R at this stage is that G may be more readily experimentally determined. For very brittle materials R, the fracture resistance, appears to be primarily the result of the energy increase due to increasing the surface area of the crack. Therefore R is relatively independent of the size of the crack in that case. It is experimentally found to be independent of crack size when the sample is in a plane strain condition. For more ductile materials, the energy to deform the material around the crack tip predominates and may increase with increasing crack size, since larger deformation regions are generated. Under plane strain conditions, the extent of this deformation is independent of the crack size, and R can also be considered crack length independent. R can then be considered a property of the sample, and in terms of this physical parameter, the criterion for a crack to propagate is

$$G > R \tag{7.23}$$

The crack will be stable as long as the energy fed to the crack from the outside work and strain energy release is less than the fracture resistance of the material. The strain energy release rate G increases with increasing crack size for the sample being considered above (central crack in an infinite plate), and when G reaches a critical value, the crack no longer is stable and extends catastrophically.

7.5. GRIFFITH'S ANALYSIS OF GLASS FIBERS

This approach was originally taken by Griffith in studying the strength of glass fibers. The change in potential energy of an infinitely large plate under plane stress when a crack of length $2c$ develops within it is

$$\Delta U = \frac{\pi \sigma^2 c^2 t}{E} \tag{7.24}$$

Griffith (1921) investigated the wide variation in strength observed in glass samples and proposed that all materials contain small imperfections and cracks. These cracks are a result of manufacturing procedures, a result of the internal morphology of the material, or a result of surface oxidation and degradation. The exact cause of the defect is of less importance in the analysis of failure than the size and location of the defect within the sample. In his analysis of the failure of such cracked samples, Griffith concluded that although an existing crack will increase in length slightly under an applied stress in a brittle material, no sudden crack propagation and resultant failure will occur if the increase in surface energy due to the formation of the surface at the newly exposed crack edges is balanced by the change in potential energy due to the extension of the sample and the reduction in stored strain energy in the sample. The result of the analysis for a sample with a very sharp crack is that if the applied stress on a cracked sample is below a critical value, the crack is stable (the increase in surface area energy balances the decrease in potential energy). If the critical stress is exceeded, the crack becomes unstable and propagates.

Griffith equated the energy/area of new surface (R) of the crack generated as being equal to the surface energy γ of the material, since glass is so brittle that negligible plastic deformation occurs. Since two new surfaces are generated as the crack propagates, the criterion for crack instability (Eq. 7.23) is

$$\frac{-d\{\pi \sigma^2 c^2 t / E\}}{dc} > \frac{d\{4ct\gamma\}}{dc} \tag{7.25}$$

yielding

$$\frac{2\pi \sigma^2 ct}{E} > 4t\gamma \tag{7.26}$$

or

$$\sigma > \left(\frac{2E\gamma}{\pi c}\right)^{0.5} \tag{7.27}$$

This was derived for plane stress conditions. For plane strain, the modificaation is

$$\sigma > \left(\frac{1}{1-v^2}\right)\left(\frac{2E\gamma}{\pi c}\right)^{0.5} \tag{7.28}$$

Griffith introduced cracks of varying sizes into glass samples and obtained good agreement with this equation. Glass is extremely brittle, with little ability to deform plastically. Irwin (1949) recognized that for materials capable of plastic deformation, the deformation of the material around the crack tip requires considerably energy. Irwin introduced the concept of R and first suggested that γ should be replaced by R for the total energy required to form the new surfaces as the crack expands.

The Griffith equation does not include the radius at the crack tip. However, inherent in the calculation is the assumption that the crack tip is of the order of atomic dimensions. The equation was derived assuming that the material is completely brittle and that no plastic or permanent deformation occurred to absorb energy during the fracture process. This assumption is excellent for material such as glass, with which Griffith was originally concerned, but inaccurate for most materials and specifically for most plastics. To expand these concepts to such materials, Orowan (1950) developed the concept of incorporating the energy absorbed by plastic deformation in this equation and replacing the surface energy term by the sum of the energy required to form the new surfaces and the energy required to deform the material.

7.6. CRITICAL STRAIN ENERGY RELEASE RATE

G_c is the value of G at which failure will occur. Then Eq. 7.27 becomes for plane stress:

$$\sigma > \left(\frac{EG_c}{\pi c}\right)^{0.5} \tag{7.29}$$

and for plane strain:

$$\sigma > \left(\frac{1}{1-v^2}\right)\left(\frac{EG_c}{\pi c}\right)^{0.5} \tag{7.30}$$

Note that G_c refers to growth of the crack, so the factor 2 in Eq. 7.27, which refers to two new surfaces being generated, is no longer present in Eq. 7.29.

G_c is commonly called the *critical strain energy release rate*, since the derivation indicates that the crack growth requires strain energy to be transferred to the surface energy and deformation processes of the growing crack. Note that by inverting this equation, knowledge of the stress at which a cracked sample fails permits us to calculate G_c. For the centrally cracked plate in plane strain the energy release rate can be calculated from Eq. 7.30 to be

$$G = \frac{(1 - v^2)^2 \sigma^2 \pi c}{E} \tag{7.31}$$

The relation between G and R can be visualized by plotting these values vs. the crack length. The fracture resistance R must depend on the energy required to produce the deformation about the crack tip and the energy to develop microvoids around the crack tip. The fracture resistance can be approximated as independent of the crack length in plane strain conditions, since the size of the plastically deformed region around the crack tip in this case is independent of the crack length. It is therefore shown in Fig. 7.4

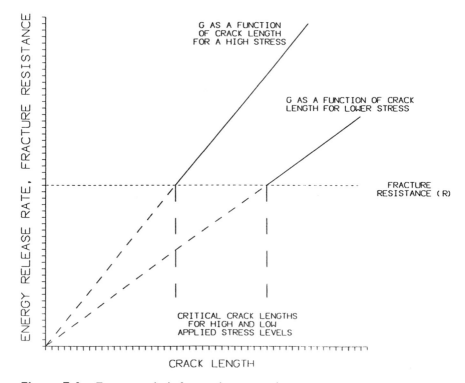

Figure 7.4. Energy analysis for crack propagation.

as a horizontal line. The energy release rate G for two different stress levels as a function of the crack length is shown as two straight lines, with greater slope for greater stress level. When the crack is sufficiently large that the energy release rate is greater than the fracture resistance, the crack will become unstable and will propagate. This will occur at the length shown in the figure. The higher the stress, the smaller the crack length at which the energy release rate exceeds the fracture resistance of the material.

For plane stress conditions, R must increase with increasing crack length, since increasing stress must be applied to continue growth of a crack. The necessity to increase the stress must be due to internal mechanisms within the sample which increase the resistance to fracture. The experimental data suggest that qualitatively, the R vs. crack length curve looks somewhat logarithmic (Fig. 7.5). At low stresses the applied G value is too small to produce any measurable crack extensions. As the stress increases, the crack grows until R reaches the value necessary to balance the applied energy release rate. Finally, if the stress is continually increased the value of G

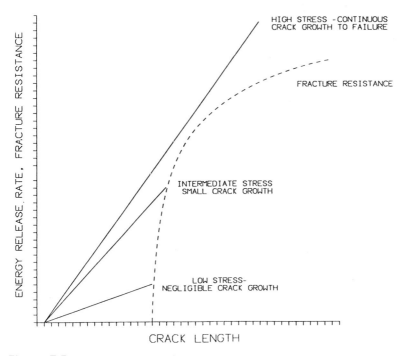

Figure 7.5. R-curve analysis for crack propagation.

exceeds the value of R possible, and the crack continually grows to cause failure (see Fig. 7.5).

Determination of G_c

The question now remains how best to determine G_c experimentally. This determination is based on the recognition that the area under the load-extension curve is the strain energy stored in the sample. The analysis proceeds by evaluating:

1. The strain energy change when a crack propagates
2. The work the external load performs on the sample

If we neglect the kinetic energy, the strain energy release rate may be experimentally found from the determination of the crack length and the specimen extension as a function of the applied load. Consider a specimen with a crack present (with length c) under an applied load. If the crack elongates by an amount dc, then the change in energy of the system will be a result of the energy transferred from the outside load, the change in strain energy of the sample, and the energy absorbed in the sample in creating the additional crack area (R).

Since the specimen is considered elastic, the variation of load with elongation can be taken as linear. If the elongation of the sample is δ, the work produced by an applied load of P causing an extension of the sample of $d\delta$ is $Pd\delta$.

Since in LEFM the $P - \delta$ plot is linear, the strain energy increase in the sample during an elongation δ is

$$\Delta U = \frac{P\delta}{2} \tag{7.32}$$

For an infinitesimal change in load:

$$dU = \frac{d(P\delta)}{2} = \frac{Pd\delta}{2} + \frac{\delta dP}{2} \tag{7.33}$$

If during this elongation a crack present in the sample grows, part of this energy must be associated with the crack development. From the definition of G_c, the energy required to create a new crack of area dA is $G_c dA$. Equating the energy input from the external load to the increase in strain energy and crack energy:

$$Pd\delta = \frac{Pd\delta}{2} + \frac{\delta dP}{2} + G_c dA \tag{7.34}$$

Therefore:

$$G_c dA = -\frac{(\delta dP - Pd\delta)}{2} \tag{7.35}$$

Multiply by P^{-2} to convert to a perfect differential:

$$\frac{G_c dA}{P^2} = -\frac{(\delta dP - Pd\delta)}{2P^2} = -\frac{d(\delta/P)}{2} \tag{7.36}$$

Therefore:

$$G_c = -\frac{P^2 d(\delta/P)}{2dA} \tag{7.37}$$

but δ/P is the compliance of part. Note that compliance in fracture mechanics is the reciprocal of the slope of the load-extension curve, rather than the reciprocal of the slope of the stress-strain diagram, as defined in other areas of mechanics. In principle G_c can be found from the compliance change of a sample as the crack extends.

A difference in sign will appear depending on the constraints imposed on the system. Under constant load, the strain energy of the system will increase and the energy being transferred to permit crack growth will also come from the external load. Under fixed grip conditions, if the crack grows, the energy to drive the crack growth comes from the strain energy, which will decrease. The sign in Eq. 7.37 will therefore depend on the particular conditions of the test.

It is possible either to use an experimentally obtained compliance vs. crack area curve obtained for different crack sizes or to obtain an analytical expression for this dependence. For example, consider a split test sample (double cantilever beam) as shown in Fig. 7.6.

For a cantilever beam, the deflection $\Delta = PL^3/3EI$, where $I = th^3/12$. Since for the split sample, there are two cantilever beams,

$$\delta = \frac{2PL^3}{3EI} \tag{7.38}$$

$$\frac{\delta}{P} = \frac{2L^3}{3EI} \tag{7.39}$$

When the crack advances a length dc, the length of the cantilever beam increases from L to $L + dL$. Therefore:

$$dA = tdc = tdL \tag{7.40}$$

$$\frac{d\{\delta/P\}}{dA} = \frac{d[2L^3/3EI]}{tdL} = \frac{2L^2}{EIt} \tag{7.41}$$

Figure 7.6. DCB (double cantilever beam) sample.

Therefore at the point where the crack begins to propagate (the critical or maximum load):

$$G_c = P^2 \frac{d\{\delta/P\}}{2dA} = \frac{P_{max}^2 L^2}{EIt} \tag{7.42}$$

and G_c can be found by a determination of the load at which crack propagation occurs. However, specific experimental details must be followed to obtain an acceptable result. These are discussed later.

Graphical Presentation of *R*

For a typical split beam sample (double cantilever beam), a load-deflection diagram is shown in Fig. 7.7. The reciprocal of the slope is the compliance required in Eq. 7.37. The longer the initial crack, the more compliant the specimen and the lower the slope of the curve. Thus, it is apparent that the slope of this curve for cracked samples is not directly related to the elastic modulus of the material. The area under the curve to a specified elongation is the strain energy stored in the sample. If the crack length is c_i, then the strain energy within the sample is the area $OA\delta_i$, where δ is the total deflection of the ends of the beams. For simplicity, consider the crack propagating at a constant load until the crack increases in length to c_f. As a result of the increased crack length, the compliance of the sample has decreased to the line OB, and the stored strain energy has decreased and is given by the area $OB\delta_f$. However, the external load has done work $P(\delta_f - \delta_i)$. The reduction in strain energy plus the external work must be

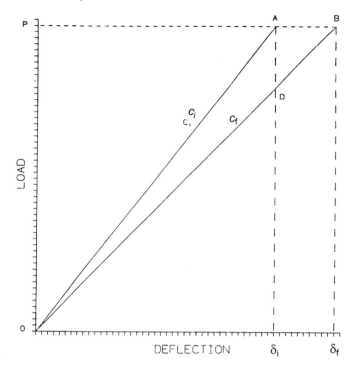

Figure 7.7. Load extension diagram for a sample with two different crack lengths. Increasing crack length lowers the compliance of the sample.

transferred to the energy for the propagation of the crack, and their sum must be equal to the work required to extend the crack. Therefore, the total crack extension energy is

$$R(A_f - A_i)$$ (7.43)

where R = energy per unit area of crack
A_f = crack area after extension
A_i = crack area before extension

$$R(A_f - A_i) = \text{area}(OA\delta_i) + \text{area}[AB(\delta_f - \delta_i)] - \text{area}(OB\delta_f) = \text{area}(OAB)$$

Therefore:

$$R = \frac{\text{area}(OAB)}{(A_f - A_i)}$$ (7.44)

This expression has been derived for a sample under a fixed load. If the specimen is maintained in rigid grips, so that the total deflection remains fixed, the load will drop from A to D under the crack extension previously considered. The work to extend the crack in this case is OADA. The difference in the case of constant load and constant extension is only ADB, which approaches zero for an infinitesimal crack extension.

Rubbery Materials

Rivlin and Thomas (1953) first applied these concepts to a polymer, studying cross-linked rubber materials. Rubbery materials are elastic but extremely nonlinear in their general behavior. They studied samples with various types of configurations and, in line with rubber technology terminology, called functions similar to G_c the tearing energy T. The sample type they used was the single-edge crack specimen. This contains short edge crack in a sheet sample. If the sample contains no crack, the strain energy is S_0. It can be shown that the introduction of the crack reduces the strain energy by $\pi c^2 b S_0 / \lambda^{0.5}$, where b is the thickness of the sample and λ is the extension ratio (final length/initial length). Differentiating to obtain the change in energy with crack length:

$$\frac{dU}{dc} = -\frac{2\pi c b S_0}{\lambda^{0.5}} \tag{7.45}$$

Since $G_c = dU/dA$, and $dA = -bdc$,

$$G_c = \frac{2\pi c (S_0)_{\text{crit}}}{\lambda^{0.5}} \tag{7.46}$$

7.7. STRESS INTENSITY

Considerable effort has been expended on studying the stresses developed at the crack tip. For materials which deform elastically around the crack, and in the vicinity of the crack tip, analysis can be performed quite accurately. For materials which exhibit some plastic deformation, the analysis is more complex, and some approximations are necessary. The procedure in general is to add some relatively simple corrections to the solutions found for the elastic case. Viscoelastic effects in plastics add to the complexity, to such an extent that they are frequently ignored. When not neglected, they are also handled by making modifications to the linear model.

Irwin (1964) modified Westergaards (1939) calculation of the stress and strain near the tip of a crack. The general form of this equation for all three

Cartesian coordinate components is

$$\sigma_{x,y,z} = \frac{K}{\sqrt{2\pi r}} f(\theta) \tag{7.47}$$

where r = distance from the crack tip
θ = angular direction from the crack
K = (the stress intensity) a factor that contains the dependence on the applied stress and the geometry of the sample and the crack

This equation for the stress intensity near a crack tip applies only close to the tip, since the equation degenerates to a zero stress as r approaches infinity. Other terms exist but are not important for our purposes.

If two samples in plane strain of the same material in the same environment have the same stress intensity (K) at the crack tip, the cracks will behave in an identical fashion, either being stable or propagating at the same rate. For plane stress, identical K values and *identical thickness* will give similar crack behavior. The reason for this difference is that for K to completely control the crack's behavior, the plastic deformation zone must be very small, since the derivation assumed elastic behavior. If there is significant plastic deformation, then K alone does not determine the stress distribution. This is the condition in which plane stress occurs, as the thickness also affects the stress distribution. When extensive plasticity occurs, other approaches (the J-integral) must be used.

From the Griffith equation and the elastic analysis of Westergaard (1939), it can be concluded that the magnitude of the stress intensity is of the form

$$K = Y\sigma\sqrt{\pi c} \tag{7.48}$$

where the value of Y depends on the specific geometry of the part and the location of the crack within the part and its relation to the direction of applied stress. This geometric factor has been evaluated for many common configurations expected for cracks in materials. In addition, many special configurations have been developed for testing to permit experimental determination of K from crack propagation studies. For a crack located centrally in an infinitely wide plate, the equation takes an especially simple form, since $Y = 1$. This special case therefore is often used as a simplifying example in discussions of material behavior.

7.8. FRACTURE TOUGHNESS

Equation 7.48 gives the stress intensity values for a particular crack and applied stress. Materials differ in their ability to blunt the crack tip and to provide alternate mechanisms for energy absorption which will prevent the crack from propagating at low stresses. When the applied stress and crack size are so great that the K value (stress intensity) exceeds a critical value of K_{Ic} for a mode I crack type, the material will fail by rapid crack propagation. Therefore K_{Ic} is a material property and a measure of the material's ability to withstand the propagation of a crack already present within the material. Fracture is considered to occur when the K_c value of the material is exceeded as a result of the applied load causing the preexisting crack to run through the material.

7.9. EFFECTIVE CRACK SIZE

Irwin (1958) stated that the plastic zone at the crack tip increases the effective size of the crack, because the crack blunts as a result of this plastic deformation. A larger size crack would be required to extrapolate the shape to a crack of zero width. More importantly, the stress in the plastic zone is limited to σ_{yield}, whereas in an idealized crack the stress would go to infinity at the crack tip. Assuming a circular plastic zone, Irwin concluded that the effective crack length should be:

$$c_{eff} = c + r_p \tag{7.49}$$

where r_p = radius of the plastic zone.

Plastic Constraint

The stress required to cause yielding at the crack tip can be computed for plane stress and plane strain by using a yield criteria. For the Von–Mises criteria:

$$(\sigma_1 - \sigma_2)^2 + (\sigma_2 - \sigma_3)^2 + (\sigma_3 - \sigma_1)^2 = \sigma_y^2 \tag{7.50}$$

From the equation of the stress field around the crack tip for plane strain:

$$\sigma_1 = \frac{K}{\sqrt{2\pi r}} \cos(\theta/2)[1 + \sin(\theta/2)] \tag{7.51}$$

$$\sigma_2 = \frac{K}{\sqrt{2\pi r}} \cos(\theta/2)[1 - \sin(\theta/2)] \tag{7.52}$$

$$\sigma_3 = v(\sigma_1 + \sigma_2) \tag{7.53}$$

For the direction along the crack $\theta = 0$

$$\sigma_1 = \frac{K}{\sqrt{2\pi r}} \qquad (7.54)$$

$$\sigma_2 = \frac{K}{\sqrt{2\pi r}} \qquad (7.55)$$

$$\sigma_3 = 2v\sigma_1 \qquad (7.56)$$

Typically for glassy polymers $v = 0.33$. Substituting these values into the yield criteria results in

$$\sigma = 3\sigma_y \qquad (7.57)$$

or the actual stress required to cause yielding is three times that required in a uniaxial tensile test.

For plane stress along the crack direction:

$$\sigma_1 = \frac{K}{\sqrt{2\pi r}} \qquad (7.58)$$

$$\sigma_2 = \frac{K}{\sqrt{2\pi r}} \qquad (7.59)$$

$$\sigma_3 = 0 \qquad (7.60)$$

and substituting into the yield criteria:

$$\sigma = \sigma_y \qquad (7.61)$$

Using these values for the size of the plastic zone r_p for plane strain

$$3\sigma_y = \frac{K}{\sqrt{2\pi r_p}}$$

or

$$r_p = \frac{K^2}{18\pi\sigma_y^2} \qquad (7.62)$$

and for plane stress

$$\sigma_y = \frac{K}{\sqrt{2\pi r_p}}$$

$$r_p = \frac{K^2}{2\pi\sigma_y^2} \qquad (7.63)$$

Then:

$$\frac{(r_\text{p})_\text{plane stress}}{(r_\text{p})_\text{plane strain}} = 9$$

The plastic zone size in plane stress is much larger than in plane strain. Although plane strain cannot exist through the whole sample and the surface must be in plane stress, the value of the average constraint suggested by Irwin (1960) which is most appropriate is close to that for plane strain. Dugdale (1960) concluded that the sample could be considered to have an extended crack running through the plastic zone, but part of the extended crack is closed by stresses equal to the yield stress being applied. These are compensated by a wedge force tending to open the crack. He arrived at

$$r_\text{p} = \frac{\pi K^2}{8\sigma^2}$$

These derivations made simplifying assumptions about the redistribution of stresses outside the boundary region close to the crack. More accurate analyses have been performed, requiring much more complex calculations. Tuba (1966) derived a stress distribution which rather than circular, exhibited the maximum extent of the plastic zone to be at a 69° angle to the plane of the crack. Experimental observations of the shape of the plastic zone and the fracture surfaces emanating from a fatigue crack by Hahn and Rosenfield (1965) indicate that the shape of the plastic zone calculated by Tuba is closest to the observed shape. However, the calculated values from Eq. 7.62 and Eq. 7.63 for the diameter of the plastic zone come reasonably close to predicting the maximum extent of the plastic zone. As a result, these simpler expressions are most commonly used. We account for the change in effective crack length by replacing the crack length c by $c + r$. Th is produces some complications, as K is a function of the crack length for most geometries, and iteration must be used in its evaluation.

The simplest geometry is that of an infinite plate with a central crack. The Y value is 1 for this case, and no iteration is required:

$$K = \sigma\sqrt{\pi c}$$

or with the inclusion of the plastic zone:

$$K = \sigma\sqrt{\pi(c + r)} \tag{7.64}$$

Substituting Eq. 7.63 into Eq. 7.64:

$$K = \sigma \sqrt{\pi \left(c + \frac{K^2}{2\pi\sigma_y^2} \right)}$$

and

$$K = \sigma \sqrt{\frac{\pi c}{1 - 0.5(\sigma/\sigma_y)^2}} \qquad (7.65)$$

When the applied stress is small compared to the yield strength, the plastic zone scarcely changes the effective size of the crack and the stress intensity. This would be the condition for a brittle material, where the yield and the failure stress are very close, and the applied stress must be kept well below the failure stress to avoid propagation of any cracks present in the sample. When the applied stress equals the yield stress, the plastic zone increases the stress intensity by 41%. Of course, under these conditions general yielding would occur, and the fundamental postulate of elastic behavior would be violated. For the fracture mechanics formulae to be applicable, the plastic zone must be small compared to the crack size.

7.10. PLANE STRAIN AND PLANE STRESS

The surfaces of a plate sample perpendicular to the crack front are necessarily stress free, and for a thin sample it can be assumed with reasonable accuracy that the stress in the direction through the plate (the z axis) is uniformly zero. This is the plane stress condition, under which the stress variation is restricted to the plane of the plate.

For the other extreme of a thick plate, the stress in the z-direction must still be zero at the surface. However, the plastic deformation of the material at the crack tip within the sample in the direction of the applied stress leads to elastic contraction in the transverse directions, calculable from Poisson's ratio. This contraction is resisted by the unstrained material near the surface. This is termed plastic constraint of the material, as is the constraint to deformation in the necked region of a tensile specimen or in a notched tensile specimen. This constraint is sufficiently severe in thick samples that it is reasonable to consider that no deformation occurs along the z-axis parallel to the crack face, and all deformations are restricted to the plane of the plate, producing a plane strain condition. This constraint generates stresses along the z-axis, producing a triaxial stress condition.

Plane Strain–Plane Stress Determination

Since there must be plane stress at the surface (no stress can exist perpendicular to a free surface), every sample has such a plane stress region. This is experimentally observable, as thin plate samples fail in tension with shear lips (at 45° to the tensile axis) running in from the surface. As the specimen width gets larger, the thickness of the shear lip reaches a maximum value, and the center of the sample after failure can be observed to contain an essentially flat region perpendicular to the applied tensile stress, with the shear lips existing on either side. For every thick samples, the major part of the sample fails with flat fracture surfaces, but the shear lips on both surfaces still exist. Figure 7.8 shows this behavior schematically as a function of thickness.

The thickness of the sample at which plane strain begins to be developed depends on the relative size of the plastically deformed region at the crack tip which generates the triaxial stress and the sample thickness. If the diameter of the plastically deformed zone is larger than the thickness of the plate, then lateral contraction is not hindered, and plane stress is developed since $\sigma_3 = 0$. The ratio of the specimen thickness to the diameter of the plastic zone therefore determines the transition from plane stress to plane

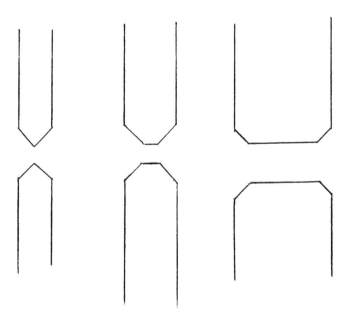

Figure 7.8. Shear lips and flat fracture for samples of different thicknesses.

strain. Since the radius of the plastic zone r_p is proportional to K_I^2/σ_y^2, to obtain plane strain the sample thickness B must satisfy a relation of the type

$$\frac{B}{K_{Ic}^2/\sigma_y^2} > \text{some value} \tag{7.66}$$

This value has been experimentally studied by Brown and Strawley (1966), among others, and it was found that for samples with thickness B satisfying the condition

$$B > 2.5\left(\frac{K_{Ic}}{\sigma_y}\right)^2 \tag{7.67}$$

then plane strain will exist through most of the sample. Similar reasoning applies to the crack size to have the plastic region small compared to the crack itself.

Practically, we cannot conclude directly from the dimensions of the sample whether plane strain or plane stress exists, but must use Eq. 7.67 to compare the size of the plastic zone around the crack tip to the constraint developed as a result of the thickness of the sample. Similarly, if the ratio in Eq. 7.66 is less than some value, plane stress can be considered to exist throughout the sample, and the fracture will be exclusively of the slant type. For intermediate values (the transition zone), a combination of conditions will exist, with significant regions of plane strain at the central portion and plane stress at the surfaces.

Figure 7.9 shows the typical variation of the experimentally determined K_c values for samples of different thicknesses. The fracture toughness, or ability to resist crack propagation, decreases with decreasing volume of material capable of deforming and therefore decreases as the sample thickness increases. Samples sufficiently thin can be considered to be in plane stress; the radius of the plastic zone is that specified by Eq. 7.63, and the toughness is high. As the thickness of the sample increases, plastic constraint begins to reduce the volume of plastically deforming material, and the fracture toughness decreases with increasing thickness until the plane strain conditions apply and plastic constraint has reached its maximum value. It is of interest to note that the variation in fracture toughness measured between thin and thick pieces may vary by as much as a factor of 3. Since the fracture toughness values measured with samples sufficiently thick are independent of sample thickness, this fracture toughness value can be considered as a true material property in any design calculation. The procedures for measuring fracture toughness which are well established therefore involve utilizing samples of sufficient thickness that the

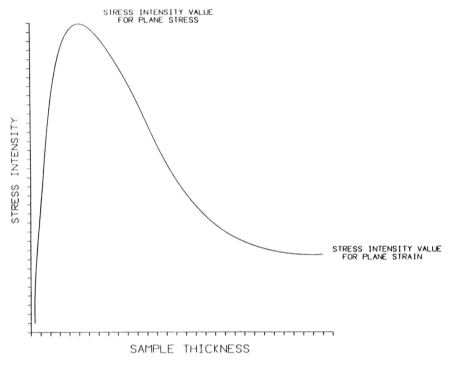

Figure 7.9.   Effect of specimen thickness on the calculated K_{Ic} values for samples of various thicknesses.

result is independent of thickness, and the results are so reported as K_{Ic} values.

Effect of Part Size

It should be noted that large structures are capable of storing large amounts of strain energy. Whereas this energy absorption ability may prevent failure in a large part under sudden load application, if flaws are present, under impact loading this elastic stored energy may be of such a large magnitude that explosive types of failures may be observed. Groshans and Takemori (1987) simulated large compliant structures by putting steel springs in series with compact tension fracture toughness samples of rubber toughened polybutylene terephthalate/polyvinyl chloride (PBT/PC) blends. They compared the behavior of normal test samples with those with the energy-absorbing springs in the system. They observed that at room temperature at low testing rates the standard samples fractured in a ductile and stable manner. The energy-absorbing system, however, exhibited initial ductile crack growth which was followed by ductile but unstable failure. This was attributed to the inability of the crack to absorb all the stored elastic energy within the system which was released by the initial crack growth.

7.11. CRAZE INITIATION

Williams and Marshall (1975) used the equation for the size of the region at the tip of a crack which experienced large stresses (the plastic zone) and equated this region to the region with sufficient stress to initiate craze growth. Replacing the yield stress for elastic materials by the craze initiation stress in the Dugdale equation the radius of the volume of material at the crack tip in which crazing occurs is

$$R = \frac{\pi}{8} \left(\frac{K}{\sigma_{craze}} \right)^2 \qquad (7.68)$$

where σ_{craze} is the craze initiation stress.

Since the craze initiation stress decreases with time with an exponential decay with an equation of the form $\sigma_{init} = ct^{-n}$, then:

$$R = c'K^2 t^{2n}$$

Excellent agreement with this equation has been found, signifying that stress intensity and viscoelastic stress relaxation both affect craze growth.

7.12. DETERMINATION OF K_{Ic}

The specification for the determination of K_{Ic} has been established in ASTM test E399. This test method was developed for metals, but specific test procedures have not been specified for plastics. To develop plane strain, as required for the determination of K_{Ic}, thick samples are required. For many plastics, the specimen sizes and procedures specified are difficult to obtain, and much of the published data has been produced by procedures which do not strictly conform to this standard. However, a significant body of data has been obtained on the fracture toughness of plastics, and the consensus is that such information is of value in design of plastic components which tend to be relatively brittle.

In the ASTM test, a crack is machined into the sample and then sharpened by fatiguing the sample until a fatigue crack propagates from the tip of the machined crack. This assures a sharp, reproducible crack tip. One common shape utilized for metal samples is that of the compact tension sample. A gage is attached at the machined edge of the crack to measure the opening dimensions as the load is increased. One of the difficulties in fracture toughness measurements is the requirement that the sample be sufficiently thick to assure that true plane strain conditions are met during the test. In plastics, the conditions necessary for this to be true have not been stringently tested. However, it is difficult to obtain samples sufficiently thick for the compact tension sample geometry to be utilized. The very general condition of a sample with a carefully defined crack is frequently used. The crack is often generated by pressing a sharp razor blade into the sample. With respect to the specification of sample thickness, the condition which has been established for metals is considered to be a satisfactory one for plastics by Brown and Strawley (1966) (Eq. 7.67)

$$\text{Thickness and crack length } (B, c) \geq 2.5\left(\frac{K_{\text{Ic}}}{\sigma_{\text{yield}}}\right)^2$$

Comparing these dimensional requirements to the size of the plastically deformed region at the crack tip, it can be seen that this corresponds to both thickness and crack length being more than 50 times the radius of the plastically deformed region around the crack tip, certainly suggesting a thick sample, which should eliminate strain in the z-axial direction.

Expected load-displacement data are shown in Fig. 7.10 for materials with limited ductility as described in the ASTM standard. The procedure for determining K_{Ic} is summarized below.

1. Determine the tangent to the load-displacement curve at the origin.
2. Draw a line from the origin with a slope of 5% less than that determined in step 1, until it intersects the curve at point P_q.
3. Choose the load for further computation as P_q or any higher value of load recorded at a smaller displacement.
4. Discard the run if the higher value of load is 10% greater than P_q.
5. Compute $K = Y\sigma\sqrt{\pi c}$, with the Y value for the sample configuration.
6. With this K value, ascertain whether the initial thickness and crack length are sufficient to assure plane strain during the test by Eq. 7.67.
7. If the conditions for thickness and crack length to produce plane strain are satisfied, this resultant K value is reported as K_{Ic}. If not, thicker samples must be used and the test reported.

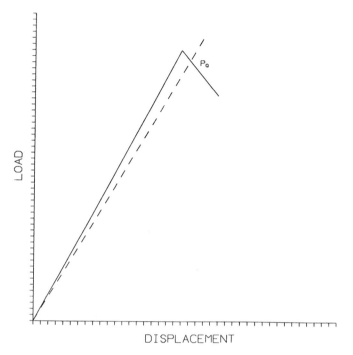

Figure 7.10.   Determination of K_{Ic} for samples with low ductility.

7.13. EMPLOYMENT OF FRACTURE MECHANICS IN DESIGN

As stated previously, the fundamental axiom of fracture mechanics is that all brittle materials inherently contain defects or cracks of various sizes. When the applied load and crack length produce a stress intensity K greater than the fracture toughness of the material, the material will fail in a brittle manner by rapid crack propagation. The maximum crack size inherent in the material can be determined by inspection. If inspection procedures are capable of observing only imperfections greater than c, then it should be assumed that cracks of this size exist in the sample. The assumption of a smaller crack size would be unsafe. The design load would then be set to avoid brittle failure:

$$\sigma_{\text{design max}} < \frac{K_{Ic}}{Y\sqrt{\pi c}} \tag{7.69}$$

where c would be determined by the inspection criteria.

For these concepts to apply in the design process, the material must be brittle and crack propagation a likely cause of failure. The majority of polymeric materials are very ductile, and it is not customary to incorporate fracture mechanics concepts in most designs. Nevertheless, for many plastics such as PS and PMMA, brittle failure is possible, and these considerations should be included in the design process.

7.14. THE *J*-INTEGRAL

Both K_{Ic} and G_{Ic} have been used extensively for glassy polymers, since the requirements for the size of samples in which the plane strain conditions necessary for evaluation can be readily met. In such relatively brittle plastics the energy dissipation during failure is primarily within the region localized around the crack tip. The remainder of the sample is elastic, and the energy storage within that part is purely elastic strain energy. When the sample behaves primarily in an elastic manner, the energy release rate G and the fracture toughness K can be computed from the elastic stress field. However, for the tougher plastics, sample thicknesses become so large that the required testing is impractical. Furthermore, the deformation is a result of strongly nonlinear behavior, and the behavior of the sample as a whole may be nonelastic. An additional condition which prevents the application of G or K is the total extent of deformation throughout the sample. If there is plastic deformation of the sample, rather than simply at the crack tip, then the criteria for the use of these factors are not satisfied. Because the energy to propagate the crack is also affected by the energy required for plastic deformation, both at the crack tip and throughout the sample, the difficulty in evaluating such materials has led to consideration of the *J*-integral developed by Rice. Although there has been study and use of the *J*-integral for metals, it is being increasingly considered for plastics. The *J*-integral is formally defined as a line integral, but the form most useful for our analysis is based on the fundamental concept that J can be considered as the change in potential energy of a cracked sample as a result of an infinitesimal increase in the crack length per unit thickness of sample:

$$J = -\frac{1}{B}\frac{\partial U}{\partial c} \tag{7.70}$$

where B is the sample thickness, U the total internal energy, and c the crack length. Whether the differential is evaluated at constant load ($P = \text{constant}$) or fixed grip condition ($u = \text{constant}$) is a second-order effect, as can be seen in Fig. 7.11. The load-deflection diagram of a nonlinear material with two cracks of length c and $c + dc$ is shown. The difference in the area under the

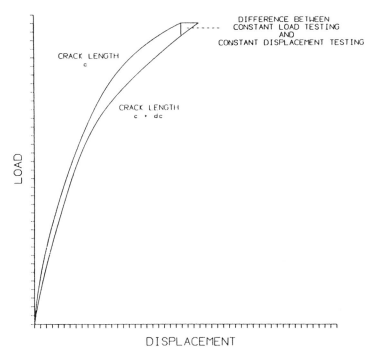

Figure 7.11. Evaluation of the *J*-integral as change in area under the load-displacement curve. The difference between constant load and constant strain conditions is negligible.

curves is *dU* (analogous to the evaluation for an elastic material). A restriction to constant total deformation *u* is not required, but it results in the work done during crack growth being completely a result of the increase in energy of the crack, rather than including the mechanical work performed in elongating the sample.

This equation therefore corresponds to the definition of *G* for a linear plastic material per unit thickness of sample. The fundamental difference is that the total internal energy being considered in the evaluation of *J* will include plastic deformation, nonlinear effects, and viscoelastic deformation energies as well as elastic strain energy. The value of *J* and the equations developed for its evaluation all therefore extrapolate to the values of the equations for *G* for purely elastic behavior. To emphasize the presence of both elastic and plastic components of the total internal energy, *J* can be written as

$$J = J_e + J_p$$

where J_e is the elastic component and J_p is the plastic component. As for the other fracture parameters, failure is considered to occur when the value of J reaches a critical value J_{Ic}.

Evaluation of the J-Integral

Based on this definition, J could be calculated from Eq. 7.70. Such a procedure would require load-displacement curves for a series of samples with increasing crack sizes.

For polymeric samples, the crack is usually initiated by pressing a razor blade into the sample, since sharp crazes tend to run directly from the small initiated crack. Some polymers do not develop this sharp craze, and fatigue loading to generate a sharp crack may be necessary. The value of the total strain energy of a cracked sample is equal to the area under a load-deflection plot. If two experimental curves are obtained for samples with slightly different crack lengths, then the difference in area under the curves would correspond to the total loss of potential energy produced by the extension of the crack, corresponding to dU, and the J value calculated as $-dU/dc$. The J-values can then be plotted as a function of crack length for fixed grip displacement. The value at which propagation of the crack occurs is J_{Ic}. However, the ASTM procedure is based on the following analysis. Since:

$$J = -\frac{1}{B}\frac{dU}{dc}\bigg|_{u=\text{const}} = -\frac{1}{B}\frac{d}{dc}\left(\int P\,du\right) = \frac{-1}{B}\int\left(\frac{dP}{dc}\right)du \qquad (7.71)$$

During the three-point bend test, for a sample which exhibits large amounts of plasticity, when the sample is loaded the entire region behind the crack becomes plastically deformed, the specimen bends in a manner similar to a hinged one, and the area directly behind the crack in this test is called a *plastic hinge*. For rotation around the hinge, the applied load to cause bending must be of the form

$$P = b^2 f$$

where $b =$ length of the uncracked sample
$f =$ some function of the modulus, the total deflection of the sample, the length of the sample, and the stress-strain characteristics of the material

Therefore:

$$\frac{\delta P}{\delta c} = -\frac{\delta P}{\delta b} = 2fb = \frac{2P}{b}$$

Substituting into Eq. 7.71 above

$$J = \frac{2\int P\,du}{Bb} \qquad (7.72)$$

The $\int P \, du$ is the area under the load-deflection curve. Therefore, with A being the area under the load-deflection curve:

$$J = \frac{2A}{Bb} = \frac{2A}{B(W - c)} \tag{7.73}$$

where W = sample width. Thus the value of the J-integral can be directly determined from the area under the load-deflection diagram of a cracked sample.

The exact test method for determining J_{Ic} is specified by ASTM E813. As with all fracture toughness testing, the test procedure requires rigorous adherence to the details to produce a result which meets all the criteria of the standard. Either tension or bend samples may be used, and the dimensions for both are given in the standard. The objective of the test procedure is to determine the value of J near the point where crack growth initiates. To obtain this point, load-displacement curves are obtained for samples with different crack lengths and the J-integral (the energy required to extend the crack) is found from Eq. 7.73 (Fig. 7.12a). The J values are then plotted vs. crack length (Fig. 7.12b). The curve of J vs. crack length is fit by a power law equation, and the point on this curve corresponding to a crack extension by propagation of 0.2 mm is chosen. This point is found by drawing the *crack blunting curve*, which approximates the extent of opening of the crack by stretching on the same graph. The crack is initially sharp, but plastic deformation during loading causes some blunting and a small extension before crack propagation occurs. From experimental data and some rather arbitrary decisions, the equation for this relation between J and the increase in crack length Δc and the yield strength of the material σ_y due to this blunting is given by

$$J = 2\sigma_y \Delta c \tag{7.74}$$

Figure 7.12b shows the procedure. A line is drawn parallel to the crack blunting curve, but at a crack opening of 0.2 mm greater. The intersection of the J vs. crack length curve and this curve is considered the value of J at which true crack extension begins. This extension in polymeric materials is not the catastrophic one associated with crack extension in elastic materials in plane strain. Rather, it is generally a slow stable increase in length requiring additional load to continue the elongation. In materials with large plastic deformations, this behavior is frequently called tearing. This value of J is therefore J_{Ic}, if all other requirements of the test are satisfied. A simplified summary of the test procedure for compact tension (CT) samples is given below.

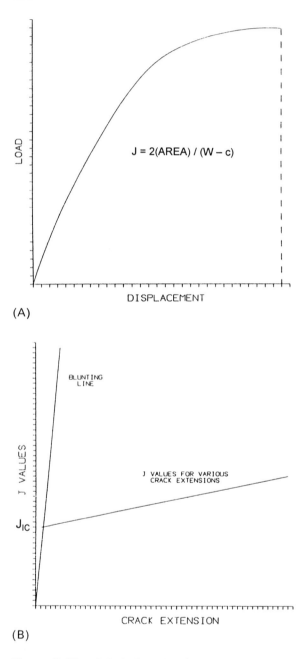

$$J = 2(\text{AREA}) / (W - c)$$

(A)

BLUNTING
LINE

J VALUES FOR VARIOUS
CRACK EXTENSIONS

J_{IC}

(B)

Figure 7.12. Calculation of J_{Ic} from load-displacement data. (A) Evaluation of J from load-displacement curve. (B) Determination of J_{Ic}.

1. The samples (a minimum of five for the multiple-sample technique is recommended) are machined, and a fatigue crack is introduced to extend a machined notch at a load based on the yield strength and dimensions of the sample given in the standard.
2. The sample is loaded until the crack extends, the load and deflection of the sample being measured. For metals the length of the fatigue crack is measured by indirect techniques such as heat tinting. For plastics, which are transparent, the crack length can usually be measured directly.
3. The area under the load-deflection curve is the energy required to extend the crack and is the value of J for that crack length.
4. In the multiple-sample technique, this procedure is repeated with different samples until at least four data points are taken, with crack extensions running between 0.15 and 1.5 mm greater than that given by the crack blunting curve.
5. The J values obtained are then plotted vs. the crack extension values.
6. The crack blunting curve is drawn on the same graph with the equation $J = 2\sigma_{yield}\Delta c$. The intersection of the J curve and a line drawn parallel to the blunting curve but for 0.2 mm greater crack length is the estimate of the J_{lc} value. If the conditions of the sample size and number of samples falling within the acceptable crack extension limits are satisfied, this is accepted as the experimental value.

Application of the J-Integral

The determination of J_{lc} requires careful and extensive testing. This value corresponds to the condition at which a crack present in the tough material will begin to grow stably, but further extension will require higher loads and additional energy input. The test does not permit determination of a J value or any other criteria for evaluating the loading conditions under which the crack has grown to such an extent that the sample will then fail! As a result, although extensive J testing is being performed on polymers, these data are rarely, if at all, used in design of polymeric parts.

The computational procedure for J analysis of a component is a part of most elastic-plastic finite element programs. The program performs an analysis of J from the load-displacement diagram and crack size with a stepwise increment in load, the calculation being repeated with increasing crack size. When the computed J value equals J_{lc}, this identifes the load at which the crack will begin to propagate in the component part. As discussed in the previous paragraph, it is uncertain at what load actual failure of the part would occur. Various procedures have been proposed, but all are at this time subject to considerable uncertainty.

Application of the *J*-Integral to Viscoelastic Materials

To begin to apply these concepts to a viscoelastic material involves the recognition that the variation of load with time, the dependence of the modulus or compliance on time, and rate of crack propagation must be specified. In principle, the *J* value can be computed as function of time, indicative of the physical reality that for a viscoelastic material the resistance to crack propagation will vary with time under load. The *J*-integral can be calculated by the following steps:

1. The hereditary integral discussed in Chapter 8 is used to find the displacement or load from the specified loading or grip displacements during the test based on the time dependence of the creep compliance or the relaxation modulus, respectively. Consider a test at constant strain rate e' for a material with a relaxation modulus given by

$$M(t) = At^{-n}$$

2. The equation for the particular specimen configuration is employed. For a double cantilever beam (DCB) sample the equation relating displacement and load is given below. Since the load P and displacement u will vary with time,

$$u(t) = \frac{8P(t)c^3}{EBh^3}$$

Substituting the modulus variation with time and solving for the load yields

$$P(t) = \frac{Bh^3 M(t)u(t)}{8c^3}$$

The load $P(t)$ which will vary with time can be calculated by the hereditary integral:

$$P(t) = \frac{Bh^3}{8c^3} \int M(t - \tau)\left(\frac{du}{d\tau}\right) d\tau \qquad (7.75)$$

Since the strain rate is constant, $du/dt = e'$, and can be taken outside the integral:

$$P(t) = \frac{Ae'Bh^3}{8c^3} \int (t - \tau)^{-n} d\tau = \frac{Ae'Bh^3 t^{-(n+1)}}{8c^3(n + 1)} \qquad (7.76)$$

3. The load can be converted into a function of the grip displacement by substituting $u = e't$ into Eq. 7.76

$$P(u) = \frac{ABh^3ut^{-(n+2)}}{8c^3(n+1)}$$ (7.77)

4. The energy input into the specimen can be computed by

$$U = \int P \, du = \frac{Ah^3u^2Bt^{-(n+2)}}{16c^3(n+1)}$$ (7.78)

5. J can be computed from the relation

$$J = \frac{1}{B} \frac{d\{\int P \, du\}}{dc} = \frac{-3Ah^3u^2Bt^{-(n+2)}}{16c^4(n+1)}$$ (7.79)

7.15. FATIGUE

Fatigue is failure due to a repeated variation in stress. All materials will fail at a lower stress under such varied stress than under a single constant application of the load. In general, the higher the applied stress, the fewer cycles to failure. Such failures usually start at some stress concentrator on the surface of the part, produced either by a sudden change in cross-sectional area designed into the part, some imperfection produced during molding, or a defect inherent in the material itself. Whatever the reason for the stress concentration, under repeated load variation a crack eventually develops at that location. As cyclic loading continues, the crack grows stepwise, reducing the area of specimen withstanding the applied load. Eventually the sound material is so reduced that it can no longer support the applied load, and the sample fails under the direct loading. If the sample is a ductile one, this part of the fracture surface exhibits a ductile failure.

As a result of this dual process of crack propagation and sudden failure, fatigue failures can often readily be identified by observing the fracture surface, although this identification is not as clear in plastics as it is in metals. As a result of a crack growing through repeated cycling, part of the sample may exhibit a series of curved lines following the path of the crack as it grows through the sample. These lines can sometimes be seen with the naked eye. Under sufficient magnification, the visible lines can be identified as being composed of a series of finely separated lines, each produced by the crack in its growth at each repeated load application. Whereas in metals, the number of cycles the sample has endured before

failure may be experimentally determined by counting these fatigue stria-
tions, in plastics they are generally not sufficienty well defined to perform
such an analysis. In addition, most amorphous polymers do not exhibit crack
growth each cycle, but at intermediate stresses multiple cycles occur without
observable growth, followed by large discontinuous crack growth. The stress
range over which such discontinuous crack growth occurs varies with
material but has been reported to be within the stress intensity range from
0.3 to 1.3 MPa m$^{0.5}$ (Elinck et al., 1971; Skibo et al., 1977). At higher stress
intensity ranges the incremental crack growth occurs at each cycle.

Stress Concentrations

Fatigue failure cracks initiate at surface imperfections. Locations at which
stress concentrations exist as a result of machining or handling defects or designed sharp changes in dimensions are frequently where such fatigue cracks
develop. Tables of stress concentration factors k_t have been compiled, where

$$k_t = \frac{\text{stress at the location}}{\text{average applied stress}}$$

These tables are available for many common configurations such as holes,
fillets, and reductions in diameter. These concentration factors are derived
from elasticity theory and therefore assume completely elastic behavior in
the sample. If fatigue strains were all truly elastic, then the fatigue life of a
part with a stress concentrator should be calculable from the stress-cycles to
failure curve at the stress computed from

$$\sigma = \sigma_{\text{applied}} \cdot k_t$$

For all materials, the actual effect of a stress concentrator on the fatigue life
is less than that computed from k_t due to plastic and viscoelastic behavior
in the area near the tip of the crack. A different factor k_f, the effective stress
concentration factor, is defined to relate the elastic stress concentration
to its effect on the fatigue life. To emphasize the significance of a stress
concentrator for fatigue life, the notch sensitivity factor q is defined as

$$q = \frac{(k_f - 1)}{(k_t - 1)}$$

This particular form for the definition of q was chosen so that $q = 0$ if the
configuration being considered has no effect on the fatigue life ($k_f = 1$), and
$q = 1$ if the sample configuration affects the fatigue life to the extent expected
for a completely elastic material ($k_f = k_t$).

For metals the notch sensitivity increases with increase in material
strength (and resultant decrease in ductility and plasticity). The notch

sensitivity has been observed in metals to decrease with decreasing radius of the defect or configuration being considered, presumably because a small volume of material is under the increased stress. Note that as the tip radius of the defect decreases, the elastic stress concentration k_t and the fatigue stress concentration factor k_f both increase, but k_f increases more slowly than k_t.

Terminology for Applied Stress

Fatigue in all materials involves similar technology regarding the applied stresses. Sinusoidally varying stress is easily obtainable experimentally in bend testing of rotating cylinders and conforms to the expected loading in automotive axles and other rotary members. It therefore is the most common type of applied load, although step functions and ramp loading tests are also performed. Referring to Fig. 7.13, average stress is defined as the mean stress

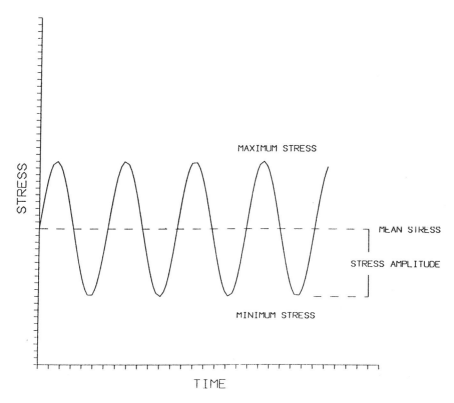

Figure 7.13. Fatigue loading stress definitions.

σ_m, and half the stress range is the stress amplitude σ_a. These are defined mathematically as

$$\sigma_m = \frac{(\sigma_{max} + \sigma_{min})}{2} \tag{7.80}$$

$$\sigma_a = \frac{(\sigma_{max} - \sigma_{min})}{2} \tag{7.81}$$

$$R = \frac{\sigma_{min}}{\sigma_{max}} \tag{7.82}$$

The load ratio $R = -1$ for a completely reversed load with $\sigma_{mean} = 0$, but can be any value. $R > 0$ signifies that the maximum and minimum stresses are both tensile.

Effect of Mean Stress

A very limited number of studies have been performed on plastics to evaluate the effect of mean stress on the fatigue life. The general approach developed for metals is the best that can be used. Clearly, as the mean stress increases, the stress amplitude must decrease for failure to be avoided, since the maximum stress applied $= \sigma_a + \sigma_m$. The Goodman relation considers that the permissible σ_a decreases linearly from the maximum possible value when $\sigma_m = 0$ to zero when the mean stress equals the tensile strength of the material. The equation for this linear variation is

$$\sigma_a = \frac{\sigma_a}{1 - \sigma_m/\sigma_{\text{tensile strength}}} \tag{7.83}$$

While this relation has been shown to provide a quite conservative estimate of the permissible applied stress for metals, and other equations have been developed which more closely approach the experimentally observed data, for plastics the lack of data suggests that this conservative model be the one employed.

Effect of Residual Stresses

Surface residual stresses can greatly affect the fatigue lifetime of specimens. Compressive stresses increase fatigue life, increasing the time for crack initiation and decreasing the growth rate of cracks. Conversely tensile stresses are harmful. Because fatigue cracks usually develop at the surface, the surface residual stresses are generally the most important. Hornberger and Devries (1987) increased the fatigue life of PC significantly by introducing compressive surface residual stresses by quenching samples in liquid nitrogen and ice water from above T_g. Ice water quenching generated stresses of about 15 MPa (2500 psi). This quenching increased the fatigue life from 39,000 cycles to 226,000 cycles. Liquid nitrogen quenching produced compressive stresses of 31 MPa (4500 psi) and increased the fatigue life to 418,000 cycles, more than an order of magnitude greater than that of the untreated samples.

Fatigue Testing Procedures

In the common rotating cylindrical beam technique the sample is subjected to an alternating load by applying a bending moment by attached weights. The bending moment increases with distance from the load, and failure occurs at the fillet. This test is accepted for metals and plastics but is rarely performed as described for plastics. In the specialized test for plastics (ASTM D671), the test is flexural fatigue of a cantilever beam. The sample is a sheet machined to have increasing width towards the fixed end. The width is designed so that a constant bending moment exists along the specimen's length, so the entire volume of sample is being fatigued at constant stress and failure may occur anywhere within the gage area of the test sample. The force applied is maintained constant throughout the test, and the frequency is specified to be 30 Hz.

The sheet thickness is to be specified in the report, and the standard specifies that comparisons of fatigue strength of different materials shall be made from test results on samples with the same thickness. Besides this warning, the standard goes on to specify conditions which make the test results really unusable in practice. The standard indicates that the results are suitable for direct application in design only when all test factors are the same as those expected in the application. These include the stress, size and shape of the part, temperature and other environmental factors, heat transfer rates and the loading frequency. The possibility of all these conditions being the same is nil, indicating the essential qualitative nature of the results of this type of testing.

Sample Size Effects

In a bending test, the stress is a maximum at the surface and decreases to zero at the neutral axis or center of the sample. The fraction of the total volume of the sample being subjected to the maximum load is fairly small. Sample size therefore affects the test results: the larger the sample, the greater the chance of a defect or crack already existing in the sample and the lower the fatigue life of the sample.

In axially loaded samples, the stress is uniform throughout the sample, and a comparison with the results of bending test samples reveals that axially loaded samples have lower fatigue lives at the same (maximum) stress level. Research activities during the past decade have emphasized axially loaded samples to avoid problems associated with the varying stresses inherent in the bend test.

Conditioning

The mechanical properties of plastics are very dependent on temperature, and moisture in the air is capable of diffusing into the sample and acting as a very effective plasticizer. In an attempt to control these factors, sample conditioning is a standard for all mechanical testing. The specifications for conditioning is given in ASTM D618. Basically, the ASTM standard requires the specimen to be held at 23°C and 50% relative humidity for a fixed length of time (40 hours) prior to testing. Although the temperature of the sample can be equilibrated readily within this time frame, water diffuses rather slowly through most plastics, and the actual water content of samples of different thicknesses will be quite disparate after this processing. Scatter in data of samples which are affected by moisture content can still therefore be expected after such conditioning.

Stress to Failure-Cycle Curves

Fatigue data obtained from this testing procedure are usually presented on graphs showing the applied stress as the number of cycles at which failure occurs. Since the number of cycles required for failures is normally quite large, it is customary to plot the applied stress vs. the log of the number of cycles. Typical data for polyethylene is shown in Fig. 7.14a. Many metals exhibit curves which tend to flatten and become parallel to the x-axis (number of cycles) at low stresses, indicating that the material is immune to fatigue failure if the applied load is kept below a value referred to as the fatigue limit. Similarly, some polymeric materials appear to exhibit such a characteristic, but others tend to exhibit curves with continuously decreasing slope, without clear evidence of a horizontal region. Consequently, for plastics without a fatigue limit, the number of cycles expected during the normal lifetime must be specified to permit determination of the safe applied stress under fatigue loading. No general separation can be performed between the class of polymers which have sharply varying fatigue curves and those with more rapidly declining slopes. The data in Fig. 7.14a does not indicate that polyethylene exhibits a fatigue limit. Figure 7.14b indicates that polystyrene has a fatigue limit. Among other plastics which have curves that do not approach a fatigue limit are Nylon, which is crystalline, and epoxy, which is cross-linked. Among the plastics which approach having a fatigue limit are crystalline polypropylene, amorphous polycarbonate, and cross linked phenolic.

Cumulative Damage

Many applications produce fluctuations in load in which the maximum amplitude is not the same for every cycle. Almost no data are available for plastics concerning the effect of such loading. Experimental data indicate that stress variations greatly affect the fatigue life. The introduction of a cycle with a higher stress amplitude retards the growth of the crack significantly for many cycles (which may run into thousands of cycles in some cases). The high stress cycle generates a larger plastic zone at the crack tip under the tensile part of the cycle. On reduction of the stress, the surrounding material applies a compressive stress to this expanded material. This generates a residual compressive stress which reduces the growth rate of the crack. The crack must grow through this material, retarding the growth rate. Once the crack has penetrated through this region into virgin material, the crack growth rate increases to that of the original material. However, for metals it is known that in carefully prepared samples with no detectable cracks, high initial loads tend to generate cracks rapidly, resulting in failure at lower cycles than specified in this equation.

For lack of better procedures, the concepts developed for metals must be utilized, but with caution. The approach pf Palgren and Miner (1954) states that the fatigue damage caused at each stress level can be calculated as the fraction of the number of cycles which will produce faulure at that loading. Failure will occur when

$$\sum \frac{n_i}{N_i} = 1 \qquad\qquad (7.84)$$

where n_i = number of cycles at stress condition i
         N_i = number of cycles to cause failure under condition i

The order in which the loads are applied does not appear in this equation. Note that this formula does not take into account the effect of overloading discussed above.

This equation is therefore unfortunately not positively a conservative one, and even more unfortunately, no conservative law has yet been established. The application of this law must therefore be carefully considered, and it must be employed with a factor of safety.

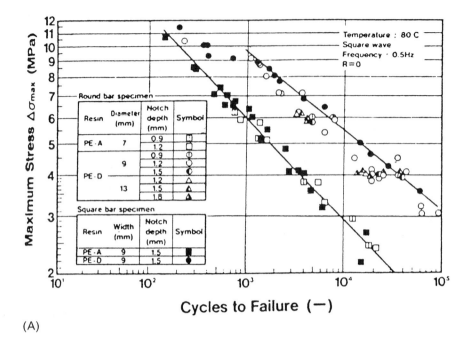

(A)

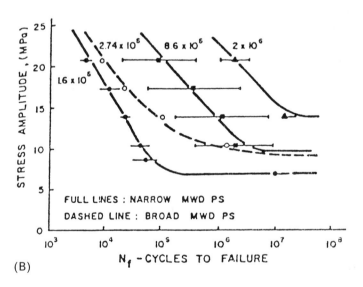

(B)

Figure 7.14. Cycles to failure as a function of maximum applied stress. (A) PE: No endurance limit is observed, the lifetime increases continuously as the maximum stress is reduced. [From H. Nishimura and I. Narisawa, *PES 31*, 399 (1991). Reprinted by permission of the publisher.]

Effect of Temperature

Fatigue properties are quite sensitive to temperature, but the temperature dependence varies markedly between polymers. The most common behavior observed is an increase in crack growth rate dc/dN with increasing temperature. Among the polymers exhibiting this behavior are acrylonitrile-butadiene-polystyrene (ABS), PMMA, PS, polysulfone (PSF), PVC.

Fatigue Failure Mechanism Changes

However, PC exhibits an interesting fatigue lifetime temperature dependence. At elevated temperatures between 75 and 125°C typical $S - N$ curves are observed, with a suggested endurance limit of the order of 5 MPa, which are not very temperature dependent in this temperature range. The entire curves shift toward longer lifetimes at a fixed stress as the temperature is lowered, as expected from the increase in ultimate strength with temperature. However, below 75°C, while at very high stresses increasing stress lowers the lifetime, an S-shaped fatigue $S - N$ curve is observed (Fig. 7.15), as a result of the samples exhibiting a larger lifetime at higher stresses in the intermediate stress region (Takemori, 1987, 1988; Matsumoto and Gifford, 1985).

This behavior indicates that there are two distinct $S - N$ curves as a result of the predominance of one of the two possible failure modes. The lifetime inversion apparently occurs when there is a transition from one branch to another, when the stress is such that a different failure mode predominates. The two distinct types of deformation that have been discussed which may ultimately lead to failure are crazing, which develops into cracks, and shear failure. In fatigue loading both shearing of the polymer and crazes occur simultaneously at the crack tip. Which is more important depends on the temperature, load, and time of loading. In PC shear fracture cracking occurs at 45° to the applied stress. The resultant crack growth rate is low. Craze growth and cracking occur at a much higher rate. The data indicate that high temperatures and high stress tend to favor shear fatigue failure. Conversely, low temperatures and low stress result in craze growth predominating. Both crazes and shear regions exist at the crack tip and failure is determined by the dominant mechanism. Because fracture by shear plane development is quite slow compared to craze and crack growth, the inversion with longer lifetimes at higher stresses (where shear failure is dominant) results.

Kim and Mai (1933) observed a fatigue failure mechanism change in

---

(B) PS: A distinct endurance limit occurs. The fatigue characteristics improve with increasing molecular weight. [From J. A. Sauer and G. C. Richardson, *Int. J. Fract.* *16*, 499 (1980). Reprinted by permission of Kluwer Academic Publisher.].

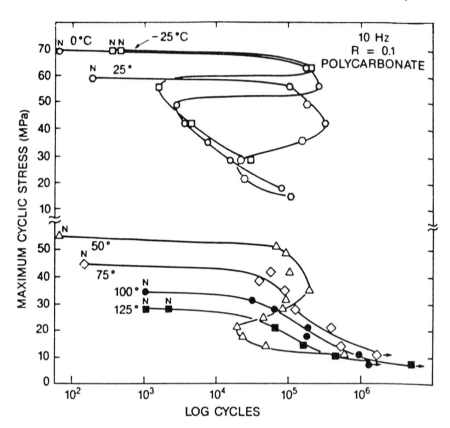

Figure 7.15. Fatigue failure lifetimes at various temperatures for PC. Data at various temperatures showing the inversion in fatigue life. The letter N indicates failure by necking; arrows indicate samples which remain unbroken. [From M. T. Takemori, *PES 28*, 641 (1988). Reprinted by permission of the publisher.]

γPVC. In the lower temperature range from -30 to $-10°C$ shear yielding was the predominant mechanism. In the high temperature range of 23 to 60°C multiple crazing at the crack tip occurred, and crack growth occurred by coalescence of the crazes. This coalescence produces a much rougher surface than that observed when shear yielding is the crack growth mechanism, a distinct change in fracture surface characteristics occurring between 0 to 20°C.

Takemori (1988) studied the effect of these variables to obtain a fatigue fracture map, indicating the conditions which would cause a specific failure mechanism. The resultant diagram in Fig. 7.16. He plotted his results as

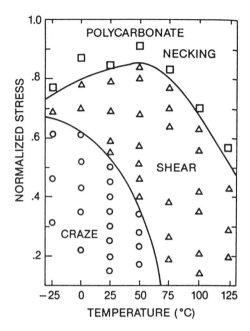

Figure 7.16. Fatigue fracture diagram for PC. Normalized stress is the applied stress/yield stress. Data obtained at 10 Hz and at $R = 0.1$. The different mechanisms for deformation and failure at different conditions are shown. [From M. T. Takemori, *PES 28*, 641 (1988). Reprinted by permission of the publisher.]

normalized stress (the applied stress/yield stress) vs. temperature. The diagram shows the distinct change of failure mode from crazing at low temperatures and lower stresses to shear failure at higher temperature and stress. Necking occurs at stresses close to the yield stress at all temperatures.

Effect of Creep on Fatigue Failure

This discussion just concluded would tend to lead to the conclusion that fatigue loading always produces failure by different mechanisms and under different conditions than by continually applied load. This is indeed the case for metallic samples. However, for many plastics, creep occurs extensively under tensile loading.

During fatigue loading in which the stress is cycled between a large and a small value (R is positive), creep occurs simultaneously with fatigue. If creep is extensive, its effect overshadows any effect of fatigue crack initiation and growth. This is apparently the case for nylon and polyester fibers, which produce an interesting variation from the expected fatigue behavior of metals. Nylon and polyesters can easily be produced as thin fibers, which are then wound into cord or rope. These are used extensively in tire cords and marine applications, where resistance to fatigue loading is one of the more important mechanical properties. The various fibers tend to rub against each other during fatigue loading, but data for single fibers and ropes of various interlacing methods all exhibit the same fatigue resistance, indicating that such rubbing is unimportant.

Under tensile fatigue in which the load is cycled between a higher and a lower tensile stress, creep occurs as a result of the load. In such testing tensile strain is accumulated during each cycle. Failure is observed when a critical strain is reached. The critical strain is the strain at which the sample fails in a direct stress-strain test. This indicates that the cyclic fatigue lifetime is controlled solely by creep rupture, and failure is affected only by the time under load, rather than the number of cycles of loading! This is shown in Fig. 7.17 (Kenney et al., 1985), which shows the applied load vs. the total testing time for nylon. The frequency of application of the load, which controls the total number of cycles, has no effect on the fatigue lifetime. Similar data for polyester also indicate that creep is the effective cause of failure.

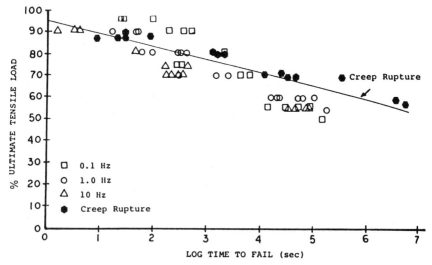

Figure 7.17. Interrelation of creep and fatigue for nylon yarns. [From M. C. Kenney, J. F. Mandell, and F. J. McGarry, *J. Mater. Sci. 20*, 2045 (1985). Reprinted by permission of Chapman & Hall.]

7.16. THERMAL FAILURE

All polymeric materials have losses which vary with loading frequency. The energy is stored in the material, increasing its temperature during loading, unless dissipated through heat transfer to the surroundings. Since the thermal conductivity of polymers is low, heat transfer is small, and the temperature rise during cyclic loading can be substantial. Fatigue failure of plastics is as likely to be a result of excessive temperature rise and local softening or melting of the material as of fatigue crack initiation and propagation. As internal mechanisms for heat generation increase with increasing temperature, the temperature rise of the specimen accelerates rapidly and may reach the melting point. At local discontinuities or internal cracks where larger strains occur, the temperature rise may be far higher than in the bulk of the material. ASTM D671 states that thermal failure is a normal result of fatigue, and artificial means of cooling should not be used to prevent such failures when they would occur in the actual application. Although this is a valuable testing procedure for full-size samples of an actual product, it is necessary to separate thermal failures from fatigue crack growth failures in experimental studies of smaller standardized specimen which will produce design data.

Crawford and Benham (1975) determined the effect of the testing frequency on the $S - N$ curves of polyacetal. The data demonstrate the large reduction of the fatigue life of polyacetal with increasing cycling rate, presumably due to this heating. The fatigue limit is decreased by about 125% by increasing the frequency from 0.167 to 10 Hz. An interesting effect is that the fatigue limit is affected to as great an extent as the fatigue strength at low lifetimes where the larger stresses would be expected to produce a higher temperature rise. Riddell et al. (1966) determined the surface temperature rise during cycling of polytetrafluoroethylene, as shown in Fig. 7.18. Note that the temperature rise measured before failure increases with decreasing stress level, probably because the highly stressed samples failed so rapidly that the surface temperature had insufficient time to rise to its equilibrium value. At 30 Hz the temperature rise for a stress of 6.3 MPa (a value corresponding to the fatigue limit) was reported to be 60°C. The temperature rise depends on the heat transfer through the sample and heat loss from the surface of the sample. In practice, the sample would need to be designed to reduce the temperature rise. In most applications, sufficient heat transfer occurs to keep the temperature down, and most concern arises during testing, where rapid accumulation of data at large cycles tends to cause investigators to use as large a cycling rate as possible.

The difficulty in analyzing fatigue data obtained at high frequencies is that these stress-cycle curves correspond to data obtained at varying and unknown temperatures. We might expect the temperature of the sample at

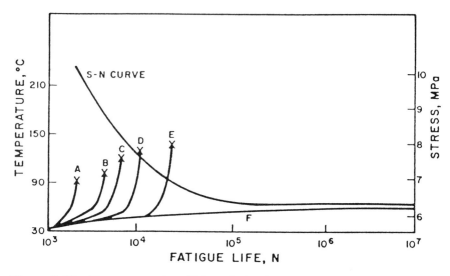

Figure 7.18. Temperature rise of PTFE during fatigue loading. (A) 10.3 MPA. (B) 9.0 MPa. (C) 8.3 MPa. (D) 7.6 MPa. (E) 6.9 MPa. (F) 6.3 MPa (no failure). [From M. N. Riddel, G. P. Koo, and J. L. O'Toole, *PES 6*, 363 (1966). Reprinted by permission of the publisher.]

equilibrium to decrease monotonically with decreasing stress. Even this simplifying condition is not absolutely correct, however, as it would pertain only to the equilibrium temperature after steady-state heat transfer is reached. For high stress levels the specimen will fail in a short time, before the maximum temperature is reached.

Another factor which further reduces the usefulness of $S - N$ curves at high cycling rates obtained in bending tests is the variation of temperature through the sample. Since the stress generated in bending varies from a maximum value at the surface to zero at the center, the heat generation also varies, being greatest at the surface. Energy is dissipated through surface conduction. The larger the surface area/volume ratio, the greater the heat dissipation, and the $S - N$ curves are raised to higher stress values before failure. The exact temperature distribution will therefore depend strongly on

1. The frequency of loading
2. The diameter of the specimen or the volume/surface ratio
3. Surface smoothness
4. Air circulation around the sample
5. Rest periods during cycling

Application of such data to any design must therefore be doubtful. The general success of plastic parts in fatigue applications must be attributed primarily to the use of relatively low stress levels, low frequencies, or sufficient surface area to dissipate the heat generated. Reinforced composite plastic parts have been successfully utilized in high stress fatigue applications. Since the reinforcements are very stiff, the internal friction tends to be low, the amount of the deformation in these applications is limited, and temperature rises are small.

Despite the importance of thermal heating resulting in failure, if such heating is controlled, normal fatigue crack initiation and propagation will occur. Figure 7.19 gives the $S - N$ curves for polyacetal at various

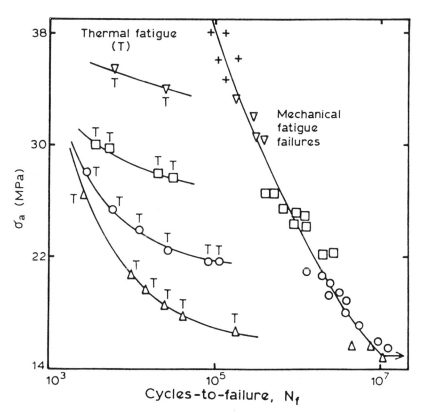

Figure 7.19. Stress amplitude vs. cycles to failure for polyacetal. Both thermal and mechanical fatigue failures occur, with different frequency dependence. [From R. J. Crawford and P. P. Benham, *Polymer 16*, 908 (1975).]

frequencies. The $S - N$ curves all exhibit two distinct regions. At high stress levels failure is by overheating, whereas at lower stress levels the failure is by cracking. The two different mechanisms have two different stress dependences, and whichever causes failure first controls the properties of the sample. As expected, the lower the cycling rate, the lower the heating, and crack propagation becomes the more dominant mechanism. Even at a cycling rate of 0.5 Hz, thermal failure still controls the properties at high stress levels. It can be seen that thermal effects are strongly frequency dependent, but crack propagation is nearly frequency independent in the range studied, as all the crack propagation curves fall on one smooth curve.

7.17. STRESS INTENSITY ANALYSIS OF FATIGUE

Although fatigue cracks may be initiated in smooth, apparently undamaged samples by repeated cycling, all materials may have inherent or induced cracks present at the time the specimen is placed in service. The safest design procedure is therefore to make the same assumption as is used in fracture mechanics analysis: the material initially contains some cracks of specified size. A crack then grows during cycling until it reaches the critical size required for failure. To perform an analysis predicting the lifetime of the part, the crack growth rate must be known.

A significant body of data now exists for the crack growth rate of plastics. These data are obtained by preparing samples with known initial crack length and crack geometry, usually by machining defects of accurately known dimensions into plastic samples. A variety of techniques are available for monitoring the growth of the crack during cycling in electrically conducting samples, but direct observation of the crack with a microscope is the only reliable technique available for these investigations in plastics. These observations are therefore tedious and time consuming.

For most geometries, crack growth rates increase with increase in crack size, since the stress intensity K at the tip of the crack increases. Early investigators developed a variety of relations between the applied stress and the crack growth rates. Because the stress intensity at the crack tip should control the driving force for crack propagation, relationships have been developed between the applied change in stress intensity and the crack growth rates. Swanson et al. (1967) demonstrated that the stress intensity controls the crack growth rate by decreasing the load during crack growth at various rates. Only when the load was decreased so that ΔK remained constant was the crack growth rate also held constant. The general relation is the Paris power law:

$$\text{Crack growth rate } = \frac{dc}{dN} = A(\Delta K)^n \tag{7.85}$$

where c = crack length
N = number of cycles
$\Delta K = K_{max} - K_{min}$
A, n = experimentally determined constants

Although many polymers follow this type of behavior, for others the rate of fatigue cracking also depends on the average value of the stress or the stress intensity. Therefore for these polymers both the average and the range of stress intensity must be included. The general form of the equation would then be

$$\frac{dc}{dN} = A(\Delta K)^n (K_{mean})^m \qquad (7.86)$$

and the objective of most crack growth studies is to determine the values of A, n and m for the particular materials. This equation yields a straight line plot of $\log(dc/DN)$ vs. $\log(\Delta K)$. This straight line relation is valid over a wide range of ΔK values. However, at the extremes of low and high stress intensity values, the crack growth rate changes, and the general overall shape is sigmoidal (Fig. 7.20). This sigmoidal shape tends to be independent of the initial crack length.

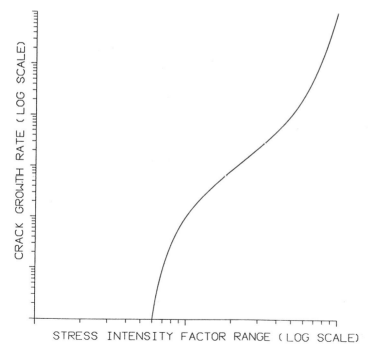

Figure 7.20. Crack growth rate vs. stress intensity factor.

Lifetime Calculations

From such data the practical lifetime of a part can be calculated if the fatigue load is constant. Direct integration may be performed by finding an analytical expression for the dc/dN curve, but the following computational scheme is simple to program without the necessity of approximating data by such an expression:

1. The initial crack is assumed to be that which is the smallest that can be detected
2. The K_{max} and K_{min} values are calculated from the applied stress and the known crack length
3. From the dc/dN data and this calculated ΔK (and K_{mean} if required) the increase in crack size is determined
4. From the new crack size new K values are computed
5. Steps 3 and 4 are repeated until the crack size is sufficiently large to cause failure

7.18. FACTORS AFFECTING FATIGUE LIFE

Effect of Temperature

The Williams–Ferry–Landau (WFL) equation (see section 8.2) is most commonly used to predict the effect of temperature on properties, and shift factors have been used for fatigue crack growth. As in other areas of mechanical property analysis, difficulty arises in using the WFL method below the glass transition temperature, since it was derived for rubbery materials. Another approach is that of Kim and Mai (1993) who used the Arrhenius equation for μPVC below T_g. The Arrhenius equation expresses the rate of molecular motion and therefore the rate of any chemical process as depending exponentially on temperature:

$$k = A \exp\left(\frac{-\Delta H}{RT}\right)$$

where k = rate of the process under consideration
  ΔH = activation energy for the process
  R = universal gas constant
  T = absolute temperature
  A = constant depending on the process and material

Kim and Mai considered that the activation energy for fatigue cracking decreased with increasing stress intensity factor variation during the loading:

$$\Delta H = \Delta H_{th} - \gamma \log \Delta K$$

where ΔH_{th} is the activation energy in the absence of mechanical stresses. The specific reason for this dependence on ΔK was presumably based on experimental data analysis. With this dependence, crack growth dependence on temperature could be given by

$$\frac{da}{dN} = \frac{B \exp([\Delta H_{th} - \gamma \log \Delta K])}{RT} \tag{7.87}$$

This equation has three constants (B, γ, ΔH_{th}) which much be experimentally determined. Once these are calculated, values for any other temperature may be obtained.

Effect of Loading Conditions on Fatigue Testing Results

Unless the loading conditions of a test are carefully noted, the results can lead to false interpretations of the ability of a polymer to withstand fatigue loading. In constant stress or load testing, degradation of the stiffness of the polymer results in increasingly rapid failure, as the deformation extremes become greater as the test proceeds. Conversely, the same polymer tested under constant strain conditions would behave admirably, since the test would be applied under continually decreasing load, and failure would be unlikely to occur.

Molecular Changes During Fatigue

Although no apparent changes occur in the external appearance of a sample during fatigue loading prior to craze or crack appearance, there is substantial evidence that continuous changes in internal structure are occurring. The major irreversible effect is the breaking of the covalent bonds that form the backbone of the molecular structure, and it is observed utilizing electron spin resonance (ESR) (Zhurkov and Tomashevskii, 1966). Such bond breakage occurs before and after crack and craze formation. Once a crack is initiated, the number of bonds broken greatly exceeds those necessary simply to extend the crack. Rather, the material surrounding the crack must also be affected. The number of unpaired electrons which are generated as a result of the breakage of covalent bonds during fatigue increases steadily during testing and crack propagation.

Effect of Molecular Structure on Fatigue Resistance

1. Higher crystallinity increases fatigue resistance
2. Finer and more uniform spherulites exhibit improved fatigue resistance
3. Both of these factors tend to decrease the amount of hysteresis heating, increasing fatigue life

Examples of these effects are:

Heating nylon at 180°C produces more uniform spherulitic structure and greater percent crystallinity, therefore improving fatigue life (Stinkas et al., 1972).

Quenching PTFE from elevated temperature forming decreases the percent crystallinity, decreasing fatigue resistance (Riddell et al., 1966).

Increasing the percentage of plasticizers in PVC increases fatigue life during strain control, again because the total load applied decreases as the test proceeds due to strain softening (VanGaut, 1965).

Constant deflection test of polyamide improved fatigue life with increasing percent moisture. This lowered the elastic modulus and lowered the applied stress during the test. Conversely, under constant load fatigue testing, heating occurred, driving off the moisture and hardening the materials. (Mafhyulis et al., 1966).

Localized Heating at Crack Tip

Some localized heating at the crack tip should occur due to the release of strain energy as the crack advances. However, the heating is negligible for crack advance rates of less than 1 cm/sec (Kambour and Barker, 1966).

Effect of Environment

Plasticization of the polymer lowers the modulus. As discussed above, under load control this results in a decrease in the fatigue life of the sample. As the solubility parameter of the environment approaches that of the polymer, the fatigue life decreases dramatically, a typical value being a decrease of a factor of 3 from that of the sample tested in air. If the sample is not cross-linked, long-term holding in this environment may eventually lead to complete solution. However, even short-term contact will plasticize the surface and lead to these reductions in fatigue life. The solubility parameter of the environmental liquid and the polymer must be within ± 5 of each other for any significant effect to occur.

Weathering

Few data are available on the important topic of the effect of weathering on fatigue characteristics. Data from Japan for 1 year of testing indicate that PS, ABS, and polyoxymethylene (POM) have fatigue lives degraded by weathering, whereas PVC and PMMA were not affected (Suzuki and Tsurue, 1976). Much more data are needed in this area.

7.19. RELATION BETWEEN STATIC AND DYNAMIC FATIGUE

While holding some plastic specimens under constant load results in a continuous increase in crack length and the application of a varying load results in discontinuous crack propagation on each cycle, the similarity in failure due to cracking has led some investigators to attempt to correlate the two. Some polymers which exhibit rapid failure under static loadings are very resistant to dynamic loading. Others exhibit the reverse type of behavior. However, evidence indicates that for a particular polymer, a correlation exists between the lifetimes in both conditions.

Zhou et al. (1991) have compared the dynamic and static lifetimes for ethylene-hexene copolymers. This copolymerization produces a random copolymer with side chains of four carbon atoms. The correlation of the time for fracture under dynamic loading and under constant load is shown in Fig. 7.21. These tests were performed for copolymers manufactured by different suppliers and produced over the time span from 1970 to 1989. Resistance to

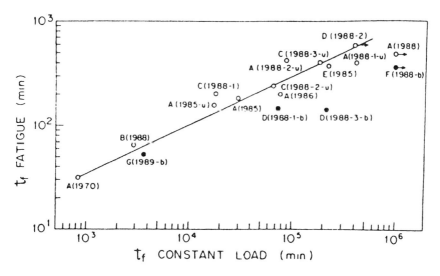

Figure 7.21. Relation between time for failure in fatigue and under constant loading. [From Y. Zhou, X. Lu, and N. Brown, *PES 31*, 711 (1991). Reprinted by permission of the publisher.]

crack propagation is provided by the tie molecules which prduce the interconnections between the various crystallites within the sample. These tie
molecules are the molecular chains with a high degree of polymerization
(DP). Increasing the average molecular weight should therefore increase the
crack resistance of the sample but also increase the difficulty of processing.
The processing parameter is the melt flow index, and all the samples had
melt flow index values close to 0.2.

8
Viscoelastic Effects

8.1. EXTRAPOLATION TECHNIQUES

Creep data determinations are time consuming, and if long-term creep data are required, the testing may take longer than the life of the project, or the investigator. Procedures for extrapolation of short-term data are invaluable and permit data to be obtained over feasible time intervals. However, as with any extrapolation technique, unexpected behavior may be observed in the extrapolated time or temperature region, such as sudden rupture which is not predicted. Such extrapolations should therefore be treated with caution.

8.2. WILLIAMS–FERRY–LANDAU EQUATION

This is one of the most powerful extrapolation techniques, with a wide range of applications, and it has been applied to virtually every mechanical property and every type of plastic. Its original derivation was for the prediction of the viscosity of liquids supercooled into the rubbery region and above the glass transition temperature, but its use has been widely extended from this beginning. The fundamental approach is that the viscosity (and by extension other mechanical properties) is dependent on the free volume within the sample. The decrease in viscosity on increasing the temperature is a result of the greater free volume into which the molecules can move during kinking and general motion.

For temperatures above T_g, the free volume increases with temperature. At T_g, the free volume becomes fixed at a value of v_g. Since we can consider that the free volume in the glassy state is independent of temperature, the volume expansion with temperature of a glass (a_g) is a result only of increases in C—C bond distances within the chains. In the rubbery range, the volume thermal expansion (a_r) is a result of this C—C bond length increase (a_g) and also an increase in the free volume ($a_r - a_g$). The free volume at any temperature in the rubber state can therefore be written

$$v_T = v_g + (a_r - a_g)(T - T_g) \tag{8.1}$$

where T = temperature
T_g = glass transition temperature
v_T = free volume at temperature T
v_g = free volume at T_g
a_r = coefficient of thermal expansion in the rubbery state
a_g = coefficient of thermal expansion in the glassy state

Williams, Ferry, and Landau (WFL) postulated that the viscosity of a liquid varies exponentially in an inverse manner with the free volume. This postulate is based on the concept that the ease of flow of a liquid depends on the available space for kink movement and uncoiling. The exponential dependence was assumed primarily to correlate with the observed dependence of viscosity with temperature. Utilizing this assumption

$$\eta_T = \exp\left(\frac{1}{v_T}\right) \tag{8.2}$$

and similarly

$$\eta_g = \exp\left(\frac{1}{v_g}\right) \tag{8.3}$$

where η_T = viscosity at temperature T
η_g = viscosity at the glass transition temperature

Then:

$$\ln\left(\frac{\eta_T}{\eta_g}\right) = \left(\frac{1}{v_T} - \frac{1}{v_g}\right) = \frac{1}{v_g + (a_r - a_g)(T - T_g)} - \frac{1}{v_g}$$

$$\ln\left(\frac{\eta_T}{\eta_g}\right) = -\frac{[(a_r - a_g)\{T - T_g\}]}{(v_g)^2 + (a_r - a_g)v_g(T - T_g)}$$

$$\log\left(\frac{\eta_T}{\eta_g}\right) = \frac{(-1/2.303)(a_r - a_g)(T - T_g)}{(v_g)^2 + v_g(a_r - a_g)(T - T_g)} \tag{8.4}$$

For many polymers, Williams, Ferry, and Landau observed that the values of $a_r - a_g$ and v_g were rather similar. The average values of many polymers were taken to be

$$a_r - a_g = 4.8 \cdot 10^{-4} \, {}^\circ C^{-1}$$

$$v_g = 0.025 \text{ cm}^3/\text{g}$$

Ferry (1961) lists values for v_g ranging from 0.016 to 0.028 and values for the expansion coefficient differences from 1 to $5 \cdot 10^{-4}$. These average values should not be considered exact for any polymer.

Substituting these average values into Eq. 8.4 yields

$$\log\left(\frac{\eta_T}{\eta_g}\right) = \frac{17.44 \, (T - T_g)}{51.6 + (T - T_g)} \tag{8.5}$$

The viscosity of many polymers extrapolated to T_g is 10^{13} poise. Using this value

$$\log(\eta_T) = 13 - \frac{17.44 \, (T - T_g)}{51.6 + (T - T_g)} \tag{8.6}$$

The WFL equation relating the viscosity of a plastic at any temperature to its viscosity at the glass transition has been derived. The problem at this point is to extend this equation for other mechanical properties. This extension requires some thought. The viscosity of a material is a measure of the ease with which molecules can move relative to each other: the greater the ease of motion, the lower the viscosity. The viscosity is therefore directly related to the relaxation time for molecular motion. However, the relaxation time is a measure of the time for the molecules to return to their original position after a external applied force dislocates them, producing kinks. This equation can therefore be generalized to other mechanical properties by replacing the viscosity with the relaxation time for a mechanical process

$$\log\left(\frac{\eta \text{ at temperature } T}{\eta \text{ at } T_g}\right) = \log\left(\frac{\text{relaxation time at temperature } T}{\text{relaxation time at } T_g}\right)$$

The WFL equation in its general form therefore does not permit computation of a property directly but relates the change in temperature of a plastic with the change in the relaxation time for mechanical properties. To emphasize this time-temperature interdependence, the ratio of relaxation times at different temperatures is denoted the *shift factor* a_T. For the modulus of a plastic, therefore

$$a_T = \frac{\text{time for } E \text{ to decay to a specified value at temperature } T}{\text{time for } E \text{ to decay to the same value at } T_g}$$

and from Eq. 8.5

$$\log a_T = \frac{17.4\,(T - T_g)}{(51.6 + T - T_g)} \tag{8.7}$$

Therefore, if the modulus vs. time curve is known at one temperature, it can be computed at any other temperature. Again, this equation was derived for a plastic above the glass transition and is applicable only under that restraint.

8.3. WFL SHIFTING PROCEDURE

The WFL equation states that a shift in temperature can exactly correspond with a shift in time. The constants in the equation given above were obtained from the average thermal expansion coefficients and free volume at the glass transition temperature of polymers. This shifting equation is very powerful if information is available for only one temperature and information must be computed for some other temperature. If information is available for several temperatures, then more precise extrapolations or interpolations can be obtained by directly shifting the experimental data. The shifting is performed by shifting high-temperature data to longer times to correspond to lower-temperature data. Such a shifting process is shown in Fig. 8.1. Extrapolation of modulus data obtained from creep data for relatively short times at a variety of temperatures to longer times can therefore be performed. Since the modulus of a rubbery plastic decreases linearly with temperature and is also affected by the density of the material, the modulus is corrected to a constant temperature by multiplying by

$$\left(\frac{T_g}{T}\right)\left(\frac{\rho_g}{\rho}\right)$$

where ρ_g and ρ are the density at T_g and T, respectively. This is a small correction and can be neglected in many cases with little error.

Although this approach is directly obtained from the WFL equation and should therefore be used only above the glass transition temperature, the shifting has been found to be effective and accurate at temperatures below T_g as well. The theoretical basis for such shifting is dubious, but it

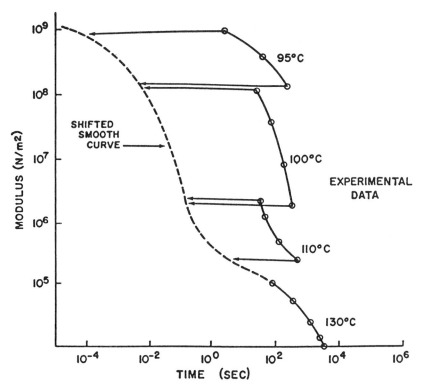

Figure 8.1. Time-temperature superposition of stress relaxation data for PS. (From *Plastics Products Design Handbook*, Part A, E. Miller, ed., Marcel Dekker, New York, 1981.)

is frequently performed. Below the glass transition the correction for temperature is unnecessary.

To perform the shifting, a reference temperature is chosen, and the modulus curves are then individually shifted horizontally until one smooth curve is obtained. The resultant curve should exhibit the expected shape, indicating the glassy region where the modulus changes only slightly with time, the transition region where the modulus changes more rapidly, and the rubbery region where the modulus again changes more slowly. For thermoplastic materials the region where the material approaches the liquid state and the modulus becomes very small should also be apparent. Note that the shape of the curve showing the dependence of the mechanical property as a function of time is the same as the shape of the curve as a function of temperature. This emphasizes the identity that a change toward longer times is equivalent to a lowering of temperature.

The completed smooth compiled curve is called the *master curve*. The shift factor a_T is now defined as

$$a_T = \frac{\text{time for a molecular response at temperature } T}{\text{time at the reference temperature}}$$

The procedure therefore permits determination of the modulus vs. time at temperatures different from that experimentally determined and extrapolation of data to much longer times.

If the reference temperature is T_g, then the shift factors should be numerically close to those computed from the WFL equation. If some other temperature is chosen as the reference, then the shift factors are arbitrary numbers, applicable to the particular master curve. An example of the shifting process is shown in Fig. 8.1 for polystyrene; the shift factor curve is shown in Fig. 8.2.

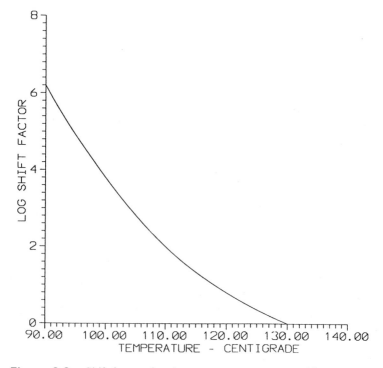

Figure 8.2. Shift factors for time-temperature superposition.

The WFL equation was derived for the rubbery state and therefore is truly applicable only for such materials. However, the concept of shifting data is so enticing that it has been applied to other states and, as shown in Fig. 8.3, works well in the crystalline region also. When applied to crystalline materials, a vertical shift of the data is also often required, to correspond to changes in modulus with changes in crystallinity with temperature and heat treatment. These vertical shifts are performed to obtain the best possible fit of the data into one smooth master curve. It is important to note that this shifting procedure is based on experimental data for the particular polymer under consideration, and the result obtained is therefore much more accurate than any value obtained from the WFL equation directly.

Example 8.1. For polystyrene with data given in Fig. 8.1, determine the time for the modulus to decay to 10^6 at 60°C.

Solution:

1. a_T at 90°C $= 10^{6.2} = 1.58 \cdot 10^6$
2. Time for modulus to decay to 10^6 at 130°C from the master curve $= 0.9$ seconds
3. Time at 90°C $= (0.9)(1.58 \cdot 10^6) = 1.42 \cdot 10^6$ seconds

Horizontal-Vertical Shifting

The concept of shifting data to produce a master curve is such a powerful one that modifications have been made to permit its use under conditions that do not necessarily satisfy all the fundamental specifications under which the WFL equation was developed. An excellent example of the extension of the WFL method is to polyethylene (PE) piping. Such piping for natural gas distribution has an expected desired lifetime of the order of 50 years. Long-time failures are usually generated by long-term creep producing rupture or slow crack growth under the applied load. Obviously, it is impossible to develop a test program which will run for the actual lifetime, and accelerated testing is required.

Shifting of high-temperature data to service temperatures by a WFL method is clearly indicated. However, analysis of experimental data by Popelar et al. (1990) shown in Fig. 8.3 indicates that the PE used in pipe production has nonlinear mechanical properties and that the WFL horizontal shifting of creep data did not produce an acceptable master curve for creep compliance. Their analysis indicated that this was due to the fact that the volume of the sample increased by 1% during a creep test

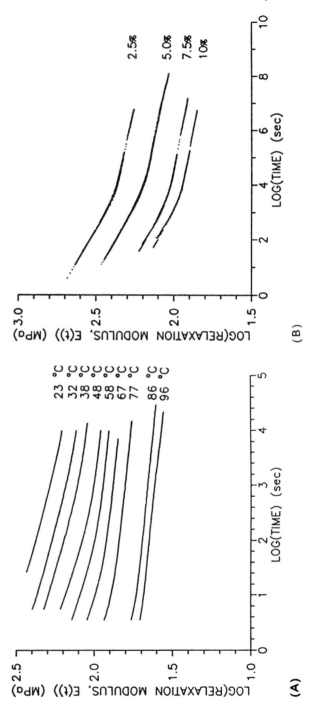

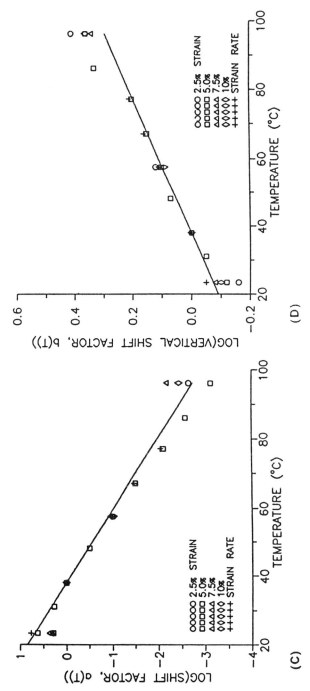

Figure 8.3. Master curves for relaxation modulus of PE at 23°C. (A) Relaxation of PE after a 5% strain at various temperatures. To obtain a master curve for this crystalline material, both horizontal and vertical shifts are necessary. (B) Master curves for various strain levels. (C) Horizontal shift factors for master curve. (D) Vertical shift factor for master curve. [From C. F. Popelar, C. H. Popelar, and V. H. Kenner, *PES 30*, 577 (1990). Reprinted by permission of the publisher.]

at 9.48 MPa. The extent of crystallinity of the sample also changed during the testing. Such changes in internal structure require adjustment along the vertical axis as well as the horizontal axis. The data could be brought into alignment to produce a master curve only by utilizing both horizontal and vertical shifting. Such shifting still could only yield master curves for each applied strain value. The failure to generate one master curve indicates the nonlinearity in mechanical properties of the PE samples.

Such vertical and horizontal shifting was applied to data on the burst strength versus time to failure for high-density PE pipe, shown in Fig. 8.4. The sudden change in slope is due to a change from a ductile type of failure at high stresses (short times) to brittle failure at low stresses (long failure times). Shifting must cause the knees of the various curves to coincide, as shown in the master curve in Fig. 8.5. From such a master curve lifetimes at various applied stresses can be predicted.

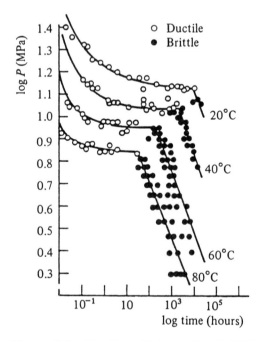

Figure 8.4. Ductile and brittle failure in HDPE pipe (data kindly supplied by BP Chemicals Limited.)

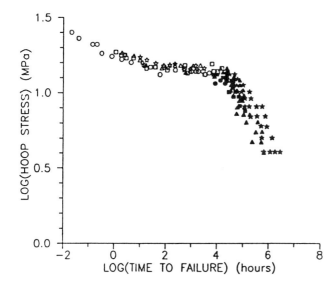

Figure 8.5. Master curve of time to failure at 20°C for the PE data of Fig. 8.4. The data are referenced to 20°C. The shift required for relaxation data, maximum stress vs. strain rate, and time to failure for PE was found to be the same and relative to the type of PE tested. [From C. F. Popelar, C. H. Popelar, and V. H. Kenner, *PES 30*, 577 (1990). Reprinted by permission of the publisher.]

8.4. EMPIRICAL EQUATIONS

When experimental data are available over a limited time or temperature range, empirical expressions are frequently utilized for the representation of the primary and secondary stages of creep. A variety of expressions have been proposed and tested for different polymer products. Table 8.1 lists some of those proposed. Due to its simplicity, the most commonly utilized expression is the first, which when written to include the dependence of the modulus on stress is

$$e = a + bt^n S^k \tag{8.8}$$

where e = creep strain
t = time
S = applied stress

The constants a, b, n, and k are determined experimentally. It has been suggested by Conway (1967) that this is a special case of the Andrade equation (Eq. 11 in Table 8.1), since in practice the exponent of time n in the power law equation is close to 1/3, and the exponent c in the Andrade equation is near zero.

Table 8.1. Expressions for creep strain as a function of time

1.	$\varepsilon = a + bt^c$	Parabolic
2.	$\varepsilon = a + bt^{1/3} + ct$	Power
3.	$\varepsilon = \dfrac{at}{1 + bt}$	Hyperbolic
4.	$\varepsilon = \dfrac{1}{1 + at^b}$	Hyperbolic
5.	$\varepsilon = a + \ln(b + t)$	Logarithmic
6.	$\varepsilon = \dfrac{a(b + \ln t)}{c + d \ln t}$	Logarithmic
7.	$\varepsilon = a + b(1 - e^{-ct})$	Exponential
8.	$\varepsilon = \sinh at$	Hyperbolic sine
9.	$\varepsilon = a + b \sinh ct$	Hyperbolic sine
10.	$\varepsilon = a + b \sinh ct^{1/3}$	Hyperbolic sine
11.	$\varepsilon = a(1 + bt^{1/3})e^{-ct}$	Power/exponential

From E. Miller and T. Sterrett, *J. Elast. and Plast. 20*: 346 (1988).

8.5. THE PSEUDOELASTIC APPROXIMATION

Design of plastic parts is commonly performed utilizing the standard elastic formula available in reference texts for various types of loadings applied to simply supported beams, cantilever beams, columns, gear teeth, pressure vessels, etc. This use of elastic formulae for viscoelastic polymers is called the pseudoelastic design procedure.

These elastic formulae were derived under the standard assumptions applicable to elastic materials. These assumptions include:

1. The strain is a linear function of stress
2. The material is homogeneous and isotropic
3. The strain is sufficiently small that second-order terms can be neglected in some derivations

Polymers are not elastic materials and do not satisfy these requirements for a variety of reasons:

1. Polymers have moduli which vary with stress level
2. As a result, the strain cannot be considered as varying linearly with stress

3. Due to their low moduli, the strains may be large. Strains of 5% or more are often acceptable for polymeric parts
4. As a result of flow during molding, polymeric parts are not isotropic. Strengths along and perpendicular to the flow direction are significantly different. Different cooling rates throughout the part also result in different mechanical properties throughout the part
5. Polymers are not homogeneous. Additives, fillers, and rubbery impact modifiers all create two-phase inhomogeneous materials

Although the differences between polymeric and elastic materials are significant, metals also do not truly obey these criteria. For example, in mechanically deformed products produced by forging, rolling, swaging, etc., the properties are strongly directionally dependent relative to the deformation direction. The complications produced by these anisotropic effects are often ignored. In addition, metals are usually not single phase. However, the various phases are all elastic. The multiphase polymeric materials have phases which are all viscoelastic, with time dependences which differ among the phases, producing more complex reactions to applied stresses. Despite these difficulties, the pseudoelastic approximation works reasonably well for most designs.

Despite these differences between the actual behavior and internal structure of polymers and the idealized behavior expected of an elastic material, it is commonly accepted that design of plastic parts can proceed utilizing these elastic equations. The modification in procedure is to replace the elastic modulus in the formula with the creep modulus of the material at the desired lifetime of the part.

8.6. TYPES OF FAILURES

Excessive Instantaneous Deformation

Polymers have stress-strain diagrams which are distinctly curved in the elastic region, resulting in the American Society for Testing and Materials (ASTM) and British Standards (BS) definitions of the yield strength differing by 15% commonly. Since the modulus of polymers is so much smaller than that of metals, the yield strength corresponds to strains of the order of 5 to 15%, as compared to that of metals, which is in the range of 0.1 to 0.3%. Instantaneous excessive deformation may occur on direct loading, often well below the yield strength of the polymer. A specification of the extent of permissible deflection must be part of the plastic product design. Since the modulus is a strong function of temperature, the maximum deflection must be computed from the modulus and applied load at the maximum expected temperature of operation.

Excessive Creep Deformation

Under the applied load, continuous creep occurs. The design stress is very often limited by the requirement that deformation of the product during its lifetime does not exceed a critical value. This creep deformation is computed from the creep modulus of the material at the expected lifetime of the product.

Creep Necking

In creep, elongation proceeds slowly and uniformly through the first and second stages, and the creep strain may be readily predicted from the equations previously described. This flow is a bulk phenomenon, and the material deforms uniformly under the applied stress. When a critical strain (5–15% depending on the polymer) is reached, large-scale yielding (the third stage of creep) occurs. Shear banding, a highly localized shear deformation within a small section of the sample, may also occur. This localized deformation may lead to failure by reduction of cross-sectional area. This large-scale deformation is generally unacceptable and is considered a design limit. These data are normally compiled in graphs of applied stress vs. time for necking initiation.

Crazing and Crack Propagation

Crazing occurs only in glassy polymers. As discussed in Chapter 7 crazing is localized, usually initiated at some small surface defect or as a result of a stress concentrator in the design. The crazes are thin sheets, approximately 0.1 to 10 μm in thickness, which appear as thin cracklike defects in the plastic. The crazes are not empty but contain elongated molecules with a density of approximately 50% of the bulk material. The appearance of a craze is not actual failure. However, the craze is unsightly and indicates a reduction in the strength of the material along the craze. Once a craze has formed, it grows steadily at constant stress. A crack develops along the craze plane, and the crack propagates through the plastic until failure results. The time for crazing to occur at a fixed stress level is a strong function of temperature and decreases as the temperature is raised to T_g.

Competition Between Shear and Cracking Failure

Whether failure occurs by general yielding or by crack propagation is a function of the applied stress and service temperature. Lower stresses tend to produce a crack propagation type of failure. The stress dependence for crack propagation is much greater than for ductile failure in many materials, and a transition therefore takes place, failure occurring by crack propagation at low stresses and by shear deformation at higher stresses. The stress at which the transition occurs increases with decreasing temperature. Crack propagation rates increase with temperature, and the yield strength decreases with increasing temperature, shifting both branches of the failure curve to shorter times at higher temperatures.

Effect of Molecular Weight

The tendency to neck is much less dependent on internal structure than are crazing, crack initiation, and crack propagation. High molecular weight polymers tend to be more resistant to crazing and cracking. As noted before, higher molecular weight improves mechanical properties in general but increases melt viscosity and increases difficulties in melt processing and mold filling.

Environmental Effects

The environment can have two major effects on properties:

1. Moisture or organic materials may diffuse into the polymer and plasticize the material, lowering the modulus, yield strength, and ultimate strength. The properties under the environmental conditions of testing should be known.
2. Environmentally assisted cracking (EAC) occurs in many organic environments and some inorganic ones. EAC requires an applied or residual tensile stress and an environment which is specific to the plastic. No theoretical method for predicting the cracking effect of environments is known, the specific environments which produce EAC must be determined experimentally. For example, the time for stress rupture of polyvinyl chloride (PVC) in gasoline and isopropyl alcohol is significantly less than in air.

Example 8.2. Consider a beam 9 feet long and 4 inches wide. The maximum deflection permissible in 1000 hours is 0.1 inch. The beam is to be loaded uniformly with 20 lb/ft. Determine the thickness required for the beam.

Note that the specification must include the use time of the object, since creep will occur continually and the maximum deflection will increase indefinitely. The available data for the mechanical properties are as follows.

Creep Modulus Data

	Modulus at specific times (psi $\cdot 10^3$)			
Stress (psi)	1 hr	10 hr	100 hr	1000 hr
2000	465	415	345	290
3000	460	405	320	270
4000	460	405	305	245

Stress Rupture Data

Stress (psi)	Time to rupture (hr)
6500	0.1
6000	28
5500	260
5000	27,000

Crazing Data

Stress (psi)	Time for craze initiation (hr)
6000	0.1
5000	0.4
4000	3
3000	100
2000	3000

$T_g = 50°C$.

Determine the criteria for failure: By interpolation of the data, at 1000 hours

Rupture stress = 5300 psi
Craze initiation stress = 2400 psi

Therefore, the maximum stress which can be applied without failure is 2400 psi. Maximum deflection = 0.1 inch.

For a uniformly loaded beam, the elastic deflection at the center is obtainable from the beam formula

$$y = \frac{5wl^4}{384EI} \tag{8.9}$$

where y = center deflection, ft
l = length of the beam, ft
w = uniform load on the beam, lb/ft
E = modulus of elasticity, psi
I = moment of inertia = $(1/12)T^3h$
h = beam width, ft
T = beam thickness, ft

This formula is used directly in the pseudoelastic method, with replacement of E by the creep modulus $E(t)$.

The moment of inertia is

$$I = \frac{hT^3}{12} = \frac{T^3}{36}$$

The weight of the beam itself should be included. The specific gravity of most plastics is close to 1. Taking this figure

$$\text{Weight/ft} = \left(\frac{4T}{12}\right)(62.4) \text{ lb/ft}^3 = 20.8T \text{ lb/ft}$$

$$\text{Total load/ft} = 20 + 20.8T$$

Substituting into Eq. 8.9

$$\frac{0.1}{12} = \frac{5(20 + 20.8T)(9^4)}{384[E(t) \cdot 144](T^3/36)}$$

At this point, a modulus value must be chosen. The modulus at 1000 hours is a function of stress, and the stress in the beam is not known at this point in the calculation. Since the maximum stress permissible is 2400 psi (due to the possibility of crazing), the modulus at 2000 psi is selected for a first try

$$E(t) = 290 \cdot 10^3$$

Substituting, we find that $T = 0.67$ feet = 8 inches.

However, the creep modulus was chosen for a stress of 2000 psi. The actual stress should now be determined. For a simple beam

$$S = \frac{Mc}{I}$$

where S = stress in the outer fiber
        c = distance from the neutral axis
        I = moment of inertia

The moment at the center for a uniformly loaded beam is

$$M = \frac{wl^2}{8} = \frac{[20 + (20.8)(0.67)]9^2}{8} = 343.6 \text{ ft} \cdot \text{lb}$$

$$S = \frac{Mc}{I} = \frac{(343.6)(0.67/2)}{(0.67^3/36)} = 13777 \frac{\text{lb}}{\text{ft}^2} = 96 \text{ psi}$$

This is a negligible stress, and a higher modulus should be used. To extrapolate the modulus to lower stress values, the equation of the form described previously can be used

$$E(t) = AS^k$$

Substituting the values from the creep modulus table given above, the best-fit equation for data at 1000 hr is

$$E(t) = 1.8 \cdot 10^6 \cdot S^{-0.24}$$

Now assume a stress of 100 psi. The computed modulus is $596 \cdot 10^3$. Utilizing this modulus in Eq. 8.9, the required thickness is 6.1 inches. The recomputed stress is 149 psi, which is sufficiently close to the assumed stress that no further recomputation is warranted.

8.7. BOLTZMANN SUPERPOSITION THEOREM

The previous calculations all involved static loads. Extension of these calculations to the more common case of time-varying loads requires special considerations, since the modulus and creep strain both vary with time. The superposition principle is frequently applied in the solution of elastic problems. In this process, the deformation produced by a complex loading can be calculated by dividing the load into several simpler parts. The total deformation is the sum of the deformations produced by the individual loads. For the superposition principle to be applicable, the total deformation must still be within the elastic region of the material. The extension of this approach to viscoelastic materials requires that the superposition be performed in time as well as for loads. To pursue the analogous comparison with elastic materials, the total load must be such that the material always behaves as a linear viscoelastic material.

The Boltzmann superposition theorem (BST) states that if loads are applied to a linear viscoelastic material at different times, the strains due to various stresses are additive in time. The power of the Boltzmann super-position principle is that creep data obtained at constant stress can be utilized directly to compute the strain for a time-dependent stress. Because creep data are obtained under constant load conditions, their use requires that the applied load be broken into parts which are constant from the time of initial application until the time at which the strain is to be computed.

For example, consider the application of the step function loading shown in Fig. 8.6A. To compute the resultant strain from creep data, the load must be decomposed into subloading parts, which can begin at any time but must remain unchanged until the end. The decomposition shown in Fig. 8.6B satisfies this requirement, but the decomposition in Fig. 8.6C does not. For

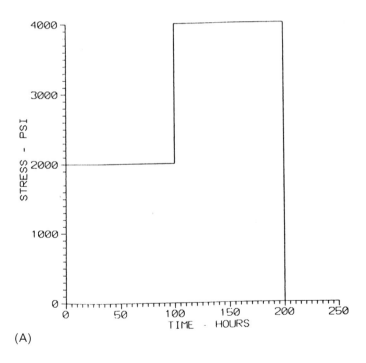

(A)

Figure 8.6. Decomposition of loading into component parts. (A) Applied load. (B1, B2) Boltzmann decomposition. (C1, C2) Incorrect decomposition—loads do not all exist at time when strain is determined.

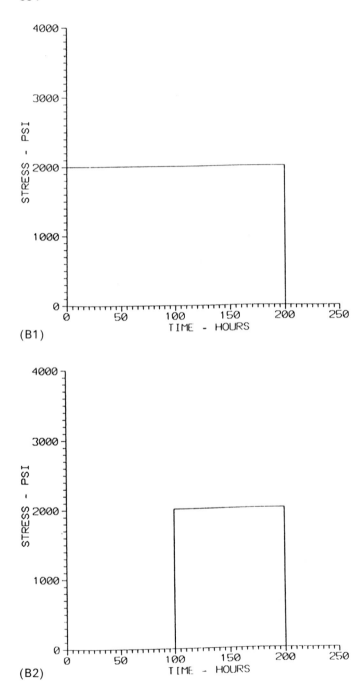

Figure 8.6. *Continued.*

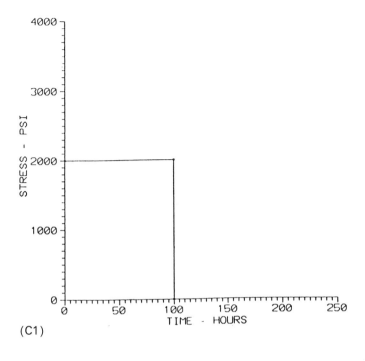

(C1)

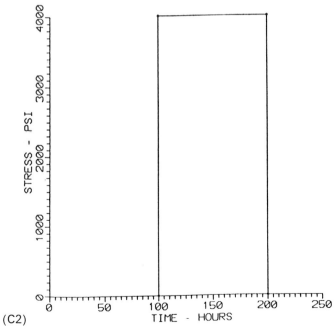

(C2)

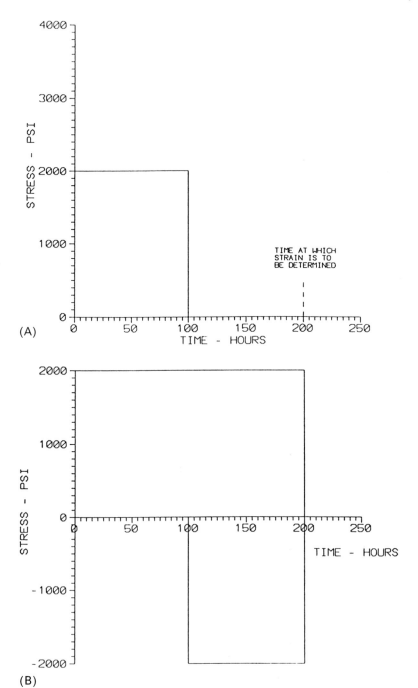

(A)

(B)

each part in Fig. 8.6B, the resultant strain can be found from the creep curve for the time of application of the individual parts of the load.

Expressed mathematically, this summation is

$$e = \sigma_1 J(t) + \sigma_2 J(t - t_2) + \sigma_3 J(t - t_3)$$

where e = total strain in the specimen at time t
  $J(t)$ = creep compliance of the material at time t
  $\sigma_1, \sigma_2, \sigma_3$ = stresses for the various subdivisions of the total applied load
  t_2, t_3 = time at which loads 2, 3 are applied

If the load is removed, the superposition can be performed as shown in Fig. 8.7, resulting in the stress relaxation curve for the material.

Experimental verification of the Boltzmann principle has been performed for several plastics. The superposition works reasonably accurately for most linear viscoelastic materials but is less accurate for loads that decrease with time than for loads that increase with time. When the loads are reversed in direction, the accuracy of the calculation is still poorer. If the load varies continually with time, the continuous function of time can be subdivided into small discrete steps (see Fig. 8.8). The subdivisions can be made with whatever accuracy is required.

Sample Problems—Boltzmann Superposition

Example 8.3.   A 20-inch-long bar is held in tension under a 2000-psi stress for 100 hours and then under a 4000-psi stress for 100 hours after that. Find the length at the end of the stress cycle for a dry sample. The creep modulus is given by

$$E(t) = 1651 \cdot 10^3 t^{-0.094} S^{-0.18}$$

with t in hrs, S in psi.

Solution: The load can be broken into a 2000-psi load for 200 hrs and another 2000-psi load for 100 hours starting at 100 hours

Modulus at 200 hours and 2000 psi = 255,000 psi

Strain due to 2000 psi load for 200 hours = $\dfrac{\text{stress}}{\text{modulus}} = \dfrac{2000}{255,000} = 0.0078$

Figure 8.7. Decomposition of interrupted loading into component parts. (A) Applied load. (B) Decomposition.

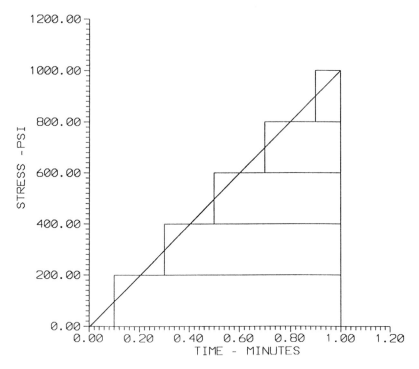

Figure 8.8. Approximation of continuous loading by small increments.

Now for the next applied stress, although the Boltzmann breakdown indicates that the stress is 2000 psi, we know that a linear viscoelastic material would deflect less than nylon, for nylon's modulus decreases with increasing stress. There are various ideas (not really theories) as to how to handle this. Probably the simplest, and as accurate as any, is to use common sense and utilize the modulus for the maximum load of the cycle at that time. Therefore we will use the modulus of nylon at a stress of 4000 psi, even if we consider that we apply a stress of 2000 psi for the second loading. The modulus at 4000 psi at 100 hr is $E(t) = 240 \cdot 10^3$ psi

$$\text{Strain due to 2000 psi load for 100 hours} = \frac{\text{stress}}{\text{modulus}} = \frac{2000}{240,000} = 0.0083$$

Total strain = 0.016
Final length = 20(1 + 0.016) = 20.32

Example 8.4. The same sample is loaded in tension at 2000 psi for 100 br and then unloaded. Find the length of a 10-ft sample after it has been unloaded for 100 hr.

Solution: Divide the load according to the BST as shown in Fig. 8.7B.

Modulus at 2000 psi for 200 hr

$$E(t) = (1651 \cdot 10^3)(200)^{-0.094}(2000)^{-0.18} = 255.4 \cdot 10^3 \text{ psi}$$

$$\text{Strain at 200 hr} = \frac{\text{stress}}{\text{modulus}} = \frac{2000}{255.4 \cdot 10^3} = 0.00783$$

Modulus at 2000 psi for 100 hr $= 272 \cdot 10^3$ psi

$$\text{Strain at 100 hr} = \frac{-2000}{272 \cdot 10^3} = -0.00734$$

Net strain $= 0.00049$

Length $= 10.0049$ ft

Note that since the stress could easily be specified to be the same in both parts of the cycle, there is no problem in citing the modulus as that at 2000 psi.

Example 8.5.   A sample of the same material is loaded at 10,000 psi for 10 hr in tension, then for 10 hr in compression. This cycle is repeated. Find the strain after the two cycles (the specimen is in compression at the end of the second cycle).

Solution: The loading condition is shown in Fig. 8.9a and the BST breakdown is shown in Fig. 8.9b. Simply looking at the BST diagram suggests that stresses of 2000 psi are being applied. However, the BST breakdown is primarily for computation, and should not overshadow the reality of the problem. The stress on the sample as applied is ± 1000 psi, and the best estimate for the creep rate should be obtained by using the compliance for that stress value. Values at various times for the creep modulus calculated from the creep modulus equation given in Example 8.3 are given below. The compliance is computed as the reciprocal of the modulus.

Time (hr)	$E(t)$ (psi)	$J(t)$ (psi^{-1})
10	$3.83 \cdot 10^5$	$2.60 \cdot 10^{-6}$
20	3.59	2.78
30	3.46	2.89
40	3.37	2.97

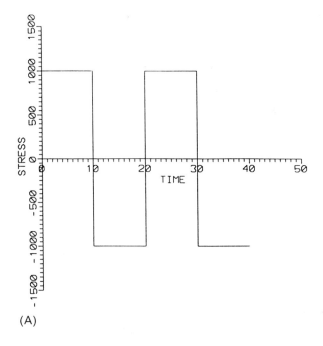

(A)

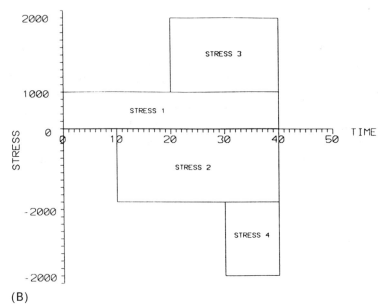

(B)

Figure 8.9. Superposition at cyclic load. (A) Load applied. (B) Decomposition.

$$e = \sum \sigma_i J(t_f - t_i)$$
$$e = 1000J(40) - 2000J(30) + 2000J(20) - 2000J(10) = 0.00245$$

Example 8.6. A polymer has a compliance given by the equation
$$J(t) = 10^{-6}t^{0.25} \ (t \text{ in minutes})$$
A load is applied which starts at zero and increases in tension until the stress is 1000 psi after 1 minute. Find the strain in the sample at that time.

Solution: Since the load is continually increasing, BST requires an approximate solution. Five steps should provide sufficient accuracy as shown in Fig. 8.8.

Mathematically, the strain is
$$e = 200\{J(1 - 0.1) + J(1 - 0.3) + J(1 - 0.5) + J(1 - 0.7) + J(1 - 0.9)\}$$
$$= 200 \cdot 10^{-6}\{0.9^{0.25} + 0.7^{0.25} + 0.5^{0.25} + 0.3^{0.25} + 0.1^{0.25}\}$$
$$= 8.06 \cdot 10^{-4}$$

Example 8.7. A beam of PVC supported at both ends has a uniform load of 12 lb/ft. The beam is 18 inches long, 1 inch wide, and 3 inches thick. The load is applied for 3 years and released for 1 year. What is the sag of the beam at its center? The creep compliance of the PVC is $J(t) = 1.9 \cdot 10^{-5}t^{0.06}$ (t in seconds). Neglect the weight of the beam.

Solution: The center deflection is given by the elastic formula
$$y = \frac{5wl^4}{384EI}$$
Replacing for the viscoelastic material
$$y = \frac{5wl^4}{384E(t)I} = \frac{5wl^4 J(t)}{384I}$$
For the specified loading
$$y = \frac{5wl^4[J(4 \text{ years}) - J(1 \text{ year})]}{384I}$$
and
$$I = \frac{h^3 b}{12} = \frac{(\frac{3}{12})^3(\frac{1}{12})}{12} = 1.08 \cdot 10^{-4}$$
where h = beam height
b = beam base dimension

Substituting

$$y = \frac{5 \cdot 12 \cdot 1.5^4 [1.9 \cdot 10^{-5} \{(1.26 \cdot 10^8)^{0.06} - (3.15 \cdot 10^7)^{0.06}\}]}{384 \cdot 8.04 \cdot 10^{-4}}$$

$$y = 4.57 \times 10^{-3} \text{ feet} = 0.055 \text{ inches}$$

8.8. HEREDITY INTEGRAL

As shown in the previous section, the Boltzmann superposition can be used to predict the behavior of a linear-viscoelastic material under a load which varies with time. Its application is quite straightforward for loads which vary only infrequently and vary discontinuously with time. The calculation may become laborious if many load fluctuations occur. However, if the load varies continously with time the BST procedure as outlined yields only an approximate solution and becomes vary tedious. This procedure involves the replacement of a continuous function by a set of discrete increments. In the limit this corresponds to the processes of calculus, and an exact solution can be found mathematically. For each increment of stress σ_0 in the BST procedure, the resultant strain at time t for a material with a creep compliance $J(t)$ is

$$e(t) = \sigma_0 J(t) \tag{8.10}$$

We wish to calculate the strain at time t_f. If the stress σ_i is applied at a time t_i later than time zero, the time of application of the load is $(t_f - t_i)$. The resultant strain at the time of interest t_f is

$$e(t_f) = \sigma_i J(t_f - t_i) \tag{8.11}$$

Summing over all the incremental stresses σ_i applied at different times t_i

$$e(t) = \sum \sigma_i J(t_f - t_i) \tag{8.12}$$

In the limit, if the stress increments are made infinitesimally small

$$e(t) = \sum J(t_f - t_i) \, d\sigma = \int J(t_f - t_i) \, d\sigma_i \tag{8.13}$$

This form of integral is called a Stieltjes-type integral. In this form the procedure for the evaluation of the strain is not very evident. To obtain a more useful form, replace $d\sigma_i$ by

$$d\sigma_i = \left(\frac{d\sigma}{dt_i}\right) dt_i$$

and assuming that $t = 0$ is counted from the first time a stress is applied to the sample, then

$$e(t) = \int_0^{t_f} J(t_f - t_i)\left(\frac{d\sigma}{dt_i}\right) dt_i \tag{8.14}$$

Analogously, if the stress is to be computed from a strain history

$$\sigma(t) = \int E(t_f - t_i)\left(\frac{de}{dt_i}\right) dt_i \tag{8.15}$$

where $E(t)$ is the creep modulus $[e(t)/\sigma]$ of the material.

If a discontinuous stress σ_0 is applied at $t = 0$ and then continuously varied from that point on

$$e(t) = \sigma_0 J(t_f) + \int J(t_f - t_i)\left(\frac{d\sigma}{dt_i}\right) dt_i \tag{8.16}$$

This equation can be put in another form by integrating by parts. Since $d(uv) = u\,dv + v\,du$, then integrating

$$\int u\,dv = uv - \int v\,du$$

Setting

$$u = J(t_f - t_i) \quad \text{and} \quad dv = \left(\frac{d\sigma}{dt_i}\right) dt_i$$

then

$$e(t_f) = \sigma(0)J(t_f) + \sigma(t)J(t_f - t)\Big|_0^{t_f} - \int_0^{t_f} \sigma(t)\,dJ(t_f - t_i)$$

$$= \sigma(0)J(t_f) + \sigma(t_f)J(0) - \sigma(0)J(t_f) - \int_0^{t_f} \sigma(t)\,dJ(t_f - t_i)$$

$$= \sigma(t_f)J(0) - \int_0^{t_f} \sigma(t)\,dJ(t_f - t_i) \tag{8.17}$$

It is customary to convert the independent variable t_i to $(t_f - t_i)$. Then

$$d(t_f - t_i) = -dt_i$$

Substituting into Eq. 8.17 we obtain the common form

$$e(t_f) = \sigma(t_f)J(0) + \int_0^{t_f} \sigma(t)\left(\frac{dJ(t_f - t_i)}{d(t_f - t_i)}\right) dt_i \tag{8.18}$$

Note that t_f is a constant, and t_i is the variable of integration. Either of these forms of Eq. 8.17 or Eq. 8.18 can be used to evaluate the total strain due to the stress history. One integral form may be more easily evaluated than another for a particular problem. If the stress varies continuously and a simple mathematical expression can be found for the stress as a function of time, then the form containing $d\sigma/dt$ may be preferred. If the stress has a discontinuity in its time variation, then the derivative in this form cannot be readily evaluated and the integral cannot be solved without the use of Dirac functions, which makes for an inconvenient solution. Therefore the form involving differentiation of the creep compliance is to be preferred, as the creep compliance is always continuous and can always be differentiated. The creep compliance must be in an analytical form. As described previously, the most representative and simplest form is

$$J(t) = at^n \tag{8.19}$$

This form is recommended for ease of usage in solving practical problems of determining the deformation of specimens by the heredity integral.

Plastic Memory

The heredity integral gives the strain at any time as a function of the past history or heredity of the specimen. It emphasizes that the behavior of a specimen which is viscoelastic is a function not only of the applied loading at the moment but also of all previous loadings. Although the effect of a load application decreases with time after the load is removed, theoretically any loading will affect the properties for all time thereafter. These effects are very evident during creep testing of samples. If a small preload is applied to a sample to test the extensometer and data acquisition systems and then removed, the sample continuously and slowly shrinks on removal of the load, making the initial length of the sample difficult to determine. Molded samples have stresses applied during the molding cycle, and sensitive measurements prior to creep testing on some samples show that the sample's length varies slowly and erratically with time as a result of the complex loading pattern during molding. If a sample has a load history of applied tension and then compression and then all load is removed, in some cases the sample will shrink, elongate, and then shrink again in delayed response to the previous load history. To demonstrate that all problems, no matter how trivial, requiring the strain produced from a stress history can be solved utilizing the heredity integral, let us first evaluate some elementary examples, which could be solved much more directly using the BST approach. However, by examining these simple problems the use of the heredity integral should be clarified.

Example 8.8. Consider first a constant stress σ applied at time $= 0$ and maintained constant until the time t_f, at which we wish to know the resultant strain. The answer must obviously be $\sigma J(t_f)$ from the definitionn of $J(t)$. The first decision must be which form of the heredity integral to use. Although the stress remains constant during the time of its application, it discontinuously rises from zero to its constant magnitude at t_0, and therefore the derivative $d\sigma/dt$ is not easily evaluated. The second form (Eq. 8.18) is therefore preferred. To evaluate the integral, the general form of the creep compliance of Eq. 8.19 will be used

$$e(t_f) = \sigma(t_f)J(0) + \int_0^{t_f} \sigma(t)\left(\frac{dJ(t_f - t_i)}{d(t_f - t_i)}\right) dt_i$$

Using

$$J(t) = at^n$$

then

$$J(t_f - t_i) = a(t_f - t_i)^n$$

$$\frac{dJ(t_f - t_i)}{d(t_f - t_i)} = an(t_f - t_i)^{n-1}$$

Since $J(0) = 0$, the first term in the evaluation of the strain is also zero. Substituting the expression given above into the integral

$$e(t_f) = \sigma an \int (t_f - t_i)^{n-1} dt_i$$

$$= \sigma a(t_f - t_i)^n \Big|_0^{t_f} - -\sigma a[(0)^n \quad t_f)^n] \blacksquare \sigma J(t_f)$$

Example 8.9. To reanalyze a simple example previously analyzed by the BST method, let us consider a stress σ applied at time t_0 and maintained until time t_1, at which time the stress drops to zero. We wish to find the resultant strain at later times. From our previous analysis we know the result is

$$e(t) = \sigma[J(t_f) - J(t_f - t_i)]$$

Utilizing the same expression for $J(t)$ as in the previous example, the heredity integral can be evaluated by adding the contributions from each of the time regions

$$e(t_f) = \sigma an \int_0^{t_1} (t_f - t_i)^{n-1} \, dt + an \int_{t_1}^{t_f} (0)(t_f - t_i)^{n-1} \, dt_i$$

$$= -\sigma a (t_f - t_i)^n \Big|_0^{t_1} = -\sigma a[(t_f - t_1)^n - t_f^n] = \sigma[J(t_f) - J(t_f - t_1)]$$

as required.

Example 8.10. Let us repeat Example 8.6 which was analyzed by the BST method. In utilizing the Boltzmann superposition principle, a continually increasing load had to be approximated by a series of discrete loads. The heredity integral can find the exact solution

$$J(t) = 10^{-6} t^{0.25} \quad (t \text{ in seconds})$$

Since the stress is varying with time, use the form of Eq. 8.16.
To emphasize the variable of integration, let us call that variable τ

$$\sigma = ct = 1000\tau$$

$$\frac{d\sigma}{d\tau} = 1000 \text{ lb/min}$$

$$J(t_f - \tau) = 10^{-6}(t_f - \tau)^{0.25}$$

substituting into the heredity integral

$$e = \int_0^{t_f} 10^{-6}(t_f - \tau)^{0.25} \times 10^3 \, d\tau = 10^{-3} \int_0^{t_f} (t_f - \tau)^{0.25} \, d\tau$$

$$= -\frac{10^{-3}}{1.25} (t_f - \tau)^{1.25} \Big|_0^{t_f}$$

$$= -\frac{10^{-3}}{1.25} [0 - (t_f)^{1.25}] = \frac{t_f^{1.25} \times 10^{-3}}{1.25} = 8.00 \times 10^{-4} \text{ as expected.}$$

It is also possible to use the other form

$$e = \sigma(t_f)J(0) + \int_0^{t_f} \sigma(\tau) \frac{\partial J(t_f - \tau)}{\partial (t_f - \tau)} \, d\tau$$

$$J(t_f - \tau) = a(t_f - \tau)^n$$

$$\frac{\partial J(t_f - \tau)}{\partial(t_f - \tau)} = an(t_f - \tau)^{n-1}$$

$$\varepsilon = anc \int_0^{t_f} \tau(t_f - \tau)^{n-1} \, d\tau$$

$$c = 10^3 \quad n = 0.25 \quad a = 10^{-6}$$

$$e = 0.25 \times 10^{-3} \int_0^{t_f} \tau(t_f - \tau)^{n-1} \, d\tau$$

$$e = 0.25 \times 10^{-3} \int_0^1 \tau(t_f - \tau)^{-0.75} \, d\tau$$

This integral is not as simple to integrate as the previous one, and is indefinite at $\tau = t_f$. Using Simpson's rule for numerical integration, an approximate solution can be found for the integral as τ approaches t_f.

For $\quad \tau = 1 - 10^{-4} \quad \varepsilon = 0.00076$

$\qquad \tau = 1 - 10^{-5} \quad \varepsilon = 0.00079$

and the integral therefore approaches 8×10^{-4} as does the other form. Clearly the choice of forms of the heredity integral depends on the ease of solution for the particular problem.

This problem is similar to a tensile test, in which the stress increases linearly with time. To solve the general expression for the strain produced in a tensile test

$$J(t) = At^n \quad \sigma = ct$$

$$e = Ac \int_0^{t_s} (t_f - \tau)^n \, d\tau = -\frac{Ac}{n+1} (t_f - \tau)^{n+1} \Big|_0^{t_f}$$

$$e(t) = \frac{Ac}{n+1} t^{n+1}$$

Note that although stress varies with time linearly, strain does not!

Example 8.11. As an example of a more complex loading pattern, let us consider a sinusoidal fatigue loading, as produced in rotating machinery. In this case the applied stress σ applied with a frequency f is given by

$$\sigma(t) = \sigma_{max} \sin(2\pi f t)$$

For a plastic with a creep compliance of the form given by Eq. 8.19, the strain at any time is given by

$$e(t_f) = an\sigma_{max} \int_0^{t_f} \sin(2\pi ft)(t_f - t)^{n-1} \, dt$$

For other than relatively simple applied loadings, the integral may become anoyingly complex. However, the integral can always be numerically integrated for various times to obtain the strain. It can be seen from the form of the equation for strain that the stress and strain will not be in phase as a result of the viscoelastic properties of the plastic, with the strain lagging behind the stress. The lag angle is largest initially, and approaches a constant value after several cycles. The magnitude of the maximum strain is also small initially, but also approaches a constant value. These effects will be described further in the section on the complex modulus of plastics, in which a different approach is taken to alternating loading. However, it should be observed that the Heredity integral provides a means of analyzing the strain produced by any type of loading.

9

Dynamic Behavior
of Polymers

9.1. INTRODUCTION

The mechanical properties of polymers under varying loads have been
discussed throughout this book, in sections under fatigue loading, the effect
of strain rate on the mechanical properties, etc. The analysis of the
deformation of the sample under either constant or varying loads has also
been discussed, and calculations have been performed employing either the
Boltzmann superposition theorem or the heredity integral. In this chapter
another form of analysis, usually reserved for rapidly varying loads of small
amplitude, is considered, resulting in the concept of the complex modulus
of materials. However, it is often useful to analyze the behavior of viscoelastic materials under both constant and varying loads using mechanical
analogues, and these analogues help understand the behavior of real
materials. These analogues are therefore discussed first.

9.2. MECHANICAL ANALOGUES OF POLYMERS

The internal atomic motions within a polymeric material produce the
viscoelastic effects we have been discussing. Additional understanding of
these properties is obtained by analyzing the behavior of mechanical models
which also behave in a viscoelastic manner. If the model faithfully reproduces
the polymer's behavior, then predictions developed from analysis of the
model will be useful for engineering extrapolations of the data. Such accurate

modeling is not easily performed unless the model has many elements. However, models with relatively few elements still produce results which are useful in helping us understand the behavior of polymers under a variety of conditions. These simple models are also the starting point for more complete models.

A polymeric material has several primary modes of response to an applied stress

1. Instantaneous stretching of the bonds within the carbon backbone of the molecule
2. Retarted (slow) coiling and uncoiling of molecular chains as a result of kink formations and kink straightening
3. Permanent deformation as a result of relative motions of chains past each other producing irreversible dimensional changes.

A spring simulates the instantaneous elastic response above of the carbon-carbon backbone atoms. Since a spring is ideally elastic, the appropriate equation is

$$\sigma = E\varepsilon$$

where E is the spring constant or modulus of the spring
σ is the applied stress
ε is the resultant strain in the spring

If the stress varies with time, the spring responds instantaneously, and

$$\sigma(t) = E\varepsilon(t) \tag{9.1}$$

Permanent flow due to chain sliding and slow kink motion are modeled utilizing dashpots containing a fluid which resists the motion of a paddle through it. The slow permanent opening of a dashpot represents chain sliding. A dashpot connected in parallel to a spring models the slow opening and formation of kinks as a result of applied loads. Upon release of the load, the system slowly retracts.

Flow is a fluid property described in terms of the viscosity. Due to the viscous effects, a shear stress along the x direction generates a velocity gradient along the y direction

$$\sigma_x = \eta \frac{dv_x}{dy}$$

where η is the viscosity of the fluid
v_x is the velocity in the x direction
σ_x is the applied shear stress

Since

$$v_x = \frac{\partial \mu_x}{\partial T}$$

where μ_x is the displacement along the x direction, then

$$\sigma = \eta \frac{\partial}{\partial y}\left(\frac{\partial \mu_x}{\partial t}\right) = \eta \frac{\partial^2 \mu_x}{\partial y \, \partial t} = \eta \frac{\partial}{\partial t}\left(\frac{\partial \mu_x}{\partial y}\right) \qquad (9.2)$$

Shear strain is defined as

$$\varepsilon_x = \frac{\partial \mu_x}{\partial y}$$

and in terms of the variables utilized in solids to describe displacements,

$$\sigma = \eta \frac{\partial \varepsilon}{\partial t} \qquad (9.3)$$

Note that although flow of polymeric fluids is in general non-Newtonian, the mechanical behavior of solid polymers is modeled with Newtonian fluid dashpots.

9.3. MAXWELL MODEL

This simple model is sometimes utilized in the analysis of the stress relaxation process. It consists of a spring and dashpot connected in series (Fig. 9.1). It is of interest in modeling the reduction in applied stress to a tightened or elongated bolt.

   For stress relaxation studies the specimen is deformed in a process which is considered instantaneous and the length of the specimen is then held constant. Internal stresses within the specimen decay with time as a result of slow kink opening and chain sliding. Therefore

$$\varepsilon_T = \text{constant}$$

where ε_T = the total strain in the model

$$\varepsilon_T = \varepsilon_{\text{spring}} + \varepsilon_{\text{dashpot}}$$

$$\frac{d\varepsilon_T}{dt} = 0 = \left(\frac{d\varepsilon_{\text{sp}}}{dT}\right) + \left(\frac{d\varepsilon_{\text{dash}}}{dt}\right) \qquad (9.4)$$

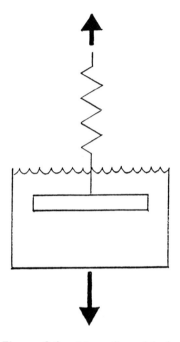

Figure 9.1.   Maxwell model of a viscoelastic material.

substituting Eq. 9.1 and 9.3

$$0 = \frac{1}{E}\frac{d\sigma}{dt} + \frac{\sigma}{\eta}$$

$$\frac{d\sigma}{\sigma} = -\frac{E}{\eta}dt$$

Integrating between σ_0, the applied stress at $t = 0$, and the stress σ at time t

$$\sigma = \sigma_0 e^{-Et/\eta} = \sigma_0 e^{-1/\tau}$$

Where $\tau = \eta/E$. This is the relaxation time, the time for the stress to decay to e^{-1} of the original stress.

The stress decays exponentially with time, asymptotically approaching zero. This behavior is observed for some thermoplastics with no permanent cross-links and no large-scale entanglements. However, either of these two restraints will prevent relative chain motion from continuing indefinitely. The curve for these types of materials will show some stress relaxation due

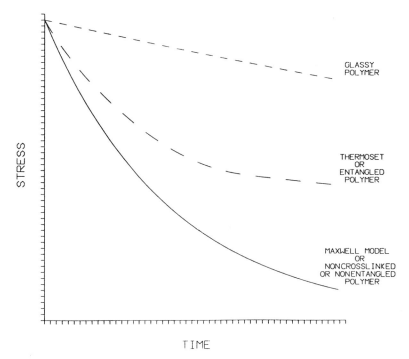

Figure 9.2. Stress relaxation data.

to kink opening, but the stress will not decrease completely to zero. The decay of stress with time is shown schematically in Fig. 9.2.

9.4. VOIGT CELL

The Maxwell model gives unsatisfactory results for creep loading at constant stress, as the spring opens instantaneously and the dashpot produces an elongation which increases linearly with time. A more interesting model for creep is the Voigt element, which produces deformations which can be qualitatively compared to those of slow kink opening under load. This model consists of a dashpot and a spring connected in parallel (Fig. 9.3). Initially the spring is at its no-load length, and the load is resisted primarily by the dashpot as a result of the rate of motion. As the element's length increases, the loading rate decreases, and an increasing proportion of the load is supported by the spring. The rate of elongation therefore decreases, reducing the resistance provided by the dashpot. Eventually, the entire load is carried

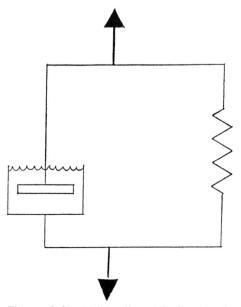

Figure 9.3.   Voigt cell model of a viscoelastic material.

by the spring and the total extension approaches a limiting value. This behavior corresponds to that expected for kink motion.

In this model

$$\varepsilon_{\text{spring}} = \varepsilon_{\text{dashpot}}$$

and the total stress σ_T is

$$\sigma_T = \sigma_{\text{spring}} + \sigma_{\text{dashpot}}$$

Substituting Eqs. 9.1 and 9.3

$$\sigma_T = E\varepsilon + \eta \frac{d\varepsilon}{dt}$$

For creep loading, the stress σ_T is constant. The equation can be solved by a variety of techniques. Rewriting

$$\frac{d\varepsilon}{dt} + \frac{E}{\eta}\varepsilon = \frac{\sigma_T}{\eta}, \qquad \sigma_T = \text{constant} \tag{9.5}$$

This equation is of the form

$$\frac{dy}{dx} + f(x)y = c$$

which can be solved by use of an integrating factor. The integrating factor (IF) is

$$IF = e^{\int f(x)\,dx}$$

If both sides of the equation are multiplied by the IF and integrated, the left-hand side always equals $(y)(IF)$ after integration. Therefore

$$(y)(IF) = \int ce^{\int f(x)}\,dx$$

As an example, let us solve

$$\frac{dy}{dx} + xy = c$$

$$IF = e^{\int x\,dx} = e^{x^2/2}$$

multiplying by the IF and setting the LHS to $(y)(IF)$,

$$y\frac{x^2}{2} = c\int e^{x^2/2}\,dx$$

and integration of the right-hand side completes the solution.

To demonstrate that the IF did convert the left term to $(y)(IF)$, differentiation returns to $(LHS)(IF)$

$$\frac{d}{dx}(ye^{x^2/2}) = e^{x^2/2}\left(\frac{dy}{dx} + xy\right)$$

Utilizing this technique for the equation for the Voigt cell

$$\frac{d\varepsilon}{dt} + \frac{E\varepsilon}{\eta} = \frac{\sigma}{\eta}, \qquad IF = e^{\int E\,dt/\eta} = e^{Et/\eta}$$

$$\varepsilon e^{Et/\eta} = \int \frac{\sigma}{\eta} e^{Et/\eta}\,dt$$

$$\varepsilon e^{Et/\eta} = \frac{\sigma}{\eta}\left(e^{Et/\eta}\frac{\eta}{E}\right) + c$$

$$\varepsilon = \frac{\sigma}{E} + ce^{-Et/\eta}$$

To obtain the constant of integration, for creep at $t = 0$ $\varepsilon = 0$. Therefore

$$\varepsilon = \frac{\sigma}{E}(1 - e^{Et/\eta}) = \frac{1}{E}(1 - e^{-t/\tau})$$

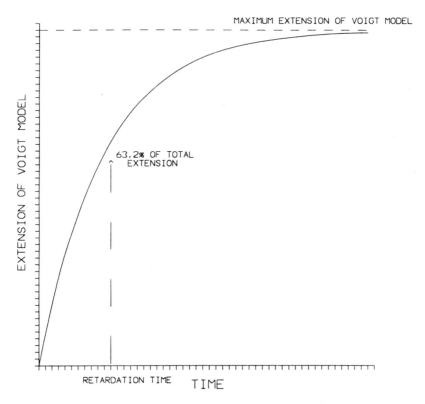

Figure 9.4.   Variation of strain with time for a Voigt cell.

where $\tau = \eta/E$ is the retardation time, the time for $(1 - e^{-1})$ or 63% of the total creep to occur. The variation of strain with time is shown in Fig. 9.4.

It is informative to replot the creep curve in logarithmic form (Fig. 9.5). In this form, it is easier to demonstrate the general behavior of the Voigt element. For times significantly smaller than τ, the element does not have sufficient time to respond, and the deformation is negligible. For times significantly larger than τ, the element has essentially opened fully, and the elongation is fixed at the maximum amount. The main movement of the element occurs at times within one order of magnitude of the retardation time. The Voigt element is taken to model kink motion. However, it can be appreciated that kinks of different sizes will respond at different rates to an applied stress; larger kinks will respond more slowly than smaller ones. One Voigt element is therefore not expected to portray accurately the behavior of a real material with various kink configurations. A Voigt element may be considered to represent the behavior of a particular size of kink, whose

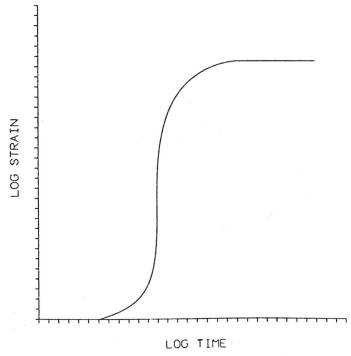

LOG TIME

Figure 9.5. Log plot of a creep curve. Note that the creep curve is almost a step function when plotted in this way.

response to applied loading is within the time τ. A complete representation should include many such elements. The behavior of each element in this array can be considered as a step function at time τ.

Retardation Time Spectrum

Since a Voigt cell has a specific retardation time, it primarily models the behavior of a plastic with kinks of one particular size, which all respond to an applied stress with the same velocity of motion. A much more accurate model would include cells of different retardation times, more closely simulating the response of a molecular mass having many different sizes of kinks. This model would contain multiple Voigt cells in series, each with a different retardation time. The equation for the deformation of such a series of n cells is then

$$e(t) = \sigma \sum_i \frac{1}{E_i} (1 - e^{-t/\tau_i})$$

and creep compliance is

$$J(t) = \sum_i \frac{1}{E_i} (1 - e^{-t/\tau_i})$$

The variation of the creep compliance with time of any plastic can be modeled accurately by taking a sufficient number of terms in this equation. Although there is a finite number of kinks in a sample, it is impossible to determine exactly which retardation times to choose, and therefore it is resonable to choose retardation times varying by one order of magnitude, yielding terms for retardation times of 0.001, 0.01, 0.1, 1, 10, etc. Computer analysis of the data is used to perform this evaluation.

In principle, any material could be modeled exactly by considering an infinite number of cells, yielding a continuous variation of retardation times, or a retardation time spectrum. In this limit

$$J(t) = \int_0^\infty \frac{1}{E_\tau} (1 - e^{-t/\tau}) \, d\tau$$

Although this seems more complex, it avoids the difficulty in choosing the retardation times and the problems with working with discrete intervals. Since the time scale for creep experiments varies so greatly, the data are commonly plotted and reported at logarithmically increasing time values. The independent variable can conveniently be converted into the logarithm of the retardation time, and the infinitesimal contribution to the compliance of the material due to the Voigt cell of retardation time with logarithms varying infinitesimally about $\log \tau$ is defined as follows:

$$J(t) = \int_0^\infty \frac{\tau}{E_\tau} (1 - e^{-t/\tau}) \frac{d\tau}{\tau}$$

and since $d \ln \tau = d\tau/\tau$, and setting $L = \tau/E$, then

$$J(t) = \int_0^\infty L_i (1 - e^{-t/\tau}) \, d \ln \tau$$

and similarly for the modulus:

$$E(t) = \int_0^\infty H e^{-t/\tau} \, d \ln \tau$$

where $H = \tau E(\tau)$.

Strain Relaxation of a Voigt Cell

If the specimen has a strain ε_0 at $t = 0$ and the load is removed, the specimen's length will slowly return toward its original value. In this case the general equation of the Voigt cell becomes

$$\frac{d\varepsilon}{dt} + \frac{E}{\eta}\varepsilon = \frac{\sigma}{\eta} = 0$$

Solving by separation of variables or by use of an integrating factor yields

$$\varepsilon = \varepsilon_0 e^{-Et/\eta} = \varepsilon_0 e^{-t/\tau}$$

and the length returns asymptotically to its initial value.

9.5. FOUR-PARAMETER MODEL

The Voigt element models kink motion. A more complete model, which includes the instantaneous deformation and permanent chain sliding, requires additional elements as shown in Fig. 9.6. Under constant stress, each of the

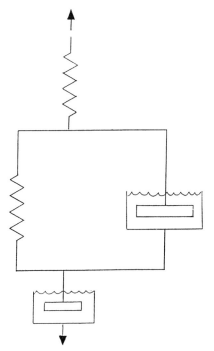

Figure 9.6. A four-parameter model of a polymer.

three components (spring, Voigt cell, dashpot) acts independently, the total deformation being the sum of all three. The summation equation for the creep compliance is

$$J(t) = \frac{\varepsilon_T}{\sigma} = \frac{1}{E_1} + \frac{1}{E_2}(1 - e^{-E_2 t/\eta_2}) + \frac{t}{\eta_3}$$

where E_1 = spring constant of separate spring
E_2 = spring constant of spring in Voigt cell
η_2 = viscosity of dashpot in Voigt cell
η_3 = viscosity of separate dashpot

The behavior of this model is shown in Fig. 9.7.

Periodic Loading

The fundamental equation for a Voigt cell under an applied periodic load is

$$\sigma_0 \cos \omega t = E\varepsilon + \eta \frac{d\varepsilon}{dt}$$

Rearranging to yield the form appropriate for solution by use of an integrating factor

$$\frac{d\varepsilon}{dt} + \frac{E\varepsilon}{\eta} = \frac{\sigma_0 \cos \omega t}{\eta}$$

Solving

$$IF = e^{\int E\,dt/\eta} = e^{Et/\eta}$$

$$\varepsilon e^{Et/\eta} = \frac{\sigma_0}{\eta} \int \cos \omega t\, e^{Et/\eta}\, dt$$

$$\varepsilon e^{Et/\eta} = \frac{\sigma_0}{\eta} \left[\frac{e^{Et/\eta}[(E/\eta) \cos \omega t + \omega \sin \omega t]}{(E^2/\eta^2) + \omega^2} \right] + c$$

If $\tau = \eta/E$, then

$$\frac{E}{E^2 + \omega^2 \eta^2} = \frac{1}{E[1 + (\omega^2\eta^2/E^2)]} = \frac{1}{E[1 + \omega^2\tau^2]}$$

and

$$\frac{\omega\eta}{E^2 + \omega^2\eta^2} = \frac{\omega\eta}{E^2[1 + (\omega^2\eta^2/E^2)]} = \frac{\omega\eta/E}{E[1 + \omega^2\tau^2]} = \frac{\omega\tau}{E[1 + \omega^2\tau^2]}$$

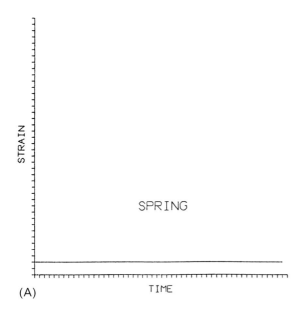

(A)

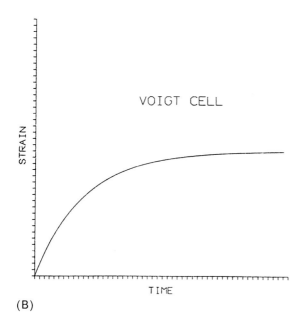

(B)

Figure 9.7. Creep response of the four-parameter model. (A) Spring (B) Voigt cell.

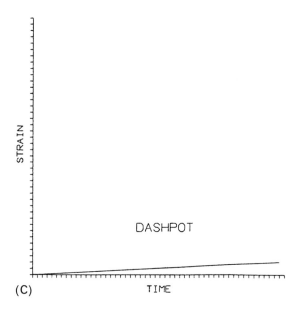

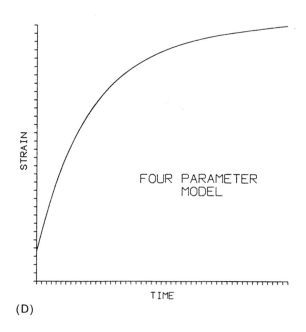

Figure 9.7. (C) Dashpot. (D) Superposition of all elements.

Then

$$\varepsilon = \sigma_0 \left[\frac{1}{E[1 + \omega^2\tau^2]} \cos \omega t + \frac{\omega\tau}{E[1 + \omega^2\tau^2]} \sin \omega t \right] + ce^{-Et/\eta} \qquad (9.6)$$

The sum of sine and cosine terms is a sinusoidally varying strain which is out of phase with the applied stress. The first term can be considered the in-phase component, or the elastic part of the response, and the second term the part 90° out of phase or the viscous part of the response. The third term is the transient response, which decays rapidly until steady state is reached.

The part in phase is symbolized by $J'(\omega)$, defined as the storage compliance at the cell

$$J'(\omega) = \frac{1}{E[1 + \omega^2\tau^2]}$$

when $\tau = 0$, $J'(\omega) = 1/E$ and the model behaves completely elastically. The part out of phase is symbolized by $J''(\omega)$, the loss compliance of the cell

$$J''(\omega) = \frac{\omega\tau}{E[1 + \omega^2\tau^2]}$$

If the applied load has a very low frequency, then the dashpot in the Voigt cell can open readily, and the strain can follow the stress accurately in time. As the frequency of applied loading increases, the viscous resistance of the dashpot prevents rapid response, and the deformation of the cell begins to lag behind the applied stress. The maximum lag angle occurs when the period of loading corresponds to the retardation time of the cell. For higher frequencies, the load is varied so rapidly that the viscous dashpot cannot begin to move in the small time frame, and the Voigt cell again begins to act as an elastic member. This is a model of the response of the kinks to an applied load. The kinks can open easily if the applied load is slowly applied and frozen at sufficiently high loading rates, behaving as elastic members.

$J'(\omega)$ and $J''(\omega)$ are plotted vs. the period of the applied loading in Fig. 9.8. Since at short times of loading the kinks cannot open, the cell behaves elastically, and $J'(t) \to 1/E$ and $J''(\omega) \to 0$. At long periods, the kinks can fully open, offering little resistance, and $J'(\omega)$ and $J''(\omega)$ both approach zero. The lag angle between the stress and the strain can also be computed by using the formula for the addition of sine and cosine waves of different amplitudes a and b

$$b \cos \omega t - a \sin \omega t = \sqrt{a^2 + b^2} \cos(\omega t + \delta)$$

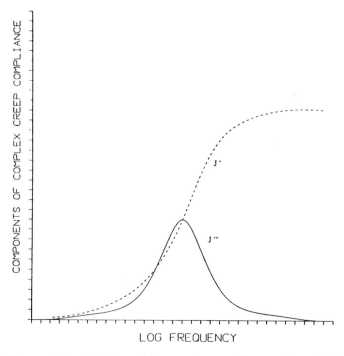

Figure 9.8. Variation of the complex compliance values with frequency.

and substituting the terms from the equation for the Voigt cell (Eq. 9.6)

$$\varepsilon = \frac{\sigma_0}{\sqrt{E^2 + \omega^2 \eta^2}} \cos[\omega t - \delta] \qquad (9.7)$$

and the lag angle between the applied stress and the resultant strain is

$$\delta = \tan^{-1}[\tau \omega] \qquad (9.8)$$

9.6. COMPLEX MODULUS

The strain lagging behind the stress produces difficulty in defining the modulus of a viscoelastic material in dynamic loading. If we consider the stress and strain shown in Fig. 9.9, the modulus can vary from zero to infinity within one cycle, depending on the time the modulus is computed as stress/strain.

The modulus for this condition can be defined by identifying the sinusoidally varying stress and strain as vectors in the complex plane

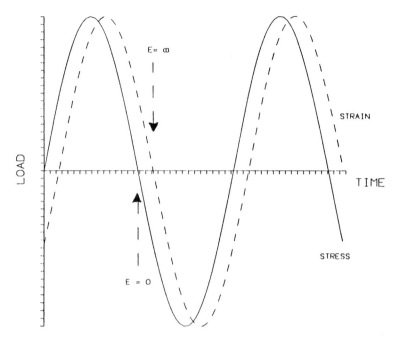

LOAD

E= ∞

STRAIN

TIME

STRESS

E = 0

Figure 9.9. Periodic loading. The modulus cannot be defined as stress/strain when the strain is not in phase with the stress.

(Fig. 9.10). Both stress and strain can be represented by vectors rotating around the origin at a fixed angular velocity. The complex strain ε^* is

$$\varepsilon^* - \varepsilon_0[\cos \omega t + i \sin \omega t]$$

or

$$\varepsilon^* = \varepsilon_0 e^{i\omega t}$$

and the complex stress which leads the strain by the angle δ is

$$\sigma^* = \sigma_0[\cos(\omega t + \delta) + i \sin(\omega t + \delta)]$$

or

$$\sigma^* = \sigma_0 e^{i[\omega t + \delta]}$$

Since there is a constant lag angle between these two vectors, the complex modulus can be defined as

$$E^* = \frac{\sigma^*}{\varepsilon^*} = \frac{\sigma_0}{\varepsilon_0} \frac{e^{i(\omega t + \delta)}}{e^{i\omega t}} = \frac{\sigma_0}{\varepsilon_0} e^{i\delta}$$

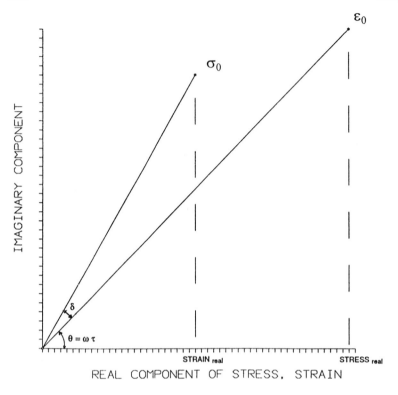

Figure 9.10. Representation of periodically varying stress and strain.

or

$$E^* = \frac{\sigma_0}{\varepsilon_0} [\cos \delta + i \sin \delta]$$

breaking the complex modulus into its components

$$E^* = E' + iE''$$

$$|E^*| = [(E')^2 + (E'')^2]^{1/2} \tag{9.9}$$

The complex modulus therefore has two components. The first one represents the component of the strain in phase with the stress. This component represents the behavior of the components of the material that produce elastic reactions to the applied stress, and consequently any energy required to deform the material in this mode will be returned on release of the applied stress, as in any elastic material. This component of the modulus is called the *storage modulus*, since the strain energy is stored within the material

$$E' = \frac{\sigma_0}{\varepsilon_0} \cos \delta$$

The other modulus represents the component of strain 90° out of phase with the stress and represents the viscous behavior of the material. Since this corresponds to an energy loss or dissipation on loading, this is the *loss modulus* of the material

$$E'' = \frac{\sigma_0}{\varepsilon_0} \sin \delta$$

and

$$\tan \delta = \frac{\sin \delta}{\cos \delta} = \frac{E''}{E'}, \qquad \delta = \tan^{-1}\left(\frac{E''}{E'}\right)$$

The loss modulus is usually several orders of magnitude smaller than the storage modulus. A polyethylene sample had the values

$$E' = 200{,}000 \text{ lb/in}^2$$
$$E'' = 2000 \text{ lb/in}^2$$
$$\delta(\text{lag angle}) = 0.6°$$

Since the lag angle is small for most polymers, the storage modulus can be closely approximated by the elastic modulus determined from the stress-strain diagram. Significant differences between the two values indicates a large loss angle.

For compliance, the definitions are similar

$$J^* = \frac{\varepsilon^*}{\sigma^*} = \frac{\varepsilon_0 e^{i\omega t}}{\sigma_0 e^{i(\omega t + \delta)}} = \frac{\varepsilon_0}{\sigma_0} e^{-i\delta}$$

$$J^* = \frac{\varepsilon_0}{\sigma_0}[\cos(-\delta) + i \sin(-\delta)] = \frac{\varepsilon_0}{\sigma_0}(\cos \delta - i \sin \delta)$$

$$J^* = J' - iJ''$$

$$|J^*| = [(J')^2 + (J'')^2]^{1/2}$$

where

$$J' = \frac{\varepsilon_0}{\sigma_0} \cos \delta$$

$$J'' = \frac{\varepsilon_0}{\sigma_0} \sin \delta$$

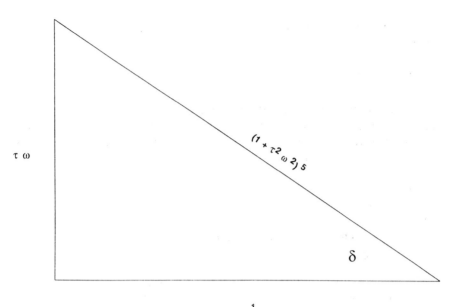

Figure 9.11.   Trigonometric relations for the lag angle for a Voigt cell.

For a Voigt cell (Eq. 9.8), $\tan \delta = \tau\omega$. The trigonometric conversion is shown in Fig. 9.11

$$\cos \delta = \frac{1}{\sqrt{1 + \tau^2\omega^2}}, \qquad \sin \delta = \frac{\omega\tau}{\sqrt{1 + \tau^2\omega^2}}$$

and

$$J' = \frac{\varepsilon_0}{\sigma_0} \frac{1}{\sqrt{1 + \omega^2\tau^2}}$$

$$\hspace{8cm} (9.10)$$

$$J'' = \frac{\varepsilon_0}{\sigma_0} \frac{\omega\tau}{\sqrt{1 + \omega^2\tau^2}}$$

These definitions agree with that calculated directly from the differential equation controlling the Voigt cell.

Relation Between Dissipation Parameters

The dissipation of energy per unit volume per cycle is

$$w = \int_0^{2\pi} \sigma \, d\varepsilon = i\omega \int_0^{2\pi} \varepsilon_0\sigma_0 e^{i\omega t} e^{i(\omega t + \delta)} \, dt = \varepsilon_0\sigma_0\pi \sin \delta$$

To compare the work dissipated to the work input, let us compute the work for a quarter cycle. This is the part of the cycle in which the tensile stress increases

$$w = \int_0^{\pi/2} \sigma \, d\varepsilon = \sigma_0 \varepsilon_0 \left(\frac{\cos \delta}{2} + \frac{\pi \sin \delta}{4} \right)$$

The total work/cycle is, therefore, four times this value. The first term is the work stored as elastic energy (stress and strain in phase) and the second term is the energy dissipated. Since

$$E' = \frac{\sigma_0}{\varepsilon_0} \cos \delta \quad \text{and} \quad J' = \frac{\varepsilon_0}{\sigma_0} \cos \delta$$

$$w_{\text{stored}} = \frac{\varepsilon_0^2 E'}{2} = \frac{\sigma_0^2 J'}{2}$$

since

$$E'' = \frac{\sigma_0}{\varepsilon} \sin \delta, \qquad w_{\text{dissipated}} = \frac{\pi \varepsilon_0^2 E''}{4} = \frac{\pi \sigma_0^2 J''}{4}$$

or the ratio of the total stored energy to that dissipated per cycle is $\pi \sigma_0^2 J''$. If we compare the energy dissipated to the maximum stored energy

$$\frac{w_{\text{diss}}}{w_{\text{stored}}} = \frac{\pi \varepsilon_0^2 E''}{\varepsilon_0^2 E'/2} = \frac{2\pi E''}{E'}$$

Since $\tan \delta = E''/E'$

$$\frac{w_{\text{diss}}}{w_{\text{stored}}} = 2\pi \tan \delta$$

9.7. EXPERIMENTAL METHODS FOR DETERMINING THE COMPLEX MODULUS

Torsion Pendulum

In an oscillatory system, the energy is proportional to the amplitude squared. If we consider two oscillations which decay with time amplitude θ_1, θ_2, when the amplitude decay is small and $\theta_1 \approx \theta_2$,

$$\frac{w_{\text{diss}}}{w_{\text{stored}}} = \frac{\theta_2^2 - \theta_1^2}{\theta_1^2}$$

Factoring:

$$\frac{w_{diss}}{w_{stored}} = \frac{(\theta_1 - \theta_2)(\theta_1 + \theta_2)}{\theta_1{}^2} \approx \frac{2\Delta\theta}{\theta}$$

so that the ratio of the energy stored in the material to that lost per cycle is proportional to the fractional amplitude decay/cycle. However, the usual measure of amplitude decay is logarithmic decrement Λ

$$\Lambda = \ln\frac{\theta_1}{\theta_2}$$

Since $\theta_1/\theta_2 \approx 1$ and $\ln(x) = 1 - x$ for x close to unity, then

$$\Lambda = 1 - \frac{\theta_1}{\theta_2} = \frac{\theta_2 - \theta_1}{\theta_2} = \frac{\Delta\theta}{\theta}$$

and as shown above

$$\frac{w_{diss}}{w_{stored}} = \frac{2\Delta\theta}{\theta} = 2\pi \tan\delta = 2\Lambda$$

or

$$\Lambda = \pi \tan\delta \qquad\qquad\qquad (9.11)$$

In the torsion pendulum experiment, the sample in the form of a thin wire is suspended from a fixed grip and caused to oscillate through a small angle (Fig. 9.12). To increase the period of oscillation, an extension arm is attached at the free end with weights at the ends to produce a large moment of inertia. The sample is set into oscillation by some mechanical means. To reduce air damping, the sample may be placed in a vacuum system and the oscillations induced by electromagnets placed close to the ends of the arm extensions. The period of oscillation and the rate of decay of the amplitude are determined, usually by observing a light reflected from a mirror attached to the sample. Since the sample is in torsion, the shear modulus G determines the motion of the sample. The general equation of motion of such an oscillating member to find G' and G'' is

$$I\ddot{\theta} + \eta\dot{\theta} + \frac{kAG^*}{l} = 0$$

where k is the spring constant and η is the viscosity of air producing damping. The solution to this equation is a damped sine wave

$$\theta = \theta_0 e^{-\beta t} e^{i\omega t}$$

$$= \theta_0 e^{+(i\omega - \beta)t}$$

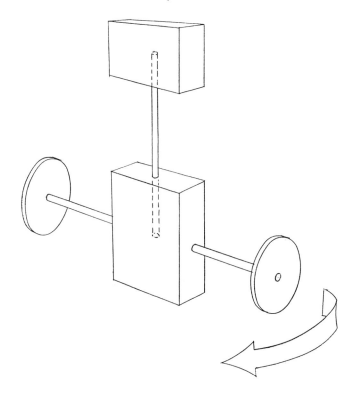

Figure 9.12. The torsion pendulum.

and for two consecutive swings with period P

$$\theta_n = \theta_0 e^{(i\omega - \beta)t_n}$$

$$\theta_{n+1} = \theta_0 e^{(i\omega - \beta)[t_n + P]}$$

$$\frac{\theta_n}{\theta_{n+1}} = e^{[\beta - i\omega]P}$$

Applying the definition of the log decrement to the real component

$$\Lambda = \ln \frac{\theta_n}{\theta_{n+1}} = \beta P$$

or if we average over m swings of the pendulum sample

$$\Lambda = \frac{1}{m} \ln \frac{\theta_n}{\theta_{n+m}}$$

To find the complex modulus

$$\beta = \frac{P}{\Lambda} \quad \text{where} \quad P = \text{period of vibration} = \frac{2\pi}{\omega}$$

Now

$$\theta = \theta_0 e^{-\beta t} e^{i\omega t}$$

$$\dot{\theta} = \theta(i\omega - \beta)$$

$$\ddot{\theta} = \theta(\beta^2 - 2i\omega\beta - \omega^2)$$

Substituting into the equation for motion and considering the real and imaginary parts separately

$$I(\beta^2 - \omega^2) - \beta\eta + \frac{kAG'}{l} = 0$$

$$\eta\omega - 2I\beta\omega + \frac{kAG''}{l} = 0$$

Substitute $\beta = P/\Lambda$

$$G' = \frac{4\pi^2 Il}{P^2 Ak}\left(1 - \frac{\Lambda^2}{4\pi^2} + \frac{\Lambda P\eta}{4\pi^2 I}\right) \tag{9.12}$$

$$G'' = \frac{4\pi Il}{P^2 Ak}\left(\Lambda - \frac{\eta P}{2I}\right) \tag{9.13}$$

Therefore, if the period P and the log decrement Λ are measured, then G' and G'' can be computed. Usually η is small. In that case only the first terms are required

$$G' = \frac{4\pi^2 Il}{P^2 Ak} \qquad \text{[note the period is determined by } G'\text{]}$$

$$G'' = \frac{4\pi Il\Lambda}{P^2 Ak} \qquad \text{[note the decrement is determined by } G''\text{]}$$

The determination of the exact length of the sample between the grips is uncertain, due to the restraint at the grips. If this uncertain additional gripped length is Δl, then

$$G' = \frac{4\pi^2 I}{P^2 Ak}(l + \Delta l)$$

A plot of P^2 vs. l for various lengths is performed. Extrapolation to $l = 0$ permits Δl to be calculated.

Application to Metals

It might be thought that losses occur only in viscoelastic materials, but there is no such thing as a completely elastic material. Metals have internal defects which also produce losses, although at much lower levels than in plastics. Any internal defect which can move and dissipate energy will result in such losses. Among the defects that are most commonly responsible for such losses in metals are interstitial atoms and dislocations. For example, in body-centered cubic (bcc) iron, which contains interstitial carbon atoms (see Fig. 9.13), the interstitial atoms are located at the center of the cube faces and/or at the center of the cube edges. The two types of positions are of the same size. Since the carbon atom is somewhat larger than the interstitial position, a stress is developed at an occupied location which tends to increase the length of the unit cell axis if an edge location is occupied.

In the unstressed state, the interstitial atoms will exist in all such sites equally, although most such positions are unoccupied. If a tensile stress is applied along the x-direction in Fig. 9.13 this will enlarge the interstitial position along this axis, and others will be compressed. The stress therefore tends to cause atoms to jump into these enlarged sites preferentially. When tested in a torsion pendulum, at frequencies too high to permit time for the atoms to perform these jumps, no such atomic motions occur, and the

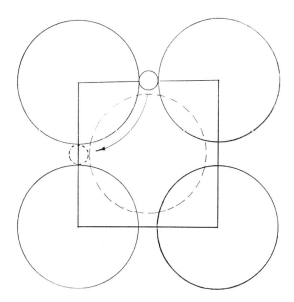

Figure 9.13. Interstitial motion to produce damping loss in metals.

material behaves elastically. At low frequencies, the atoms have sufficient time to move, and the stress and strain are in phase. Note that greater strains occur, and therefore the elastic modulus of the metal is lower. At intermediate frequencies, the strain is out of phase with the applied stress, since the interstitial atoms cannot follow, and a lag angle develops between the applied stress and the resultant strain. This behavior is identical to that of kinks in a polymeric material, and damping occurs, which is observable in a metallic torsion pendulum sample.

Hysteresis curves are used with metallic and polymeric samples to describe such behavior. In Fig. 9.14A, at frequencies too high for defects (interstitials or kinks) to follow the applied stress, the stress and strain are in phase, and increasing and decreasing stresses result in the same tensile diagram. In Fig. 9.14B, at very low frequencies, the stress and strain are still in phase, but the elastic modulus is lowered. At intermediate frequencies, where the period is of the order of the relaxation time of the defect, the strain

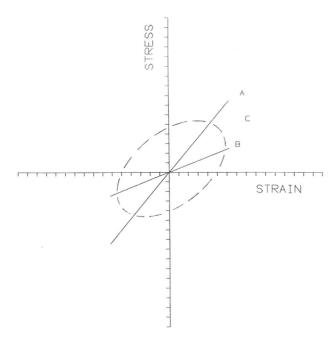

Figure 9.14. Stress-strain relations at different loading frequencies. (A) High frequency. Defects cannot follow the stress, and no hysteresis occurs. (B) Low frequency. Defects follow the stress easily, and no hysteresis occurs. The modulus is lower since the material can respond to the alternating stress. (C) Intermediate frequency. The defects lag behind the stress, resulting in hysteresis losses.

on increasing stress is not the same as during stress reduction, since the defects cannot relax sufficiently rapidly. An open hysteresis loop therefore develops (Fig. 9.14C). The area inside the loop is a measure of the energy dissipated in the cycle.

Nonresonant Forced Vibration

In this method, the forcing transducer applies a steady sinusoidal input to the sample (Fig. 9.15). Due to the viscoelastic response of the sample, the input force and the resultant displacement are out of phase by the lag angle δ. If the lag angle and the magnitude of the force and displacement are measured E^* can be directly obtained. The input force is

$$F^* = F_0 e^{i\omega t}$$

The resultant displacement is

$$\mu^* = \mu_0 e^{i(\omega t - \delta)}$$

since

$$\sigma^* = \frac{F^*}{A} \quad \text{and} \quad \varepsilon^* = \frac{\mu^*}{l_0}$$

The definition of complex modulus can be directly employed

$$\frac{F^*}{A} = E^* \frac{\mu^*}{l_0}$$

and

$$E^* = \frac{F_0}{\mu_0} \frac{l}{A} e^{i\omega\delta}$$

$$E' = \frac{F_0}{\mu_0} \frac{l}{A} \sin \delta$$

$$E'' = \frac{F_0}{\mu_0} \frac{l}{A} \sin \delta$$

The complex modulus can be found by this method over the range from about 0.1 to 100 Hz. The transducer ability to oscillate limits the lower range, and greater frequencies approach the natural frequency of the sample.

Although the principle is a simple application of the definition of the modulus, in practice the electronics are complex and the necessity to avoid the natural frequencies of the sample and the equipment add additional complexities. Corrections for grip effects on the effective length of the sample

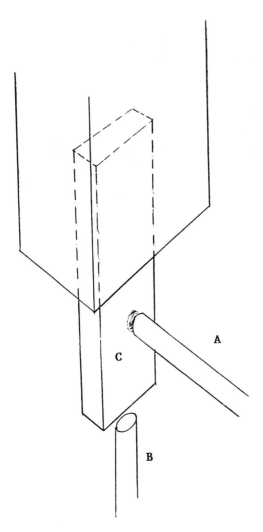

Figure 9.15. Experimental arrangement for determination of the complex modulus by the decay of the amplitude of oscillations. (A) Measuring transducer. (B) Forcing transducer. (C) Specimen.

must be made as in torsion pendulum experiments. Other corrections must be made for the compliance of the fixtures and grips.

Decay of the Amplitude of Vibration of a Normal Frequency

This technique is applicable in the frequency range of 20 Hz to 20 kHz. The equation of motion for a vibrating beam is

$$E*I \frac{\partial^4 \mu}{\partial x^4} + \rho A \frac{\partial^2 \mu}{\partial t^2} = F*$$

In the experiment the sample is put in motion and then the forcing transducer is turned off ($F* = 0$). We expect each point on the bar to oscillate with a different amplitude but to oscillate with the natural frequency of the bar. The time dependence can therefore be separated

$$\mu(x, t) = \mu(x)e^{i\omega t}$$

$$\frac{\partial \mu}{\partial t} = i\omega\mu$$

$$\frac{\partial^2 \mu}{\partial t^2} = -\omega^2\mu = -\omega^2\mu(x)e^{i\omega t}$$

Substituting to reduce the equation to only one variable

$$e^{i\omega t}E*I \frac{d^4\mu(x)}{dx^4} - \rho A\omega^2\mu(x)e^{i\omega t} = 0$$

$$E*I \frac{d^4\mu}{dx^4} - \rho A\omega^2\mu = 0$$

$$\frac{d^4\mu}{dx^2} - \frac{\alpha\mu}{l^4} = 0, \qquad \alpha = \frac{\rho A\omega^2}{EI}l^4$$

The solution (which can be verified by differentiation) is

$$\mu = a \sin \frac{\alpha x}{l} + b \cos \frac{\alpha x}{l} + c \sinh \frac{\alpha x}{l} + d \cosh \frac{\alpha x}{l}$$

The constants a to d are determined by the boundary conditions.

At the clamped end no displacement, no slope: $\mu = 0$, $d\mu/dx = 0$
At the free end, no bending moment, no shear force: $d^2\mu/dx^2 = 0$, $d^3\mu/dx^3 = 0$

The solutions, except for the trivial zero solutions, are only for values of α which satisfy $(\cos \alpha)(\cosh \alpha) = -1$. Possible α values are

n	α
1	1.875
2	4.694
3	7.854
4	10.996
5	14.37
⋮	⋮

We see that only specific frequencies of vibrations may occur. These are the natural frequencies of the member.

To find the complex modulus, these values of α can be substituted back into the equation defining α, and E^* can be found. The expression for $\tan \delta$ (where δ is the lag angle) is then found

$$\tan \delta = \frac{\ln\left(\frac{\mu(t_1)}{\mu(t_2)}\right)}{\pi f_{nat}(t_2 - t_1)} = \frac{2.303 \log_{10}\left(\frac{\mu(t_1)}{\mu(t_2)}\right)}{\pi f_{nat}(t_2 - t_1)}$$

In this field, audio terminology is used, utilizing the intensity I of the vibration

$$dB = 10 \log_{10} I/10^{-16}, \qquad I \text{ in watt/cm}^2$$

or, since intensity or energy is porportional to the amplitude squared,

$$dB_{change} = 10\left(\log_{10}\frac{I_2}{I_1}\right) = 10 \log_{10}\frac{\mu_2^{\,2}}{\mu_1^{\,2}} = 20 \log_{10}\frac{\mu_2}{\mu_1}$$

A common procedure is to permit the amplitude to decay 60 dB and record the time required. In this case

$$60 \text{ dB} = 10 \log_{10}\left(\frac{\mu_1}{\mu_2}\right)^2 = 20 \log_{10}\left(\frac{\mu_1}{\mu_2}\right)$$

Therefore

$$\log\left(\frac{\mu_1}{\mu_2}\right) = 3$$

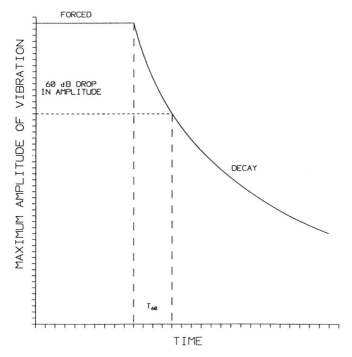

Figure 9.16.   Decay of vibrations at a natural frequency.

Substituting into the equation for tan δ

$$\tan \delta = \frac{2.303 \times 3}{\pi f_{\text{nat}} T_{60}}$$

where T_{60} = time for the amplitude to decay 60 dB, or when $\mu_2 = 10^{-3}\mu$ (see Fig. 9.16). Determination of the time for decay therefore yields the lag angle.

Forced Vibration of a Natural Frequency

As for the natural frequency method, the fundamental equation is

$$E^*I \frac{\partial^4 \mu}{\partial x^2} + \rho A \frac{\partial^2 \mu}{\partial t^2} = F^*$$

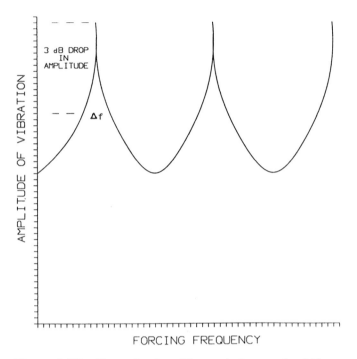

Figure 9.17. Determination of lag angle from peak width.

Continuous vibrations are induced by a transducer close to the end of the sample. In this case, $F^* = F_0 e^{i\omega t}$ at the free end and $F^* = 0$ elsewhere.

The solution to the equation can be considered to be the result of superimposing the contribution of all the natural frequencies of the sample (each with a different amplitude). A plot of amplitude vs. forcing frequency then appears as shown in Fig. 9.17. Large amplitudes occur when the forcing frequency approaches one of the natural frequencies of the sample, and much lower amplitudes occur for other frequencies. When the applied force has a frequency close to that of a natural frequency, the contribution of the other frequencies can be neglected, and the solution for E' for forced vibration at this frequency yields

$$E' = \left(4\pi \sqrt{3\rho} \times \frac{l^2}{t} \right)^2 \times \left(\frac{\omega_{nat}}{\alpha^2} \right)^2$$

where $t =$ thickness of bar sample.

The width of the peak is a function of the loss angle, and for the case where the energy of oscillation has dropped to one-half the value at the maximum, this corresponds to a 3 dB drop

$$dB = 10 \log \frac{I_2}{I_1} = 10 \log 0.5 = 3.01$$

At that point the lag angle can be determined from (see Fig. 9.17)

$$\tan \delta = \frac{\Delta f}{f}.$$

10
Composites

10.1. COMPOSITE TERMINOLOGY

(Although the general term composite material could apply to any combination of individual materials or phases, most commonly it implies the use of thin fibers of one material which have been impregnated with a plastic matrix. The fibrous materials most often used are fiberglass, Kevlar, boron, or carbon fibers. The mechanical properties of the resultant composite are a function of the type and volume fraction of the fibers, the type and volume fraction of the polymeric matrix, and the strength of the bond between the fibers and the matrix. Although three-dimensional moldings are possible, most commonly, since the fibers are usually quite thin, the resultant mixture is a thin layer, and larger, stronger specimens are built from many layers superimposed on each other during the molding or layup process.) This chapter is devoted to such laminates. The design of a structure manufactured from composites requires that the orientation of the fibers within the different layers be directed so that optimum mechanical properties are obtained in relation to the direction of the applied loads.

(The fibrous material can exist as either unidirectional individual fibers or a woven mat, with fibers running in two mutually perpendicular directions. Three-dimensional weavings are also possible. In general, the fibers have ultimate strengths and moduli several orders of magnitude greater than those of the matrix material. A reasonable approximation is to consider that the mechanical strength in the direction of the fiber axis is due to the fibers exclusively and that the principal function of the polymer is to transfer the load between the fibers. In the direction transverse to the fibers, in a unidirectional fiber layer, the stresses are controlled solely by the polymer, and the strength and modulus values are much lower.

Laminates of Carbon Fibers

Individual fibers are prepared in the range of 10 to 400 μm in diameter and are available in twisted yarn or untwisted tows of 300 to 350,000 filaments. These are then woven into fabrics of different weave designs or used directly as unidirectional tapes. The fiber bundle or woven cloth is impregnated with the resin to generate a prepreg. Final ply thickness can vary from 0.005 to 0.002 inch, the thinner plies being most costly. The prepregs are supported by a thin Mylar backing to help maintain the fiber alignment during handling. These thin layers are then placed on the tool, with the fibers aligned in the direction specified in the design. The product is then built up by adding additional layers as described in the section on molding. The Mylar backing is removed after the layer has been placed in the correct orientation, and excess resin is squeezed from the tape during molding. Each of these layers is called a lamina, and the final product is the resultant laminate. The final layer is only slightly thicker than the diameter of the fiber. Since it is virtually impossible to align all the fibers exactly, mechanical properties tend to vary widely between samples. When woven into cloth, the fibers must be intertwined, and the layers are thicker after curing than for the unidirectional prepreg tape. While the weaving process tends to bend the fibers, and therefore lowers the mechanical strength of the resultant product, the fibers are placed in tension during the weaving process. This results in the final product having more uniform alignment, resulting in less scatter in mechanical properties than for unidirectional tapes.

Composite Matrix Materials

The matrix material serves the following functions:

1. The matrix holds the material together forming a solid part, so that the product is air- and watertight
2. The matrix prevents mechanical abrasion of the fibers
3. The matrix transmits the stress from one fiber to another
4. In unidirectional material, the matrix properties control the transverse properties

Moisture absorption, temperature limitations, and outgassing characteristics are primarily determined by the matrix. Polyesters are most commonly used with fiberglass. They have good mechanical properties, are easily handled and molded, and have low cost compared to other matrix materials. Epoxy matrix materials are used when higher strength or higher chemical resistance is required. The cost is considerably higher than for polyesters. For carbon, boron, and Kevlar fibers, epoxy matrix is almost exclusively employed. Fiberglass is coated with polyester for most applications and epoxy for specialty high-strength applications. The maximum temperature of operation is of the order of 350°C. Phenolics have high strength and high heat resistance. The bonding strength to the fibers is not as great as for epoxies. Poly(ether ether ketone) (PEEK) is a thermoplastic of considerable interest as a matrix material. The use of a thermoplastic permits molding techniques which are unavailable for the other thermosetting polymers.

The failure strength of a composite is usually intermediate between the failure strength of the fibers and that of the matrix, with some fibers breaking at different times and some fiber pullout where matrix-fiber adhesion fails. Because it is the matrix polymer which transfers the load between the fibers, the matrix can affect the strength of the composite. The adhesion of the matrix to the fibers and the modulus of the matrix are the two properties that determine the ability of the matrix to transfer the stress. If the matrix modulus is very low, the fibers can exhibit almost independent motion within the composite, since the matrix can conform readily to the length each fiber takes. Since the fibers can stretch independently, the fibers break individually when they reach their breaking stress, and the failure mode is by progressive breaking of the fibers. The strength of the composite will therefore be low, as the fibers do not all withstand the load simultaneously. If the matrix has a high modulus and sufficient adhesion to the fibers, the entire composite structure will act as a single unit and failure will occur simultaneously in all fibers. This will result in a much higher failure stress. However, strong fiber-matrix adhesion is important, or else the fibers carrying the most load will be pulled out of the matrix individually, causing failure at low stresses.

Prepregs of glass, carbon, or Kevlar impregnated with partially polymerized resin are standard in the aerospace industry. The prepolymerization leaves the prepreg tacky. The impregnated fibers are supported by a Mylar backing sheet for ease of handling. Sheet molding compound (SMC) contains 15–30% glass fibers aligned within the plane of the sheet sample. Bulk molding compound (BMC) contains 15–35% glass fibers. The fibers are shorter and tend to be directed in all three directions when used in compression or other molding techniques to generate solid objects.

10.2. ADVANTAGES OF COMPOSITES

Molding of many hollow structural entities such as boats and aerospace structures is most directly performed utilizing the layup processes possible with composite materials. Fiberglass structures have been used for decades and have been completely accepted into every part of industrial and pleasure usage. Kevlar and carbon fiber products were originally developed for the aerospace industry but are finding increasing usage elsewhere, in tennis rackets, fishing rods, bicycles, etc.

In evaluating the relative strength of metals, the convention in the past has been to evaluate them in terms of the load-carrying ability per unit area. This is a comparison of the properties on an equal volume basis. For most transportation applications, it is appropriate to compare the load-carrying ability and resultant deformations of structures on the basis of equal weights of the materials. For example, in a commercial jet, approximately 3 pounds of fuel and engine are required to lift 1 pound of payload and aircraft structure. The importance of reducing the weight of any such structure cannot be overemphasized.

For applications where the weight of the component is the central factor, it is more pertinent to evaluate the mechanical properties by comparing the strength and stiffness on an equal weight basis. This evaluation can be performed by dividing the yield strength, ultimate strength, and elastic modulus by the density of the material. For such applications, carbon fiber and Kevlar fiber-epoxy composites have properties which are superior to those of the more common structural materials.

For comparison, aluminum and aluminum alloys are commonly specified for transportation applications due to their low density and therefore good specific strength properties. However, the stiffness-to-weight ratio of aluminum is less than one-sixth that of graphite-epoxy composites. Table 10.1 gives the specific strength and stiffness of common materials of construction. The true ability of a given mass of material to support a load depends on the specific configuration of the part. If there is no size or

Table 10.1. Mechanical Properties of High-Modulus Fibers

Material	Tensile strength		Elastic modulus		Specific gravity
	MPa	ksi	$MPa \times 10^3$	$ksi \times 10^2$	
Boron	3450	500	0.4	58	2.6
Kevlar 49	2750	400	0.13	19	1.45
E-glass	1510	220	0.07	10	0.55
High-modulus graphite	2050	300	0.38	55	1.94
High tensile strength graphite	2750	400	0.26	38	1.76

geometric constraint to the shape of the part, the low density of the material may be even more important. If there is no restriction on the diameter of a loaded thin-wall tube, the relative effectiveness is proportional to the ratio of the modulus to the density squared. For a structure in which torsional stiffness is critical, the relative efficiency is proportional to the modulus/density ratio cubed.

Thermal Expansion Coefficients

Another characteristic of carbon and Kevlar composites is their low thermal conductivity. This property is of major importance in applications where sun exposure or large thermal gradients may occur. As a result of such temperature differences, distortion of the structure may occur. If there are bonds between materials of different thermal expansion coefficients and constraints to expansion exist, the stresses generated by the differences in thermal expansion may cause failure or distortion. The thermal expansion coefficients of metals are all positive and are generally taken to be isotropic, with the same value in all directions. Graphite and Kevlar fibers are distinctly anisotropic, with thermal expansion coefficients along the fiber axes being negative and those perpendicular to the axis being positive. In both directions the expansion is an order of magnitude lower than that of most metals. The thermal expansion coefficients of the polymer matrices are all positive. The thermal expansion coefficient in the resultant composite therefore is anisotropic, and by proper choice of fiber direction and fiber-to-resin ratio the thermal expansion may be controlled so that it is close to zero.

Fiberglass

Fiberglass is one of the most versatile and most widely used fibrous reinforcing materials. Due to its extensive usage, it is available in a wide variety of configurations, including continuous filaments with twisted or parallel strands, short chopped fibers, woven cloth, rovings, and prepregs.

Glass fibers are produced by passing molten glass through a platinum crucible (the bushing) which contains many small holes (tips) through which the glass is drawn, forming the fibrous material. Immediately on passing through the tips, the drawn glass fibers are subjected to a carefully controlled water spray and a humidity-, and temperature-controlled air blast for rapid cooling. The glass is then coated with an appropriate size and then wound on mandrils for packaging and further handling. The sizing is usually a starch-oil emulsion which absorbs water to provide lubrication during the subsequent winding and handling operations.

Table 10.2. Glass Grades for Fiberglass

Type	Description
A	Glass of soda-lime composition similar to bottle glass. Poor thermal and chemical properties, not used for fibers.
C	Chemically resistant soda-lime-borosilicate glass used for its high corrosion and chemical attack resistance.
D	A low-density glass with high electrical resistance.
E	Pyrex composition glass. Good electrical properties and good for general-purpose application when a combination of good strength and chemical resistance is observed.
S	A high-strength, high-modular glass for specific applications. Higher in cost.

Fibers are manufactured in the range of 10 to 25 μm. The thinner-diameter fibers can be more easily wet and coated by resin and provide better drape to contour accurately a complex mold. However, they are more difficult to produce and require more care in handling. The finer the fiber, the greater the surface area for a given weight of glass, and the poorer the resistance to chemical attack.

Table 10.2 lists the various glass grades available in fibers. Glass fibers are amorphous and behave elastically until their breaking point, failing with about 3% elastic strain. Over 90% of the fiberglass used for reinforcement is of the E-glass type. This glass has excellent electrical properties and good mechanical properties and bonds well to most plastics after an appropriate coupling agent is employed. S-glass is higher in strength but difficult to manufacture and is utilized primarily in aerospace applications where strength is critical. Chemically resistant C-glass is utilized when chemical resistance is a critical requirement. The C-glass is usually employed as an outer layer with the more customary E-glass as the bulk of the fiberglass structure.

Graphite Fibers

Graphite fibers are manufactured principally from three organic starting materials: rayon, polyacrylonitrile (PAN), and pitch. Fibers of these starting materials are produced by the techniques for fiber manufacture, in which a solution of the polymer is forced through a spinneret head containing minute holes to produce fine fibers. These starting fibers are converted to pure carbon by decomposition at elevated temperatures in a series of heating steps in air and inert atmospheres to temperatures as high as 3000°C. All components except the carbon evaporate at these temperatures, and the crystal structure changes simultaneously. The fibers are stretched to align the molecules along the fiber axis, either prior to or concurrent with the heating process. The modulus and strength of the fibers tend to increase with increasing temperature and the extent of stretching.

As the degree of conversion to carbon increases with increasing heat treating temperature, the modulus also increases. The surface of the fiber is somewhat porous when heated at low temperatures. The surface becomes smoother and sealed as the heat treating temperature increases. The chemical behavior of the surface is altered, and the epoxy matrix material does not chemically bond as well to the high-modulus fibers. The graphite fibers are given various surface treatments, and coupling agents are applied to the surface to improve their compatibility with the resin matrix, as well as their handleability and abrasion resistance.

Kevlar Fibers

Kevlar is the trade name of the aramid (polyphenylene terephthalamide) manufactured by Dupont. Kevlar is manufactured by extrusion of an acid solution of the polymer through spinnerets at 50 to 100°C with hot air, followed by quenching in cold water. The fiber diameter is typically 12 μm. The tensile strength and modulus of the fiber are much higher than for other polymeric fibers. Kevlar fibers behave elastically until failure at an elongation of approximately 2%. While Kevlar is thermoplastic, the fibers are resistant to combustion and are stable to fairly high temperatures. The strength is 70% of its initial value after 1800 hours at 175°C. The fibers are resistant to attack by oils, fuels, and solvents but are susceptible to some acids. Kevlar is much less brittle than graphite and can be woven into a variety of cloth patterns which require bending of the fibers during weaving. The tensile strength is comparable to that of high-strength graphite fibers, but the compressive strength is low. The modulus is high for a polymer but not as high as that of graphite fibers.

The principal superiority of Kevlar over graphite fibers is its greater fiber toughness and hence greater resistance to impact and abrasion. Kevlar fiber composites are therefore used alone and also as outer plies of graphite laminates to protect the graphite from mechanical damage. Typical properties are given in Table 10.1.

Boron Fibers

The modulus of boron is the highest of that of the high-strength fibers, and boron is used in applications where maximum stiffness is required. The boron fibers are manufactured starting with boron trichloride (BCl_3), which is reduced with hydrogen to generate the boron. A heated tungsten wire is slowly passed through a container in which the BCl_3 gas is held. The BCl_3 decomposes on contact with the hot wire, and pure boron deposits on the wire. Due to its high cost and the difficulties associated with handling boron, this material is less commonly used than carbon or Kevlar fiber.

Tensile Data

Despite the viscoelastic properties of the polymeric matrix, the bond between the fibers and the matrix is so strong that graphite, Kevlar, and boron fiber composites behave elastically until failure. The stress-strain diagrams for carbon and boron composites are essentially linear and elastic until failure. Kevlar has very different properties in tension and compression. The compression strength is only 20% of the tensile strength, and the stress-stress diagram in compression becomes distinctly nonlinear at stress approaching the maximum values. These effects are due to buckling of the fibers.

Creep

The strong direct C—C bonds in the graphite fibers result in very small creep rates, even after long times. Kevlar having the normal bonding of a polymeric material has much higher creep strains. Both materials exhibit long secondary creep stages, in which the creep strain can be predicted accurately by an equation of the standard form

$$\varepsilon = \varepsilon_0 + at^n$$

Graphite-epoxy composites have excellent resistance to stress rupture, since the pure C—C bonds in the fibers support the load. Polymeric Kevlar fibers are more susceptible.

Impact Resistance

(Graphite fiber composites have low resistance to impact loads. Cracks propagate perpendicular to the fibers, and a result of the strong interfacial bond between the matrix and the fibers, once initiated, the crack will run through the entire structure. The impact resistance is therefore highly dependent on orientation, begin lowest perpendicular to the fiber axis. A typical value of impact energy perpendicular to the fiber axis is $20 \, kJ/m^2$.) Cracks in Kevlar-fiber composites propagate parallel to the fiber axis, and thus these materials have higher impact resistance, values as high as $130 \, kJ/m^2$ being observed.

Impact resistance can be obtained in structures by designing a hybrid of Kevlar and graphite. The Kevlar layers are placed on the outermost lamina, where the impact is to occur. The coefficients of thermal expansion of the two materials are similar, permitting polymerization during the molding process and subsequent cooldown to occur with few thermal stresses being generated. A design utilizing a Kevlar-graphite hybrid has the desirable characteristics of the high strength in both tension and compression of the graphite and the toughness of the Kevlar outer layers. The increase in toughness results with only a minor loss in strength and stiffness.

Vibration Damping

The graphite-epoxy composites are in the same range as metals. Vibration damping, such as the reduction of flutter in wing structures, is about the same as in metal wings. Similarly, the low loss factor obviates the necessity of being concerned about rapid heat and temperature buildup during fatigue or alternating cycling. Kevlar composities have loss factors an order of magnitude higher. Vibration damping by such structures is greatly improved, but some consideration needs to be given to the possibility of part heat-up during cyclic loading.

10.3. FABRICS

Fabrics are interwoven fibers, with the fibers running in two perpendicular directions. The long fibers are the *warp* direction, and the short cross-fibers are the *fill* direction or fill fibers. The use of fabrics reduces handling costs and provides a material which tends to be more amenable to molding around complex tools. The material has built-in strength in two directions, reducing the number of laminae which must be specified. Laminae cannot be as precisely specified, as the orientations are restricted to two perpendicular directions. The strength in the two directions does not necessarily need to be the same, as the fabric count, the numbers of fibers in the warp and fill direction, does not have to be the same. A bigger count in one direction will produce a fabric which has greater strength in that direction.

Types of Weaves

Plain Weave

This is the simplest type of weave, in which the long warp yarns pass over and then under the perpendicular fill fibers. The plain weave is the firmest of the common woven patterns, since the tight weaving permits the least shifting and misalignment of the fibers during manufacturing. Since the resultant woven structure has an equal density of fibers in both perpendicular directions, the strength and mechanical properties tend to be the same in both directions.

Satin Weave

In the satin weave, the fill threads are not interwoven over every alternate warp thread but cross under at more widely spaced intervals. Various possible weaving patterns have the interweaving at every fourth, fifth, eighth, or tenth fiber, generating what is called different *harness weaves*. As a result of this uneven overlap, satin fabric weaves appear differently on the two sides. The advantage of the satin-type weaves is that they are more flexible than the plain weaves due to less interleaving of the fibers. Thus, for complex shapes and compound curvature of the tool, the fabric can more easily be molded to the correct contour if a satin weave is employed. The greater the number of fibers skipped in the weaving, the more flexible is the fabric. For example, an eight harness satin is more flexible than a four harness one. In addition, since the fibers are bent less often and at a lower radius of curvature, there is less damage to the fibers, and the resultant fabric is of higher strength per fiber than for a plain weave. The fibers can be more closely packed due to the reduction in bending. This improves the strength of the fabric, but there is less open space and entrapped air within or beneath the fabric is more difficult to remove during the molding operation, and extra care must be employed to ascertain that the fabric has been completely deaerated.

10.4. LAMINA ORIENTATION

The exact orientations of the various laminae are determined by the appropriate design. The number of laminae in specific directions determines the relative strength, modulus, Poisson's ratio, and thermal expansion of the resultant product. The convention used for specifying laminae directions has been standardized, and the complete specification is in the USAF Composite Design Guide. An arbitrary direction (usually a convenient one in relation to the shape of the tool) is chosen as $0°$. The angles are specified looking toward the tool surface. The code presents the lamina direction relative to the zero axis, the number of laminae at each angle, their total number, and the sequence of the laminae proceeding from the tool surface. In the case of hybrid composites made from more than one material, the type of material in each lamina can also be specified. The standard laminae specifications for a laminate are as follows:

1.   The angular degrees of the fibers in the particular lamina from the defined reference direction. Only angles between 0 and 90° are used. Positive angles are defined as counterclockwise; negative angles are clockwise.
2.   A slash mark is usually used to indicate separate lamina orientation, with a subscript indicating the number of laminae of the same orientation. The laminae are specified with the one next to the mold surface being written first.
3.   Laminae are often placed next to each other symmetrically opposed around the reference direction. For such laminae, a + or − sign is used to specify the directions of the lamina. For example, ±30 means one lamina in the 30° direction and one in the −30° direction.
4.   Symmetric laminates which have a central plane of symmetry have specific properties which make them commonly used. For such symmetric laminates, only half the total number of laminae is specified, up to the mirror plane. The designation of a subscript S indicates that this convention is being used. If the total laminate is specified, the subscript T indicates that fact.
5.   A set is a lamina sequence which is repeated in the laminate. The number of such sets is specified by a numerical subscript preceding the subscript S or T.

Examples of laminate specifications are

$$(\pm 45/0_2/\pm 90)_T = +45/-45/0/0/+90/-90$$
$$[(0/60)_S 90]_S = 0/60/60/0/90/90/0/60/60/0$$

A *balanced* layup has a +ply for every −ply. For example

$[\pm 45]_T$ is balanced but not symmetric
$[0/\pm 30]_S$ is both balanced and symmetric
$\{0/\pm 30/30]_S$ is symmetric and not balanced

10.5.   EFFECT OF FIBER CONTENT ON THE ELASTIC MODULUS

{In an individual laminae, since the fibers are aligned in one direction, composite materials have properties which vary with direction. Equations which relate the direction of the applied stresses and resultant strains relative to the fiber direction have been developed to permit optimum use of the composite. The material properties also depend on the volume fraction of the fibers. The two principal directions in the plane of the lamina are the fiber or longitudinal direction and the transverse direction perpendicular to the fiber axis. Since the composite sheet is normally longer in the fiber direction, the transverse direction is also labeled as the width of the sheet.)

If we assume that when a stress is applied in the fiber axial direction the interfacial bond between the fibers and the matrix is not destroyed, then the total strain in the two components must be the same. Since the fibers are high-strength, high-modulus materials with strains at failure of less than 2%, they exhibit negligible viscoelastic effects. Since the total strains to failure of the composite are also of the order of 2% and the constraint of the fibers on the matrix through the interfacial bond is large, both the fibers and the matrix material can be considered as simple elastic materials.

The convention that the subscript f refers to the fibers, the subscript m refers to the polymer matrix, and the subscript comp refers to the total composite will be used. In the fiber direction, if the interfacial bond between the fibers and the matrix holds, the total strain in each of the components must be equal. The stress in each can then be computed using the assumption of perfect elasticity

$$\sigma_f = E_f \varepsilon$$

$$\sigma_m = E_m \varepsilon$$

The total force F acting on the sample is the sum of the forces in the fibers and the matrix

$$F = \sigma_{comp} A_{comp} = \sigma_f A_f + \sigma_m A_m$$

where A is the area of each constituent.

Since $\sigma_{comp} = E_{comp} \varepsilon$,

$$E_{comp} A_{comp} = E_f A_f + E_m A_m$$

Since the fraction of the total area of the cross section of the laminae is equal to the volume fraction v of the component

$$E_{comp} = E_f v_f + E_m v_m \tag{10.1}$$

This is the *rule of mixtures* and applies only in the fiber direction. This simple additive rule applies for Poisson's ratio in the fiber direction and the density of a composite. It cannot be applied for most properties in the transverse direction.

Transverse to the fibers the total strain is the sum of the strains in the two components; the stress is transferred directly, so it is considered to be the same in the two components. The stress is assumed to generate a uniform strain distribution within the matrix and the fibers. Clearly, some irregularities will occur at the interface, and some longitudinal stresses will be developed due to the differences in the Poisson's ratio of the fibers and the matrix. Neglecting these effects

$$\varepsilon_f = \frac{\sigma}{E_f} = \frac{E_{comp}\varepsilon_{comp}}{E_f}$$

$$\varepsilon_m = \frac{\sigma}{E_m} = \frac{E_{comp}\varepsilon_{comp}}{E_m}$$

If the total width of the laminae is W, the total deformation is

$$\Delta W = \varepsilon_{comp} W = \varepsilon_f v_f W + \varepsilon_m v_f W$$

Substituting the expressions for the strain in the previous equations:

$$\varepsilon_{comp} = \frac{E_{comp}\varepsilon_{comp} v_f}{E_f} + \frac{E_{comp}\varepsilon_{comp} v_m}{E_m}$$

and solving for the modulus of the composite

$$E_{comp} = \frac{E_f E_m}{v_f E_m + v_m E_f} \tag{10.2}$$

The maximum density of fibers possible in a uniaxial laminae is obtained by considering the fibers laid in contact with each other and the matrix material filling the interstices. At this maximum density of packing, if the packing produces a square array of fibers, the volume fraction of fibers is $\pi/4 = 78\%$. If the fibers are laid in the most densely packed possible face-centered array, the density of packing possible is 90%. In practice, the maximum fiber concentration does not exceed 65%. Note that at this fraction of fibers, even for the stiffest fibers, the modulus of the composite is only double that of the matrix in the transverse direction. The modulus in that direction is therefore always quite low, and other laminae must be placed with the fibers directed to support loads in that direction.

10.6. MATRIX EXPRESSIONS FOR MODULUS AND COMPLIANCE

The properties of the composite are not isotropic but are orthotropic, containing two perpendicular planes of symmetry within the sheet (the longitudinal and transverse directions). Because the stresses and strains perpendicular to the sheet can be neglected for many applications, only the properties within the plane of the sheet need to be evaluated. Since stresses in one direction produce strains in the transverse direction, the relation between stress and strain can be put into matrix form, where the subscript 1 refers to the direction along the fiber axes and the subscript 2 the direction perpendicular to the fiber axis. (See Fig. 10.1.)

The normal stress along the fiber axes is therefore σ_1, that in the transverse direction is σ_2, and the shear stress in both directions is τ_{12}. Similar nomenclature refers to the normal strains ε_1 and ε_2 and the shear strain δ_{12}.

The laminae is thin and σ_3 is assumed zero. For equilibrium $\tau_{12} = \tau_{21}$

$$\{\sigma\} = [Q]\{\varepsilon\} \tag{10.3}$$

is the shorthand notation for the matrix relation in which $[Q]$ is the stiffness matrix, to replace the complete two-dimensional matrix equation

$$\begin{Bmatrix} \sigma_1 \\ \sigma_2 \\ \tau_{12} \end{Bmatrix} = \begin{bmatrix} Q_{11} & Q_{12} & Q_{13} \\ Q_{21} & Q_{22} & Q_{23} \\ Q_{31} & Q_{32} & Q_{33} \end{bmatrix} \begin{Bmatrix} \varepsilon_1 \\ \varepsilon_2 \\ \delta_{12} \end{Bmatrix} \tag{10.4}$$

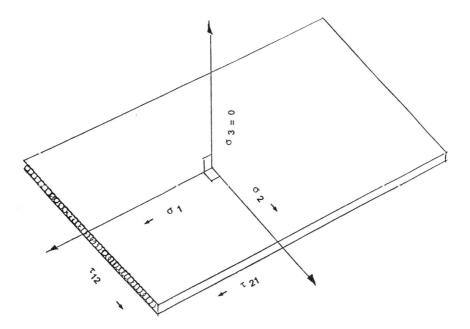

Figure 10.1.   Stresses in two dimensions of a laminae. The laminae is thin, and the stress perpendicular to the laminae (σ_3) is assumed to be zero. For equilibrium the shear stress $\tau_{12} = \tau_{21}$.

which in turn is expanded by the rule

$$\sigma_i = \sum_j Q_{ij}\varepsilon_j$$

The matrix equation, when expanded, yields three equations

$$\sigma_1 = Q_{11}\varepsilon_1 + Q_{12}\varepsilon_2 + Q_{13}\delta_{12}$$
$$\sigma_2 = Q_{22}\varepsilon_1 + Q_{22}\varepsilon_2 + Q_{23}\delta_{12}$$
$$\sigma_3 = Q_{31}\varepsilon_1 + Q_{22}\varepsilon_2 + Q_{33}\delta_{12}$$

For the orthotropic symmetry of the composite laminae, the stiffness matrix is symmetric; that is, $Q_{ij} = Q_{ji}$. For an orthotropic composite material, there is no interaction between the normal stresses and the shear strain as long as the normal stresses are directed along or perpendicular to the fiber axis. If a stress is applied in the fiber direction, no distortion or shear will be produced in the laminae. This is the behavior observed in isotropic materials and the one expected for metals. This is not necessarily correct if the stress is directed at some angle to the fibers which produces anisotropic behavior (see Fig. 10.2). The matrix equation can therefore be rewritten with the interaction terms between normal stress and shear strain being zero for these

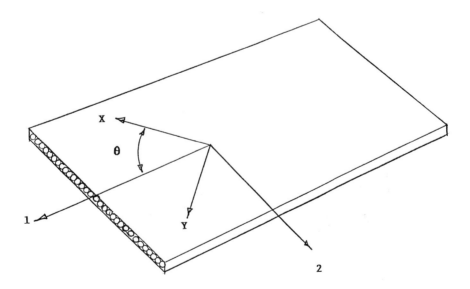

Figure 10.2. Convention for directions in a laminae.

principal stresses

$$
\begin{Bmatrix} \sigma_1 \\ \sigma_2 \\ \tau_{12} \end{Bmatrix} = \begin{bmatrix} Q_{11} & Q_{12} & 0 \\ Q_{12} & Q_{22} & 0 \\ 0 & 0 & Q_{33} \end{bmatrix} \begin{Bmatrix} \varepsilon_1 \\ \varepsilon_2 \\ \delta_{12} \end{Bmatrix}
$$

Written out completely, the full equations are

$$\sigma_1 = Q_{11}\varepsilon_1 + Q_{12}\varepsilon_2$$

$$\sigma_2 = Q_{12}\varepsilon_1 + Q_{22}\varepsilon_2$$

$$\tau_{12} = Q_{33}\delta_{12}$$

showing the lack of any shearing strain resulting from normal stresses and normal strains from shear stresses.

Similarly, the compliance matrix $[S]$ is given by the equation $\{\varepsilon\} = [S]\{\sigma\}$ or expanded

$$
\begin{Bmatrix} \varepsilon_1 \\ \varepsilon_2 \\ \delta_{12} \end{Bmatrix} = \begin{bmatrix} S_{11} & S_{12} & S_{13} \\ S_{21} & S_{22} & S_{23} \\ S_{31} & S_{32} & S_{33} \end{bmatrix} \begin{Bmatrix} \sigma_1 \\ \sigma_2 \\ \tau_{12} \end{Bmatrix}
$$

and recognizing that no normal strain is produced by a shear stress and that this matrix must also be symmetric

$$
\begin{Bmatrix} \varepsilon_1 \\ \varepsilon_2 \\ \delta_{12} \end{Bmatrix} = \begin{bmatrix} S_{11} & S_{12} & 0 \\ S_{12} & S_{22} & 0 \\ 0 & 0 & S_{33} \end{bmatrix} \begin{Bmatrix} \sigma_1 \\ \sigma_2 \\ \tau_{12} \end{Bmatrix}
\qquad (10.5)
$$

Relation of Matrix Elements and Engineering Material Constants

We recognize from Eq. 10.3 that the elastic modulus of the laminae is different in the fiber and transverse directions. Similarly, Poisson's ratio must also be measured and defined in both these directions. Poisson's ratio defines the relation between the strains produced in two perpendicular directions under an applied load

$$\nu_{12} = \frac{-\varepsilon_2}{\varepsilon_1} \qquad (10.6)$$

where ν_{12} refers to the strain produced in direction 2 as a result of an applied stress in direction 1. The negative sign is to maintain Poisson's ratio as a positive number. Similarly

$$v_{21} = \frac{-\varepsilon_1}{\varepsilon_2}$$

specifies the strain in the longitudinal fiber direction when a stress is applied in the transverse direction.

If we consider a stress applied along the fiber axis (σ_1)

$$\varepsilon_1 = \frac{\sigma_1}{E_1} \tag{10.7}$$

$$\varepsilon_2 = -\varepsilon_1 v_{12} = \frac{-\sigma_1 v_{12}}{E_1} \tag{10.8}$$

Similarly, for a stress applied in the transverse direction

$$\varepsilon_2 = \frac{\sigma_2}{E_2} \tag{10.9}$$

$$\varepsilon_1 = -\varepsilon_2 v_{21} = \frac{-\sigma_2 v_{21}}{E_2} \tag{10.10}$$

Note that although the Poisson's ratios for stress applied along the fiber and longitudinal direction are different, they are not independent physical constants but are related by

$$v_{12} E_2 = v_{21} E_1 \tag{10.11}$$

These expressions can now be related to the compliance matrix elements. Expanding the compliance matrix (Eq. 10.5) to obtain the equation for ε_1

$$\varepsilon_1 = S_{11} \sigma_1 + S_{12} \sigma_2$$

Comparing this equation to Eqs. 10.7 and 10.9 above, if a stress in direction 1 is applied, then

$$S_{11} = \frac{1}{E_1}$$

and similarly, for a uniaxial stress in direction 2

$$S_{22} = \frac{1}{E_2}$$

and from Eq. 10.10 and Eq. 10.11

$$S_{12} = \frac{-v_{21}}{E_2} = \frac{-v_{12}}{E_1}$$

Although the Poisson's ratios in the two directions parallel and perpendicular to the fiber axes are different, they are related by Eq. 10.11, and thus there are truly only four independent materials constants, S_{11}, S_{22}, S_{12}, S_{33}. Since there is no interaction between the normal and shear components

$$S_{33} = \frac{1}{G}$$

where G is the shear modulus of the composite laminae.

All the elements of the compliance matrix in terms of the familiar of engineering constants have now been computed, and the complete matrix is

$$\left\{ \begin{array}{c} \varepsilon_1 \\ \varepsilon_2 \\ \delta_{12} \end{array} \right\} = \left[\begin{array}{ccc} \dfrac{1}{E_1} & -\dfrac{v_{12}}{E_1} & 0 \\ -\dfrac{v_{21}}{E_2} & \dfrac{1}{E_2} & 0 \\ 0 & 0 & \dfrac{1}{G} \end{array} \right] \left\{ \begin{array}{c} \sigma_1 \\ \sigma_2 \\ \tau_{12} \end{array} \right\} \quad \text{with} \quad -\frac{v_{12}}{E_1} = -\frac{v_{21}}{E_2} \qquad (10.12)$$

The stiffness matrix terms can now be determined by inverting the compliance matrix

$$Q_{11} = \frac{E_1}{(1 - v_{12}v_{21})}$$

$$Q_{22} = \frac{E_2}{(1 - v_{12}v_{21})}$$

$$Q_{12} = \frac{v_{12}E_2}{(1 - v_{12}v_{21})} = \frac{v_{21}E_1}{(1 - v_{12}v_{21})}$$

$$Q_{33} = G$$

The interrelations between the material properties commonly defined and the matrix components are useful in evaluating the matrix elements from mechanical test data and in relating these elements to the more familiar properties of metals. For analysis of the behavior of the laminae and composite laminate, direct use of the matrices is preferred, and no direct physical connection between the elements and the mechanical test properties need be considered.

Test Methods

The values of the matrix components of engineering constants must be determined by experimental testing of the composities. Tensile tests to determine elastic properties are performed routinely on metal samples. However, since composites are either orthotropic or anisotropic, with four or six independent material parameters, respectively, more elaborate testing programs must be developed for the composites. All the elastic constants for an isotropic metal are determined from one test. Three independent tests are required for composites. To determine the strength of the laminate along the fiber axis, test samples with laminae having all fibers oriented in the same

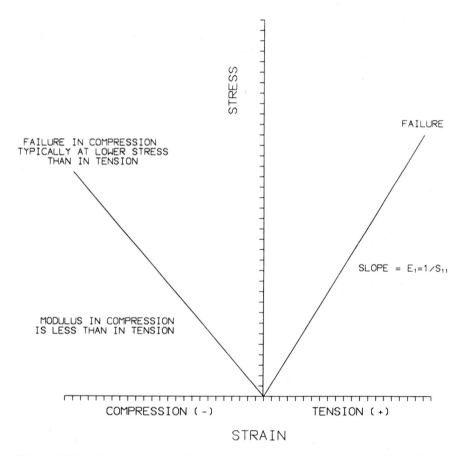

Figure 10.3. Representation of stress-strain data of a laminae tested with the load along the fiber direction.

direction are molded. The number of plies has varied widely between investigators. The test can be performed in either tension or compression. In compression the sample length must be controlled so that buckling does not occur. The critical length is determined by

$$L = \left(\frac{\pi^2 EI}{P}\right)^{1/2}$$

where P is the applied load. These laminae behave elastically until failure, and the deviations from linearity are quite small. Measurement of longitudinal and transverse strain determine S_{11} and S_{12}. Test samples in the transverse direction generate stress-strain curves which are not as linear as the test in the fiber direction, since the load is primarily supported by the viscoelastic matrix.

The shear modulus G of isotropic materials need not be determined independently, as it can be computed from the elastic modulus and Poisson's ratio

$$G = \frac{E}{2(1 - v)}$$

For orthotropic laminae, a pure shear test is required for its determination. Although pure shear can be obtained in torsion tests on tubular specimens,

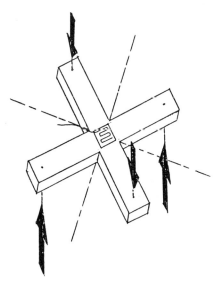

Figure 10.4.   Biaxial testing. Arrows indicate loading directions. Pure shear occurs at a 45° angle to the arms of the sample. A strain rosette at the center of the specimen measures the resultant shear.

the cost of molding such samples is high, since the ends of the sample must be strengthened to permit gripping without crushing the tube. The alternative test is to place a sample in biaxial loading with equal tensile and compressive stresses on the two axes. Figure 10.4 shows the cross-beam test.

The laminae to be tested is bonded in the center of the testing beam fixture, and biaxial loading is applied. Pure shear exists at an angle of 45° to the loaded beams. The shear stress-shear strain diagram is not linear. The slope at the origin is usually taken as the shear modulus, but it is more appropriate to take the modulus at the actual applied stress. This yields the final matrix component S_{33}.

10.7. EFFECT OF ORIENTATION OF STRESS ON COMPLIANCE AND STIFFNESS

We now have the relationship between stress and strain along the principal directions of laminae (the fiber and transverse directions) and can incorporate the usual material properties in these relationships. When the stress is not directed along or perpendicular to the fiber axis, the compliance and stiffness matrices transform to produce different element values. The matrices for an arbitrary stress direction can be obtained from those specified in Section 10.6 by replacing the actual stress by its components along and perpendicular to the fiber axes. We therefore need a conversion of the components of stress directed along the x-, y-axes (which indicate the stress directions) at an arbitrary angle to the fiber direction (1) to its components along and perpendicular to the fiber axis. Fig. 10.2 shows the relation between these directions.

Axes 1 and 2 are fixed by the fiber and transverse directions. The x- and y-axes are the directions applicable to the applied stress, θ is the angle between x- and 1-axes. The positive direction of rotation is defined from the x- to the 1-axis in a counterclockwise direction.

The fiber axis and the transverse direction are called the principal axes because there is no interaction between normal and shearing stresses only when the normal stresses are directed along these directions. The equation for the transformation of components from one set of axes to another is given in any standard strength of materials test. The transformation matrix from the directions along which the actual stress is applied to the principal axes is the transformation matrix $[T]$. These and all subsequent conversions and calculations are almost universally performed by computer programs which are readily available.

$$\{\sigma\}_{1,2,12} = [T]\{\sigma\}_{x,y,xy} \qquad (10.13)$$

with $[T]$ given by

$$[T] = \begin{bmatrix} \cos^2 \theta & \sin^2 \theta & 2\sin\theta\cos\theta \\ \sin^2 \theta & \cos^2 \theta & -2\sin\theta\cos\theta \\ -\sin\theta\cos\theta & \sin\theta\cos\theta & \cos^2\theta - \sin^2\theta \end{bmatrix} \qquad (10.14)$$

The transformation for strains can be performed similarly. However, care must be taken with the definition of strain.

Distortion as a result of shear stress also rotates the specimens. The rotation can be removed if we consider the block as being distorted equally along the x- and y-axes, so that the centerline is unchanged. The definition of the tensor strain (which is irrotational) is the pure shear strain, and is therefore $1/2$ the engineering strain. The transformation of strain follows the same matrix computation as stress if the tensor strain is used in the equations. These are designated $\varepsilon^{\text{TENS}}$. The engineering strain will be given the symbol $\{\varepsilon\}$ or $\{\varepsilon_{1,2,12}\}$ to denote the three components.

$$\{\varepsilon\}_{1,2,12}^{\text{TENS}} = [T]\{\varepsilon\}_{x,y,xy}^{\text{TENS}}$$

$$\left\{ \begin{array}{c} \varepsilon_1 \\ \varepsilon_2 \\ \dfrac{\delta_{xy}}{2} \end{array} \right\} = [T] \left\{ \begin{array}{c} \varepsilon_x \\ \varepsilon_y \\ \dfrac{\delta_{12}}{2} \end{array} \right\}$$

where $[T]$ is the transformation matrix given in Eq. 10.14.

The stiffness matrix of the laminae when the stresses are not along the fiber axis depends on the angle between the applied stress and the fibers. The stiffness matrix under these general conditions can be obtained employing the transformation matrix. However, the strain in the original stiffness matrix refers to the engineering strain. The transformation equation is applicable to the tensor strain. The conversion between the two strain definitions can be obtained from the Reuter matrix

$$\varepsilon = [R]\{\varepsilon\}^{\text{TENS}}$$

where

$$R = \begin{bmatrix} 1 & 0 & 0 \\ 0 & 1 & 0 \\ 0 & 0 & 2 \end{bmatrix} \qquad \left\{ \begin{array}{c} \varepsilon_1 \\ \varepsilon_2 \\ \delta_{12} \end{array} \right\} = [R] \left\{ \begin{array}{c} \varepsilon_1 \\ \varepsilon_2 \\ \varepsilon_{12}^{\text{TENS}} \end{array} \right\}$$

since multiplication through of the matrix yields the components

$$\varepsilon_1 = \varepsilon_1{}^{\text{TENS}}$$

$$\varepsilon_2 = \varepsilon_2{}^{\text{TENS}}$$

$$\delta_{12} = 2\varepsilon_{12}{}^{\text{TENS}}$$

The steps to affect the calculation of the stress in any arbitrary direction from the known strains in those directions can now be computed by the following steps: The conversion from principal to arbitrary axes is from Eq. 10.13

$$\{\sigma_{x,y,xy}\} = [T]^{-1}\{\sigma_{1,2,12}\} \tag{10.15}$$

Since from Eq. 10.3

$$\{\sigma_{1,2,12}\} = [Q]\{\varepsilon_{1,2,12}\}$$

Substituting into Eq. 10.15

$$\{\sigma_{x,y,xy}\} = [T]^{-1}[Q]\{\varepsilon_{1,2,12}\}$$

To transform from engineering shear strain to tensor shear strain use the [R] matrix

$$\sigma_{x,y,xy} = [T]^{-1}[Q][R]\{\varepsilon_{1,2,12}{}^{\text{TENS}}\}$$

To transform the strain to the x, y axes use the $[T]$ matrix again

$$= [T]^{-1}[Q][R][T]\{\varepsilon_{x,y,xy}{}^{\text{TENS}}\}$$

And to transform back to engineering shear strain use the inverse [R] matrix

$$= [T]^{-1}[Q][R][T][R]^{-1}\{\varepsilon_{x,y,xy}\}$$

Multiplying these matrices to generate the general stiffness matrix for a stress applied at any arbitrary direction to the fiber axis $[\bar{Q}]$

$$\{\sigma_{x,y,xy}\} = [\bar{Q}]\{\varepsilon_{x,y,xy}\}$$

where

$$[\bar{Q}] = [T]^{-1}[Q][R][T][R]^{-1}$$

Multiplying the matrices together, the terms of the general stiffness matrix are

$$\bar{Q}_{11} = Q_{11}\cos^4\theta + 2(Q_{12} + 2Q_{33})\sin^2\theta\cos^2\theta + Q_{22}\sin^4\theta$$
$$\bar{Q}_{12} = (Q_{11} + Q_{22} - 4Q_{33})\sin^2\theta\cos^2\theta + Q_{12}(\sin^4\theta + \cos^4\theta)$$
$$\bar{Q}_{22} = Q_{11}\sin^4\theta + 2(Q_{12} + 2Q_{33})\sin^2\theta\cos^2\theta + Q_{22}\cos^4\theta$$
$$\bar{Q}_{13} = (Q_{11} - Q_{12} - 2Q_{33})\sin\theta\cos^3\theta + (Q_{12} - Q_{22} + 2Q_{33})\sin^3\theta\cos\theta$$
$$\bar{Q}_{23} = (Q_{11} - Q_{12} - 2Q_{33})\sin^3\theta\cos\theta + (Q_{12} - Q_{22} + 2Q_{33})\sin\theta\cos^3\theta$$
$$\bar{Q}_{33} = (Q_{11} + Q_{22} - 2Q_{12} - 2Q_{33})\sin^2\theta\cos^2\theta + Q_{33}(\sin^4\theta + \cos^4\theta)$$

$$(10.16)$$

This matrix has no zero terms. Thus a normal stress will generate a shear strain and an angular distortion of the laminae will occur. Similarly, a shear stress will generate normal stresses and a change in the shape of the laminae. This complexity requires care in the design of the composite structure but also permits improvement in the design of intentionally twisted parts such as fan and impeller blades. The interactions between normal stresses and shearing strains also affect molding, as the thermal stresses generated on cooling produce complex warping of the product.

The $\bar{Q}_{11}$ term represents the stiffness in the direction of the stress. It is a maximum when the stress is applied along the fiber axis and a minimum in the transverse direction. The stiffness drops off rapidly, and for stresses at an angle greater than $60°$ the stiffness is close to that of the matrix material alone. The $\bar{Q}_{22}$ curve mirrors the $\bar{Q}_{11}$ curve because at a $90°$ angle the stress is applied along the transverse direction, and the stiffness $\bar{Q}_{22}$ then refers to the fiber direction. The term $\bar{Q}_{33}$ refers to the shear stiffness and reaches a maximum at $45°$. Similarly, the $\bar{Q}_{12}$ term, indicating the generation of transverse stress by a longitudinal strain, is also a maximum at $45°$. The $\bar{Q}_{13}$ term is a measure of the normal-shear stress coupling, which is greatest when the stress is applied at an angle to the fibers which is close to $30°$.

Following an identical path, the generalized compliance can be found

$$\{\varepsilon_{1,2,12}\}^{\text{TENS}} = [R]^{-1}\{\varepsilon_{1,2,12}\}$$
$$\{\varepsilon_{x,y,xy}\}^{\text{TENS}} = [T]^{-1}[R]^{-1}\{\varepsilon_{1,2,12}\}$$
$$\{\varepsilon_{x,y,xy}\} = [R][T]^{-1}[R]^{-1}\{\varepsilon_{1,2,12}\}$$

Since

$$\{\varepsilon_{1,2,12}\} = [S]\{\sigma_{1,2,12}\}$$
$$\{\varepsilon_{x,y,xy}\} = [R][T]^{-1}[R]^{-1}[S]\{\sigma_{1,2,12}\}$$

$$(10.17)$$

and multiplication of these matrices produces the general compliance matrix $[\bar{S}]$ for a load at an angle to the fibers

$$[\bar{S}] = [R][T]^{-1}[R]^{-1}[S]$$

$$\bar{S}_{11} = S_{11}\cos^4\theta + (2S_{12} + S_{33})\sin^2\theta\cos^2\theta + S_{22}\sin^4\theta$$

$$\bar{S}_{12} = S_{12}(\sin^4\theta + \cos^4\theta) + (S_{11} + S_{22} - S_{33})\sin^2\theta\cos^2\theta$$

$$\bar{S}_{22} = S_{11}\sin^4\theta + (2S_{12} + S_{33})\sin^2\theta\cos^2\theta + S_{22}\cos^4\theta$$

$$\bar{S}_{13} = (2S_{11} - 2S_{12} - S_{33})\sin\theta\cos^3\theta$$
$$\qquad - (2S_{22} - 2S_{12} - S_{33})\sin^3\theta\cos\theta$$

$$\bar{S}_{23} = (2S_{11} - 2S_{12} - S_{33})\sin^3\theta\cos\theta$$
$$\qquad - (2S_{22} - 2S_{12} - S_{33})\sin\theta\cos^3\theta$$

$$\bar{S}_{33} = 2(2S_{11} + 2S_{22} - 4S_{12} - S_{33})\sin^2\theta\cos^2\theta$$
$$\qquad + S_{33}(\sin^4\theta + \cos^4\theta) \tag{10.18}$$

The generation of shearing strains by normal stresses and normal strains by shearing stresses is evident.

10.8. GENERALIZED ENGINEERING MATERIAL CONSTANTS FOR LAMINAE

When the stress is applied off the fiber axis, the response of the composite is anisotropic, as interaction between the shearing and normal components occurs. These interactions appear in the compliance modulus as nonzero $\bar{S}_{13}$ and $\bar{S}_{23}$ terms. These have no corollary values for isotropic metals and polymers. $\bar{S}_{13}$ gives the strain in the fiber direction produced by a shearing stress, and $\bar{S}_{23}$ gives the strain in the transverse direction due to a shearing stress.

The correlation of these compliance matrix elements and the engineering material constants can be obtained by considering the standard uniaxial material strength tests. If a stress is applied only along one direction arbitrarily called the x-axis, then σ_x is given, $\sigma_y = 0$, $\tau_{xy} = 0$

$$\begin{Bmatrix} \varepsilon_x \\ \varepsilon_y \\ \delta_{xy} \end{Bmatrix} = \begin{bmatrix} \bar{S}_{11} & \bar{S}_{12} & \bar{S}_{13} \\ \bar{S}_{21} & \bar{S}_{22} & \bar{S}_{23} \\ \bar{S}_{31} & \bar{S}_{32} & \bar{S}_{33} \end{bmatrix} \begin{bmatrix} \sigma_x \\ 0 \\ 0 \end{bmatrix}$$

$$\varepsilon_x = \bar{S}_{11}\sigma_x \tag{10.19}$$

and since for a tensile test the modulus is defined by $\sigma = E\varepsilon$

$$\overline{S}_{11} = \frac{1}{E_x}$$

In the y direction

$$\varepsilon_y = \overline{S}_{21}\sigma_x$$

Defining the Poisson's ratio for this off-fiber axis test loading

$$v_{xy} = \frac{-\varepsilon_y}{\varepsilon_x}$$

$$\varepsilon_y = -v_{xy}\varepsilon_x = \frac{-v_{xy}\sigma_x}{E_x}$$

Therefore

$$\overline{S}_{21} = \frac{-v_{xy}}{E_x}$$

Shear strain exists as a result of this test

$$\delta_{xy} = \overline{S}_{31}\sigma_x \tag{10.20}$$

This coupling of shear strain and normal stress does not exist in metals or isotropic polymers or for on-fiber loading of composites. Additional material constants which are not required for isotropic materials must be defined for orthotropic materials. The coefficient of mutual influence of the second kind has been defined to relate the shear strain produced by normal stresses (in a parallel definition to Poisson's ratio for normal stresses)

$$v_{x,xy} = \frac{\delta_{xy}}{\varepsilon_x} \tag{10.21}$$

In this text, for these ratios the second subscript refers to the strain produced not in the stress direction (in this case shear in the x-y plane), and the first subscript refers to the stress direction. These coefficients have been experimentally observed to be either positive or negative in magnitude, and therefore the defining Eq. 10.21 has a positive sign for simplicity. To relate the compliance matrix terms to Eq. 10.21 substitute Eq. 10.19 and Eq. 10.20 into Eq. 10.21.

$$v_{x,xy} = \frac{\overline{S}_{31}\sigma_x}{\overline{S}_{11}\sigma_x} = \frac{\overline{S}_{31}}{\overline{S}_{11}}$$

or

$$\bar{S}_{31} = v_{x,xy}\bar{S}_{11} = \frac{v_{x,xy}}{E_x}$$

Similarly, for a uniaxial stress test in which the stress is applied along the y-axis

$$\sigma_y \neq 0, \qquad \sigma_x = \tau_{xy} = 0$$

$$\varepsilon_y = \bar{S}_{22}\sigma_y = \frac{1}{E_y}\sigma_y$$

$$\bar{S}_{22} = \frac{1}{E_y}$$

To compute S_{12}

$$\varepsilon_x = \bar{S}_{12}\sigma_y$$

$$v_{yx} = \frac{-\varepsilon_x}{\varepsilon_y}, \qquad \varepsilon_x = -v_{yx}\varepsilon_y, \qquad \varepsilon_y = \frac{\bar{S}_{12}\sigma_y}{v_{yx}}$$

$$\bar{S}_{12} = \frac{-v_{xy}}{E_x} = \frac{-v_{yx}}{E_y}$$

To find $S_{32} = \bar{S}_{23}$

$$\delta_{xy} = \bar{S}_{32}\sigma_y, \qquad v_{y,xy} = \frac{\delta_{xy}}{\varepsilon_y}, \qquad \delta_{xy} = \frac{v_{y,xy}\sigma_y}{E_y}$$

$$\bar{S}_{32} = \frac{v_{y,xy}}{E_y}$$

where $v_{y,xy}$ is defined as the coefficient of interaction of the second type; and for a shear test

$$\tau_{xy} \neq 0, \qquad \sigma_x = \sigma_y = 0$$

$$\begin{Bmatrix} \varepsilon_x \\ \varepsilon_y \\ \delta_{xy} \end{Bmatrix} = \begin{bmatrix} \bar{S}_{11} & \bar{S}_{12} & \bar{S}_{13} \\ \bar{S}_{21} & \bar{S}_{22} & \bar{S}_{23} \\ \bar{S}_{31} & \bar{S}_{32} & \bar{S}_{33} \end{bmatrix} \begin{Bmatrix} 0 \\ 0 \\ \tau_{xy} \end{Bmatrix}$$

$$\delta_{xy} = \bar{S}_{33}\tau_{xy}$$

Since for a shear test $\tau = G\delta$, $\bar{S}_{33} = 1/G$.

An interactive term between the shear stress and normal stress appears which is not present for fiber on-axis testing

$$\varepsilon_x = \bar{S}_{13}\tau_{xy}$$

The coefficient of mutual influence of the first kind is defined to identify this type of interaction

$$v_{xy, x} = \frac{\varepsilon_x}{\delta_{xy}}$$

$$\varepsilon_x = v_{xy, x}\delta_{xy}, \qquad \varepsilon_x = \frac{v_{xy, x}\tau_{xy}}{G}$$

$$\bar{S}_{13} = \frac{v_{xy, x}}{G}$$

Similarly

$$\bar{S}_{23} = \frac{v_{xy, y}}{G} = \frac{v_{y, xy}}{E_y}$$

The compliance matrix for engineering constants is now complete

$$\{\varepsilon\} = [\bar{S}]\{\sigma\}$$

$$
\begin{Bmatrix} \varepsilon_x \\ \varepsilon_y \\ \delta_{xy} \end{Bmatrix} =
\begin{bmatrix}
\dfrac{1}{E_x} & \dfrac{-v_{yx}}{E_y} & \dfrac{v_{xy, x}}{G} \\[2mm]
\dfrac{-v_{xy}}{E_x} & \dfrac{1}{E_y} & \dfrac{v_{xy, y}}{G} \\[2mm]
\dfrac{v_{x, xy}}{E_x} & \dfrac{v_{y, xy}}{E_y} & \dfrac{1}{G}
\end{bmatrix}
\begin{Bmatrix} \sigma_x \\ \sigma_y \\ \tau_{xy} \end{Bmatrix}
\qquad (10.22)
$$

This matrix is symmetric, so there are six independent constants. Note that the symmetry condition holds for the matrix but not for the engineering constants. $\bar{S}_{13} = \bar{S}_{31}$, but $v_{x, xy}$ does not equal $v_{xy, x}$. It is worth observing the increase in the number of material constants which must be specified as the symmetry of the material with respect to loading direction decreases, as shown in Table 10.3.

The variation of these material constants in the compliance modulus of Eq. 10.22 with orientation of the applied stress relative to the fiber axis can be found by the transformation of Eq. 10.17. The result of that matrix

Table 10.3. Material Constants

Type of behavior	No. of constants	Constants
Isotropic (metals, polymers)	2	E, ν
Orthotropic (on fiber axis stress)	4	E_x, E_y, G, ν_{12}
Anisotropic (off fiber axis stress)	6	$E_x, E_y, G, \nu_{xy}, \nu_{xy,x}, \nu_{xy,y}$

multiplication is

$$\frac{1}{E_x} = \frac{1}{E_1} \cos^4 \theta + \left(\frac{1}{G_{12}} - \frac{2\nu_{12}}{E_1}\right) \sin^2 \theta \cos^2 \theta + \frac{1}{E_2} \sin^4 \theta$$

$$\nu_{xy} = E_x \left[\frac{\nu_{12}}{E_1}(\sin^4 \theta + \cos^4 \theta) - \left(\frac{1}{E_1} + \frac{1}{E_2} - \frac{1}{G_{12}}\right) \sin^2 \theta \cos^2 \theta\right]$$

$$\frac{1}{E_y} = \frac{1}{E_1} \sin^4 \theta + \left(\frac{1}{G_{12}} - \frac{2\nu_{12}}{E_1}\right) \sin^2 \theta \cos^2 \theta + \frac{1}{E_2} \cos^4 \theta$$

$$\frac{1}{G_{xy}} = 2\left(\frac{2}{E_1} + \frac{2}{E_2} + \frac{4\nu_{12}}{E_1} - \frac{1}{G_{12}}\right) \sin^2 \theta \cos^2 \theta + \frac{1}{G_{12}}(\sin^4 \theta + \cos^4 \theta)$$

$$\nu_{x,xy} = E_x \left[\left(\frac{2}{E_1} + \frac{2\nu_{12}}{E_1} - \frac{1}{G_{12}}\right) \sin \theta \cos^3 \theta\right.$$
$$\left. - \left(\frac{2}{E_2} + \frac{2\nu_{12}}{E_1} - \frac{1}{G_{12}}\right) \sin^3 \theta \cos \theta\right]$$

$$\nu_{y,xy} = E_y \left[\left(\frac{2}{E_1} + \frac{2\nu_{12}}{E_1} - \frac{1}{G_{12}}\right) \sin^3 \theta \cos \theta\right.$$
$$\left. - \left(\frac{2}{E_2} + \frac{2\nu_{12}}{E_1} - \frac{1}{G_{12}}\right) \sin \theta \cos^3 \theta\right] \tag{10.23}$$

where the subscripts 1 and 2 refer to the values parallel and perpendicular to the fiber axis as before. It is possible for the material constants to reach values greater than that of the fiber axis and transverse directions. However, for the commonly used fibers and matrix materials, the fiber direction always provides the greatest stiffness.

For an isotropic material, Poisson's ratio is positive and the strain in the direction perpendicular to the applied stress is always negative. Right angles remain 90°. In shear, since the strains in the x- and y-directions are zero, the volume change of the sample is also zero. For the off-axis composite, application of a normal stress produces a distortion of the angles, and a rectangular sample no longer remains rectangular. Since $v_{x,xy}$ is usually a positive value, the application of a shear stress increases the volume of the part. If $v_{x,xy}$ is negative, the volume will be decreased by a shear stress.

10.9. STIFFNESS OF LAMINATES

Symmetric Laminates

Equation 10.23 provides the engineering constants for a lamina which has stresses applied at an angle to the fiber axis. However, since the lamina has very low strength in the transverse directions, it is rare that the product contains laminae all directed parallel to each other. The optimum properties are obtained by combining laminae of different orientations, so that the desired strength and stiffness are obtained in all directions. The combination of multiple layers of the laminae is called a laminate. Any combination of laminae is possible, but most commonly balanced symmetric laminates are designed. A laminate is symmetric if the plies are arranged symmetrically around the center of the laminate with respect to the orientation of the plies and is also symmetric with regard to the thickness of the plies. In a hybrid composite, the symmetry would include symmetrical layup of the material of the plies around the central ply. A laminate which is balanced has an equal number of + and − plies. A regular symmetric balanced laminate has plies laidup so that adjoining laminae have opposite signs of the orientation angle relative to the central ply chosen as the reference or 0° ply.

Due to this symmetry, symmetric laminates are not anisotropic but are orthotropic in their behavior. Their behavior and the analysis of their response to an applied load are greatly simplified, since this symmetry eliminates the interaction between the shearing and normal components. As a result, when an in-plane load is applied, no out-of-plane bending occurs. An additional advantage of the symmetric laminates is that the use of the angle plies produces a laminate with a reasonably large value of the shear modulus and good resistance to shear deformation.

For nonsymmetrical laminates, twisting out of the plane results from tensile or compressive loads within the plane. The matrix polymer of the laminates is polymerized at elevated temperatures. Thermal stresses are developed as the composite cools to room temperature. These stresses produce twisting and distortion, which can be reduced by the use of the symmetric balanced laminates. Since the center of the laminate is the plane of symmetry and alternative layers must have opposite angular signs, if an odd number of plies is required, the middle of the central ply is the center of symmetry and the remaining plies are laidup symmetrically around that center.

Deformation of a Laminate

As a result of the applied stresses, the laminae may change length within its plane and bend out of the plane. The analysis assume that the total thickness of the laminate is small and a perfect bond exists between adjacent plies, so that the fundamental precepts of thin-shell theory can be applied—in particular, the Kirchhoff-Love hypothesis that straight lines in the x-z and y-z planes remain straight along bending of the laminate (Fig. 10.5).

The resultant strain of the laminate can be separated into the strain of the midplane of the laminate (ε_0) and the additional strain due to bending. The smaller the radius of curvature and the greater the distance of the point under consideration from the midplane, the greater the resultant strain. Since the laminae are considered to be perfectly bonded, the strain varies linearly through the laminate. However, the different laminae have very different stiffness values, and the resultant stress changes discontinuously at the interfaces between the laminae (Fig. 10.6).

The radius of curvature resulting from the applied bending stress K is defined as

$$K_x = \frac{\partial^2 w}{\partial^2 x}$$

Considering displacement due only to bending, the elongation in the x-direction is zero at the neutral axis and increases in proportion to the distance from the neutral axis. If u is the displacement along the x-direction and w is the displacement along the z-direction

$$u = z\frac{\partial w}{\partial x}, \quad \text{since} \quad \varepsilon_x = \frac{\partial u}{\partial x}, \quad \text{then} \quad \frac{du}{dx} = z\frac{\partial^2 w}{\partial x^2}$$

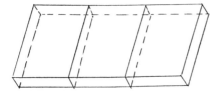 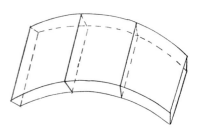

Figure 10.5.   Straight lines remain straight in bending.

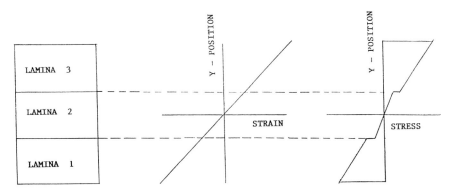

Figure 10.6.   Stress discontinuites at laminae interfaces.

Then

$$\varepsilon_x = K_x z$$

The strain in other directions depends on the radius of curvature in the particular direction. The curvature is therefore a vector and will be denoted as $\{K\}$.

The total strain at any point can therefore be given by

$$\{\varepsilon\} = \{\varepsilon_0\} + \{K\}z$$

The applied stress is related to the strain by

$$\{\sigma\} = [\bar{Q}][\varepsilon] = [\bar{Q}][\{\varepsilon_0\} + \{K\}z] \tag{10.24}$$

The stresses, therefore vary throughout the thickness of the beam, not only due to bending (through the $\{K\}z$ term), but also as a result of the variation of properties in the different laminae $[\bar{Q}]$. Because the laminae are constrained to have the same strain, the laminae which are stiffer in the direction of the applied stress will have a larger internal stress.

The force developed in the laminate in a specific direction can be calculated by integrating through the thickness of the laminate

$$F = \int_{t/2}^{-t/2} w\sigma \, dz$$

where w is the width of the sheet.

Or consider the force per unit width of the sheet

$$\{F\} = \int_{-t/2}^{t/2} \{\sigma\} \, dz$$

and the average stress within a specific laminae is

$$\sigma_{aver} = \frac{1}{t} \int_{-t/2}^{t/2} \{\sigma\} \, dz$$

and for an individual component (for example in the x-direction)

$$\sigma_x = \frac{1}{t} \int_{-t/2}^{t/2} \sigma_x \, dz$$

since

$$F_z = \int_{-t/2}^{t/2} \sigma_x \, dz, \qquad \text{then} \qquad \sigma_x = \frac{1}{t} F_x$$

We will continue to work with the force F for simplicity. Substituting Eq. 10.24

$$\{F\} = \int_{-t/2}^{t/2} [\bar{Q}][\{\varepsilon_0\} + \{K\}z] \, dz$$

The resultant force F on the entire laminate can be taken as the sum of the forces on each laminae. However, the stiffness matrix of each lamina is a constant for the lamina, and can be taken outside the integral. For N laminae

$$\{F\} = \sum_{i=1}^{N} [\bar{Q}]_i\{\varepsilon_0\} \int_{h_{i-1}}^{h_i} dz + \sum_{i=1}^{N} [\bar{Q}]_i\{K\} \int_{h_{i-1}}^{h_i} z \, dz$$

where the integral now is over the thickness of one laminae from its top dimension h_i to its bottom at h_{i-1} (see Fig. 10.7). Since the curvature K is constant throughout the thickness, it also is removed from the integral. Integrating

$$\{F\} = \sum_{i=1}^{N} [\bar{Q}]_i\{\varepsilon_0\}[h_i - h_{i-1}] + \frac{1}{2}\sum [\bar{Q}]\{K\}[h_i^2 - h_{i-1}^2] \qquad (10.25)$$

The relationship between the deformations of the entire laminate and the force on the laminate has therefore been obtained. This relationship involves the properties of all the laminae [in the $\sum [\bar{Q}]_i$]. This sum is customarily identified as a property of the laminate, and Eq. 10.25 can be written

$$\{F\} = [A]\{\varepsilon_0\} + [B]\{K\} \qquad (10.26)$$

where the matrices A and B are

$$[A] = \sum_{i=1}^{N} [\bar{Q}]_i [h_i - h_{i-1}] \qquad \text{[extensional stiffness matrix]} \quad (10.27)$$

$$[B] = \frac{1}{2} \sum_{i=1}^{N} [\bar{Q}]_i [h_i^2 - h_{i-1}^2] \qquad \text{[coupling stiffness matrix]} \quad (10.28)$$

The significance of these matrices follow directly from Eq. 10.26. The $[A]$ matrix relates extension of the laminae as a result of the applied force. Matrix $[B]$ relates the coupling of the out-of-plane bending and the in-plane forces. Each term of these matrices is found by summing the appropriate $\bar{Q}_{ij}$ values of the stiffness matrix of the individual laminates.

The fiber direction of one lamina (usually the central one of a symmetric laminate) is used to define 0°. The angles of the remaining laminae are defined relative to the reference lamina. The $\bar{Q}_{ij}$ can then be computed and summed. Arithmetic simplification of these terms is possible using trigonometric identities, but the computation is usually performed by a computer program, and the computation can be performed readily to attain the extensional stiffness of any laminate.

For any laminate in which the $[B]$ matrix is not zero, any in-plane force will therefore generate a curvature and out-of-plane bending. The resultant shape will be very complex, and requires solution of Eqs. 10.26 to 10.28. The general desirability of a symmetric material should therefore be apparent. In a symmetric matrix, the coupling matrix is always zero, an in-plane force will result in deformation only within the plane of the laminate, and the laminates therefore will remain planar. Because a single lamina is symmetric about its midplane, the $[B]$ matrix is always zero for a unidirectional laminate or a single lamina.

As an example, consider a three-layer symmetric laminae $\theta/0/-\theta$ with laminae thickness h (Fig. 10.7). The components of the $[B]$ matrix are

$$B_{11} = \frac{1}{2} [\bar{Q}_{11}(-\theta)] \left[\left(\frac{-h}{2}\right)^2 - \left(-3\frac{h}{2}\right)^2 \right] + Q(0) \left[0 - \left(-\frac{h}{2}\right)^2 \right]$$

$$+ Q(0) \left[\left(\frac{h}{2}\right)^2 - 0 \right] + \bar{Q}_{11}[\theta] \left[\left(\frac{3h}{2}\right)^2 - \left(\frac{h}{2}\right)^2 \right]$$

From Eq. 10.16 it can be seen that since the sine and cosine terms are all squared in the stiffness matrix and therefore

$$\bar{Q}_{11}{}^2(\theta) = \bar{Q}_{11}{}^2(-\theta)$$

and therefore $B_{11} = 0$. All other elements of the matrix are similarly zero.

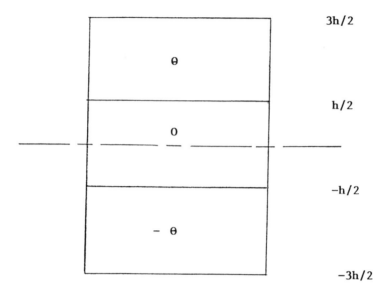

Figure 10.7. Laminae orientation and thickness dimensions.

For a symmetric laminate, the analysis is therefore greatly simplified because Eq. 10.26 reduces to $F = [A]\varepsilon_0$.

Now let us evaluate an $[A]$ matrix element for the same laminate. The $[A]$ matrix element A_{11} is

$$A_{11} = t[\bar{Q}_{11}(\theta) + \bar{Q}_{11}(0) + \bar{Q}_{11}(-\theta)]$$

$$\bar{Q}_{11}(\theta) = \bar{Q}_{11}(-\theta) \qquad \text{and} \qquad \bar{Q}_{11}(0) = Q_{11}$$

Therefore

$$A_{11} = (2\bar{Q}_{11}(\theta) + Q_{11})t$$

The other elements of the matrix can be found similiarly.

10.10. FLEXURAL STIFFNESS OF LAMINATES

The development of the flexural stiffness of a laminate follows the same general procedure as performed for plane extensional stiffness. For in-plane stiffness the stacking order has no effect on the overall stiffness, but does affect the possible coupling of in-plane loading and out-of-plane twisting. For flexural stiffness, since the loading and deflections are out of the plane of the laminate, the order of stacking is critical. The elastic constants for this type of loading are defined in the flexural modulus matrix $[D]$, to be derived below.

Since the resistance to flexure varies with distance from the neutral axis, the equations for the moment require integration over the thickness of the laminate. The sign of the moment must be carefully defined. Positive moments correspond to a positive stress in the upper half of the laminate, and the laminate is convex in an upward direction.

In a similar manner, the moment on the laminate can be obtained by integrating the applied stresses over the distance from the midplane

$$M = \int_{-t/2}^{t/2} \sigma z \, dz$$

where M is the moment per unit width of the laminate.

Substituting Eq. 10.24 for σ

$$M = \int_{-t/2}^{t/2} [\bar{Q}][\{\varepsilon_0\} + \{K\}z]z \, dz$$

and again recognizing that the stiffness matrix $[Q]$ and the curvature $\{K\}$ are constant within an individual lamina

$$M = \sum_{i=1}^{N} [\bar{Q}]_i \{\varepsilon_0\} \int_{h_{i-1}}^{h_i} z \, dz + \sum_{i=1}^{N} [\bar{Q}]_i \{K\} \int_{h_{i-1}}^{h_i} z^2 \, dz$$

$$= \left[\frac{1}{2} \sum_{i=1}^{N} [\bar{Q}]_i \{\varepsilon_0\}[h_i^2 - h_{i-1}^2] \right] + \left[\frac{1}{3} \sum_{i=1}^{N} [\bar{Q}]_i \{K\}[h_i^3 - h_{i-1}^3] \right]$$

$$(10.29)$$

The first term can be recognized as relating to the coupling stiffness matrix (Eq. 10.28). We can therefore rewrite Eq. 10.29 as

$$M = [B]\{\varepsilon_0\} + [D]\{K\}$$

where

$$[D] = \frac{1}{3} \sum_{i=1}^{N} [\bar{Q}][h_i^3 - h_{i-1}^3] \qquad \text{[flexural modulus matrix]}$$

The existence of the $[B]$ matrix emphasizes that in nonsymmetric laminates the generation of an extensional strain within the plane of the laminate will generate a moment and the application of a moment will produce strain and elongation within the plane of the laminate. This emphasizes the coupling of bending and extension in these complex laminates. The flexural modulus matrix relates the moment to the curvature developed.

To complete our example of the $\theta/0/-\theta$ symmetric laminate, the computation of the elements of the $[D]$ matrix proceeds as before, remembering that $\bar{Q}_{11}(0) = Q_{11}$.

$$D_{11} = \frac{1}{3}\left\{\left[\bar{Q}_{11}[\theta]\left(-\frac{h}{2}\right)^3 - 0\right] + Q_{11}\left[\left(\frac{h}{2}\right)^3 - \left(-\frac{h^3}{2}\right)\right]\right.$$

$$\left. + \bar{Q}_{11}[-\theta]\left[\left(\frac{3h}{2}\right)^3 - \left(\frac{h}{2}\right)^3\right]\right\}$$

$$= \frac{h^3}{3}\left[\bar{Q}_{11}[\theta]\left[-\frac{1}{8}\right] + Q_{11}\left[+\frac{1}{4}\right] + \bar{Q}_{11}[-\theta]\left[\frac{27}{8} - \frac{1}{8}\right]\right]$$

$$= \frac{h^3}{3}\left[\bar{Q}_{11}[\theta]\left\{\frac{25}{8}\right\} + \frac{Q_{11}}{4}\right]$$

$$= \frac{h^3}{12}\left[\bar{Q}_{11}[\theta][12.5] + Q_{11}\right]$$

10.11. ENGINEERING CONSTANTS FOR A LAMINATE

The extensional compliance matrix can be calculated by inverting the extensional stiffness matrix $[A]$

$$\{\varepsilon\} = [A]^{-1}\{F\} = [SL]\{\sigma \cdot t\}$$

where $[SL]$ is the compliance matrix of the laminate.
   Following the usual inversion process for a matrix,

$$SL_{11} = \left(\frac{A_{22}}{A_{11}A_{22} - A_{12}^2}\right)$$

$$SL_{12} = \frac{-A_{12}}{A_{11}A_{22} - A_{12}^2}, \qquad SL_{22} = \frac{A_{11}}{A_{11}A_{22} - A_{12}^2}, \qquad SL_{33} = \frac{1}{A_{33}}$$

As for the case of an individual lamina, these matrix terms can be related to the engineering constants by utilizing the results of mechanical testing. For example, if we apply $\sigma_x \neq 0$, $\sigma_y = \tau_{xy} = 0$, then

$$\varepsilon_x = SL_{11}\sigma_x t$$

Since

$$\varepsilon_x = \frac{1}{E_x}\sigma$$

then

$$E_x = \frac{A_{11}A_{22} - A_{12}^2}{tA_{22}}$$

Similarly, considering a uniaxial stress along the y-axis

$$E_y = \frac{A_{11}A_{22} - A_{12}^2}{tA_{11}}$$

Elastic Modulus

Graphical presentation of the stiffness can be used for common sets of laminae to give quick, reasonably accurate values of the engineering constants for preliminary design work. As an example of typical values, consider the modulus of asymmetric $0°/45°/90°$ laminate in comparison to that of a unidirectional laminate with the same total number of plies. The modulus depends primarily on the presence of plies in the $0°$ orientation, the $45°$ and $90°$ plies contributing almost negligibly. Clearly, if only properties in that one direction were of interest, a uniaxial layup would be used. Of course, the cross angle and $90°$ plies are added for strength in the y-direction, and the cross plies for resistance to shear.

Poisson's Ratio

Poisson's ratio for isotropic plastics is close to 0.3, and for rubbery polymers it is 0.5. Orthotropic laminates are observed to have a wide range of Poisson's ratio. Such large variations produce different lateral strains in adjacent laminae under loading, since the laminae at different angles contract by different amounts. These variations between laminae produce interlaminar stresses which may cause delamination in extreme cases. The variation of Poisson's ratio for different ratios of $0°/45°/90°$ plies is notable. For 50% of the laminate being in the $0°$ orientation, the v_{xy} value can vary from a tenth to almost the full value of a uniaxial laminate depending on the fraction of remaining plies in the $45°$ orientation.

Shear Modulus

Shear within the plane of the laminate is resisted almost exclusively by the $45°$ plies, and therefore the value of G_{xy} increases linearly with increasing fraction of the cross plies. Proper design must correlate the extent of cross plies with the necessity for shear resistance, as the increase in cross plies reduces the ability of the laminate to withstand the longitudinal and transverse stresses for a fixed total number of plies.

10.12. FAILURE CRITERIA

The ultimate strength of a composite in the fiber and transverse direction and the shear strength can be readily determined experimentally. For composites, failure may occur completely within the matrix as a result of transverse loading, by fracture of the fiber in tension or fiber microbuckling in compression, or by disbonding at the fiber-matrix interface due to shear. Questions analogous to those studied about the effect of stress direction on stiffness need to be asked about the strength of the composite. Specifically, how does the strength compare for loads that are not applied along the fiber and transverse directions to these direct loadings? In addition, for a laminate, how does the ultimate strength vary if loading other than simple uniaxial loads occurs?

The simplest failure criterion that could be reasonably proposed is that failure occurs when the component of the applied stress exceeds the failure stress for each of the three principal cases of fiber-directed, transverse or shear failure. This approach gives reasonable results for uniaxial loads applied close to the fiber axis direction or close to the transverse direction in tension, since these are the conditions established experimentally for the criteria. For applied loads at other angles and for compression the agreement is poor. Many failure criteria have been proposed (over 30 are readily found in the literature). The two most commonly employed are given below.

Tsai–Hill Criterion

A general form of an equation with terms including the interaction between the different stresses was developed by Hill (1950) for anisotropic materials. In its general form to include the interactions in the two dimensions of the laminate the criterion is

$$A\sigma_1{}^2 + B\sigma_2{}^2 + C\sigma_1\sigma_2 + D\tau_{12}{}^2 = 1 \tag{10.30}$$

The squares of the stresses are a result of Hill embodying the concepts of the Van Mises yield criteria, in which the distortional energy (proportional to the square of the stress) is considered to control yielding.

Tsai incorporated the strength values of the laminae in unidirectional loading into this equation. In a tensile test along the fiber axis

$$\sigma_1 = \sigma_1{}^{ult}, \qquad A = \left(\frac{1}{\sigma_1{}^{ult}}\right)^2$$

In the transverse direction

$$\sigma_2 = \sigma_2{}^{ult}, \qquad B = \left(\frac{1}{\sigma_2{}^{ult}}\right)^2$$

and similarly, as a result of a shear test,

$$D = \left(\frac{1}{\tau_{12}{}^{ult}}\right)^2$$

Evaluation in three dimensions give the value of $C = (1/\sigma_1{}^{ult})^2$ in Eq. 10.30.

Substituting these values into Eq. 10.30 gives the Tsai–Hill failure criterion:

$$\left(\frac{\sigma_1}{\sigma_1{}^{ult}}\right)^2 + \left(\frac{\sigma_2}{\sigma_2{}^{ult}}\right)^2 + \frac{\sigma_1\sigma_2}{(\sigma_1{}^{ult})^2} + \frac{\tau_{12}}{(\tau_{12}{}^{ult})^2} = 1 \qquad (10.31)$$

Failure will occur if the left-hand side of this equation exceeds unity.

In the special simple application of a unidirectional stress at an angle θ to the fiber axis, the components can be computed directly from the transformation equations as

$$\{\sigma_{1,2,12}\} = [T]\{\sigma_{x,y,xy}\}$$

or

$$\sigma_1 = \sigma_x \cos^2(\theta)$$

$$\sigma_2 = \sigma_y \sin^2(\theta)$$

$$\tau_{12} = \sigma_x \cos(\theta) \sin(\theta)$$

Substituting these values into Eq. 10.31, the failure criterion becomes

$$\sigma_x{}^2\left[\frac{\cos^4(\theta)}{(\sigma_1{}^{ult})^2} + \frac{\sin^4(\theta)}{(\sigma_2{}^{ult})^2} + \sin^2(\theta)\cos^2(\theta)\left(\frac{1}{(\tau_{12}{}^{ult})^2} - \frac{1}{(\sigma_1{}^{ult})^2}\right)\right] = 1 \quad (10.32)$$

and if σ_x has a value which makes the left-hand side (LHS) greater than one, failure will occur.

Tsai–Wu Criterion

To include all possible interactions, Tsai and Wu (1971) proposed a criterion which included the first power of all three principal stresses and all possible interactions between these stresses. The symmetry of the orthotropic material removes the terms involving interactions between shear and normal stresses. The resultant equation includes terms for the differences in strength of the material in the tensile and compressive directions. If this difference is neglected, the Tsai–Wu criterion is

$$2\left(\frac{\sigma_1}{\sigma_1{}^{ult}}\right) + 2\left(\frac{\sigma_2}{\sigma_2{}^{ult}}\right) - \left(\frac{\sigma_1}{\sigma_1{}^{ult}}\right)^2 - \left(\frac{\sigma_2}{\sigma_2{}^{ult}}\right)^2 + \left(\frac{\tau_{12}}{\tau_{12}{}^{ult}}\right)^2 + 2F_{12}\sigma_1\sigma_2 = 1$$

$$(10.33)$$

The last interaction term is difficult to compute, as it requires biaxial tensile tests. Fortunately, the value of this term does not significantly affect the accuracy of the prediction for failure, and a value of $-1/2$ with stress measured in units of 100 ksi is often chosen as typical.

Although these failure criteria reasonably accurately predict the biaxial stress interactions, they do not permit evaluation of the failure mode. Exact failure analysis is required to evaluate which of the three types of failure listed at the beginning of this section occurs.

Failure of a Laminate

The laminate comprises all the laminae, which can be examined independently for failure. The steps are:

1. The stresses acting on the individual laminae are determined
2. The stresses are transformed to the components relative to the fiber axis

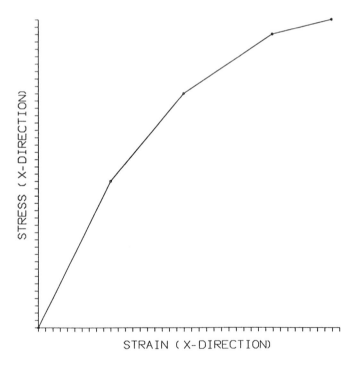

Figure 10.8. Progressive failure of a laminate. When an individual laminae fails, the stiffness decreases, but total failure may not occur until several laminae have failed successively.

3. The LHS of Eq. 10.33 is computed for each laminae
4. If the LHS exceeds 1, the laminae will fail. If all the LHS values are less than unity, the laminae with the largest value of the LHS is closest to failure
5. Failure of one laminae does not necessarily require the entire laminate to fail. If the LHS exceeds 1, we must assume that this lamina has failed and will carry no load. The computation can then be repeated for the remaining laminae with a higher stress based on one less lamina. The remaining laminae may still carry the applied load, but the stiffness will decrease (see Fig. 10.8)

These calculations can readily be programmed, and the output indicate the laminae which is closest to failure and whether the other laminae are capable of withstanding the load if the most stressed laminae fails. Graphical presentation of the failure envelope computed from the failure criterion gives a quick overview of the strength of the laminate.

References

A. S. Argon and M. M. Salama, *Phil. Mag. 35*, 1217 (1977).

A. Arzak, J. I. Equiazabal, and J. Nazabal, *PES 32*, 586 (1992).

S. Bandyopadhyay, J. R. Roseblade, and R. Muscat, *PES 33*, 3777 (1993).

C. Bauwens-Crowet, J. C. Bauwens, and G. Homes, *J. Polym. Sci. A2*, 735 (1969).

P. Beardmore and S. Rabinowitz, *J. Mater. Sci. 9*, 81 (1974).

R. Benavente and J. M. Perena, *PES 27*, 913 (1987).

P. P. Benham, *PES 13*, 398 (1973).

F. W. Billmeyer, *J. Polym. Sci. 4*, 83 (1949).

P. B. Bowden and J. A. Jukes, *J. Mater. Sci. 7*, 52 (1972).

P. W. Bridgman, *Trans. ASM 32*, 553 (1954).

L. J. Broutman and S. M. Krishnakumar, *PES 16*, 74 (1976).

N. Brown, *Creep, Stress Relaxation and Yielding in Engineering Plastics*, Vol. 2, Metals Park, Ohio: ASM, 1988, p. 728.

W. F. Brown and J. E. Strawley, *ASTM STP 410* (1966).

P. J. Cloud, F. C. McDowell, and S. Gerakaris, *Plast. Tech. 22*, 76 (1976).

J. B. Conway, *Numerical Methods for Creep and Rupture Analysis*, Gordon and Breach, 1967, Newark, New Jersey.

R. J. Crawford and P. P. Benham, *Polymer 16*, 908 (1975).

CRC Handbook of Solubility Parameters and Other Cohesion Parameters, Boca Raton, FL: CRC Press, 1983.

A. D'Amore, J. M. Kenny, and L. Nicolais, *PES 30*, 314 (1990).

D. S. Dugdale, *J. Mech. Phys. Sol. 8.* 10 (1960).

J. P. Elinck, J. C. Bauwens, and G. Homes, *Int. J. Fract. Mech. 7*, 227 (1971).

C. E. Englis, *Proc. Inst. Naval Architects 55*, 219 (1913).

J. Evans, *J. Soc. Plast. Eng. 16*, 76 (1960).

425

J. D. Ferry, *Viscoelastic Properties of Polymers*, New York: Wiley, 1961.

E. Flexman, D. D. Huang, and H. L. Snyder, *Polym. Prepr. 29*, 189 (1988).

T. G. Fox, *Bull. Am. Phys. Soc. 1*, 123 (1956).

A. Gent, *J. Mater. Sci. 5*, 925 (1970).

A. N. Gent, *J. Mater. Sci. 15*, 2884 (1980).

A. A. Griffith, *Trans. Royal Soc. London A221*, 163 (1921).

V. L. Groshans and M. T. Takemori, *PES 27*, 1010 (1987).

H. Gupta and R. Salovey, *PES 30*, 455 (1990).

G. T. Hahn and A. R. Rosenfeld, *Acta Met. 13*, 293 (1965).

S. Hashimi and J. G. Williams, *J. Mater. Sci. 20*, 4202 (1985).

R. W. Hertzberg, J. A. Manson, and M. D. Skibo, *PES 15*, 252 (1975).

R. Hill, *The Mathematical Theory of Plasticity*, Oxford University Press, London, 1950, p. 252.

L. E. Hornberger and K. L. Devries, *PES 27*, 1473 (1987).

Y. Huang and N. Brown, *J. Polym, Sci. Polym. Phys. Ed. 28*, 2007 (1990).

E. Inglis, *Proc. Inst. Naval Architects 55*, 219 (1913).

K. Inouye, *Tetsu to Hagane 41*, 593 (1954).

B. R. Irwin, *Proc. 7th Sagamore Conf.* (1960).

G. R. Irwin, *Fracture of Metals, Cleveland*: ASM, 1949.

G. R. Irwin, *Handbuch der Physik*, Vol. 1, Berlin: Springer, 1958, pp. 551–590.

G. R. Irwin, *Appl. Mat. Res 3*, 65 (1964).

R. P. Kambour, and R. E. Barker, Jr., *J. Polym. Sci. Polym. Lett. Ed. 4*, 35 (1966).

M. C. Kenney, J. F. Mandell, and F. J. McGarry, *J. Mater. Sci. 20*, 2045 (1985).

H. Kim and Y. Mai, *PES 33*, 5479 (1993).

B. D. Lauterwasser and E. J. Kramer, *Phil. Mag. A39*, 369 (1979).

A. N. Mafhyulis, M. I. Pugina, A. A. Zhechyus, V. K. Kuchinskas, and A. P. Stasyunas, *Mekh. Polim. 2* (1), 60 (1966).

S. S. Manson, *NACA Tech Note 2933* (1954).

D. S. Matsumoto and S. K. Gifford, *J. Mater. Sci. 20*, 4610 (1985).

E. Miller, *Properties Modification by Use of Additives in Engineering Plastics*, Vol. 2, Materials Park, OH: ASM, 1988, p. 493.

N. J. Mills, *J. Mater. Sci. 11*, 363 (1976).

M. A. Miner, *J. Appl. Mech. 12*, 159 (1954).

J. Mitz, J. DiBeneditto, and S. Petrie, *J. Mater. Sci. 13*, 1427 (1978).

I. Mondragon and J. Nazabal, *PES 25*, 178 (1985).

A. Nadai, *Theory of Flow and Fracture of Metals*, New York: McGraw-Hill, 1950, p. 973.

L. E. Nielsen, *Predicting the Properties of Mixtures*, New York: Marcel Dekker, 1978, p. 22.

H. X. Nguyen and H. Ishida, *Polym. Compos. 8*, 57 (1987).

A. E. Oberth and R. S. Brenner, *Trans. Rheol. Soc. 9*, 165 (1965).

R. H. Olley, D. C. Bassett, and D. J. Blundell, *Polymer 27*, 344 (1986).

E. Orowan, *Fatigue and Fracture of Metals*, Cambridge, MIT Press, 1950.

R. J. Oxborough and P. B. Bowden, *Phil. Mag. 28*, 547 (1973).

R. E. Peterson, *Stress Concentration Design Factors*, Wiley, New York, 1974.

C. F. Popelar, C. H. Popelar, and V. H. Kenner, *PES 30*, 577 (1990).

J. Puglisi and M. A. Chaudhari, *Epoxies in Engineering Plastics*, Vol. 2, Materials Park, OH: ASM, 1988, p. 240.

R. Regel, *Tech. Phys. USSR*, 7, 353 (1956).

M. N. Riddell, G. P. Koo, and J. L. O'Toole, *PES 6*, 363 (1966).

R. S. Rivlin and A. G. Thomas, *J. Polym. Sci. 10*, 291 (1953).

R. E. Robertson, in *Polymer Science*, A. D. Jenkins, ed., North Holland, Amsterdam, 1972.

H. Sasabe and S. Saito, *J. Polym. Sci. A26*, 1401 (1968).

J. A. Sauer and G. C. Richardson, *Int. J. Fract. Mech. 17*, 499 (1980)

A. Siegmann, S. Kenig, and A. Buchman, *PES 27*, 1069 (1987).

M. D. Skibo, R. W. Hertzberg, J. A. Manson, and S. L. Kim, *J. Mater. Sci. 12*, 531 (1977).

P. So, and L. J. Broutman, *PES 16*, 785 (1976).

A. P. Stasyunas, *Meck. Polim, 2(1)*, 60 (1966).

R. T. Steinbuch, *Br. Plast. 37*, 678 (1964).

R. T. Sternstein and L. Ongchin, *Am. Chem. Soc. Polym Repr. 10*, 1117 (1969).

A. V. Stinkas, Y. P. Bhaushis, and I. P. Bareishis, *Mekh. Polim, 8*, 59 (1972).

H.-J. Sue, *PES 31*, 275 (1991).

S. Suzuki and T. Tsurue, *Int. Conf. Mech. on Mater.*, 2nd, Boston, ASM, Materials Park, OH, (1976).

S. R. Swanson, F. Cicci, and W. Hoppe, *ASTM STP 415*, 312 (1967).

M. T. Takemori, *PES 27*, 46 (1987).

M. T. Takemori, *PES 28*, 641 (1988).

K. R. Tate, A. R. Perrin, and R. T. Woodhams, *PES 28*, 740 (1988).

B. S. Thakkar and L. J. Broutman, *PES 21* 155 (1981).

S. W. Tsai and E. M. Wu, *J. Comp. Mat. 5*, 58 (1971).

S. Tuba *J. Strain Analysis 1*, 115 (1966).

S. Turner, *PES 6*, 303 (1966).

Y. N. VanGaut, *Mekh. Polim. 1(3)*, 151 (1965).

T. T. Wang, M. Matsuo, and T. K. Kwei, *J. Appl.* Phys. *42*, 4188 (1971).

H. M. Westergaard, *J. Appl. Mech. 61*, 49 (1939).

J. G. Williams and G. P. Marshall, *Proc. Roy. Soc. A342*, 55 (1975)

J. S. Yu, D. M. Lim, and D. M. Kalyon, *PES 31*, 145 (1991).

Y. Zhou, X. Lu, and N. Brown *PES 31*, 711 (1991).

S. N. Zhurkov and E. E. Tomashevskii, *Conf. Ser. I Inst. Phys. Physical Soc. Oxford*, "Physical Basis of Yield and Fracture," 1966.

Index

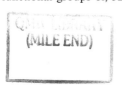